The human species faces a new threat to its health – perhaps to its survival. Our burgeoning numbers, technology and consumption are overloading Earth's capacity to absorb, replenish and repair. These global environmental problems pose health risks, not just from localised pollution, but from damaged life-support systems. Might we, too, become an 'endangered species'?

The risks of skin cancer and cataracts from ozone destruction are evident enough: likewise the health hazards from greenhouse-related heatwaves and hurricanes. More insidious, though, may be the indirect health consequences of climate change on food production and the spread of infections; likewise, immune suppression by ultraviolet radiation, erosion of overworked soils, depletion of freshwater, and loss of genetic and biological resources for producing food and medicines. Rapid urban growth poses other health hazards. Population health cannot be sustained without intact ecosystems.

Professor McMichael assesses these problems, and their scientific uncertainties, within a broad biological, historical and social context. Relations between the world's rich and poor, and their environmental consequences, are explored. Above all, the health of *Homo sapiens* is examined within an ecological framework.

Tony McMichael, previously Professor of Environmental Health at the University of Adelaide, Australia, is now Professor of Epidemiology at the London School of Hygiene and Tropical Medicine, UK. He is an advisor to the Environmental Health Division of WHO, and is coordinating the scientific assessment of the health impact of climate change for the Geneva-based Intergovernmental Panel on Climate Change (IPCC). He has travelled and taught in many countries.

PLANETARY OVERLOAD

PLANETARY OVERLOAD

Global environmental change and
the health of the human species

A. J. McMICHAEL
Department of Community Medicine
University of Adelaide
Adelaide
South Australia

CAMBRIDGE
UNIVERSITY PRESS

Published by the Press Syndicate of the University of Cambridge
The Pitt Building, Trumpington Street, Cambridge CB2 1RP
40 West 20th Street, New York, NY 10011–4211, USA
10 Stamford Road, Oakleigh, Melbourne 3166, Australia

First published 1993
Reprinted 1994 (twice)

Printed in Great Britain at the University Press, Cambridge

A catalogue record for this book is available from the British Library

Library of Congress cataloguing in publication data

McMichael, Anthony J.
Planetary overload: Global environmental change and the health of the human species/
A. J. McMichael.
p. cm.
Includes bibliographical references and index.
1. Environmental health. 2. Environmental policy. I. Title.
RA565.M384   1993
304.2′8–dc20   92-38292   CIP

ISBN 0 521 44138 2 hardback
ISBN 0 521 45759 9 paperback

TAG

To Anna, Celia, and other stewards of the next millennium

Contents

Contents

Preface

The most serious potential consequence of global environmental change is the erosion of Earth's life-support systems. Yet, curiously, the nature of this threat to the health and survival of the world's living species – including our own – has received little attention.

Over aeons, the evolution of life has gradually transformed the environment that clothes the planet's surface. The lower atmosphere's composition has changed; stratospheric ozone has formed from oxygen emitted by plants; soil has been created by oxidation, plants and microbes; and forests speed the recirculation of rainwater. Life's genetic diversity confers a capacity for adaptive change. However, this fabric of life-supporting mechanisms is now starting to unravel, in a brief geological moment, as the cumulative global impact of human activity escalates.

We fret about the more easily understood effects of environmental damage upon national economies, property values, amenities and pristine nature. In its 1992 Report, the World Bank says: 'Soils that are degraded, aquifers that are depleted, and ecosystems that are destroyed in the name of raising incomes today can jeopardize the prospects for earning income tomorrow.'[1] The report also notes, *en passant*, that local environmental pollution by toxic chemicals may impose costs to human health that retard economic development. This exemplifies how we typically overlook the more fundamental fact that Earth's natural systems provide the essential life-support services that enable organisms to remain healthy and to breed. Today's unprecedented global environmental changes – particularly climate change, ozone layer depletion, land degradation and loss of biodiversity – may therefore have profound effects upon the health of human populations.

This is unfamiliar territory. Overloading the biosphere can affect population health in ways that differ fundamentally from the local, direct-acting, toxicity of environmental pollutants such as sulphurous fumes in air and heavy metals in food. Rather, such overload reduces the stability and productivity of the natural systems that support life. We have created environmental agencies to address the familiar type of toxic pollutant problem highlighted at the 1972 UN Environment Conference – such as those due to air pollutants, contaminated drinking water, toxic waste sites and garbage disposal. Today, however, we face environmental problems that reflect ecological disruption, transcend national and regional boundaries, and pose a more profound, albeit longer-term, threat to health.

The political consequences are as complex as the science. For example, we may soon live in a world in which global warming, caused mainly by industrialised and industrialising countries, causes inundation of coastal communities in Bangladesh and increased spread of malaria to highland communities in Latin America and Africa. For such reasons, some governments have begun examining the health impacts of greenhouse-induced climate change and ozone layer depletion. The World Health Organization submitted to the 1992 UN Earth Summit a report on the health impact of current patterns of energy use, urbanisation, industry and agriculture. The UN's Food and Agricultural Organization now acknowledges that the combination of soil erosion, desertification, climate change and population growth portends more food shortages.

Overall, however, our response has been tentative. I think this is largely because we still have a shallow understanding of the ultimate dependence of our health upon the integrity of ecosystems. We talk about 'life-support' systems, but, frankly, the idea that the survival of *Homo sapiens* depends upon the sustaining of ecosystems still seems a bit far-fetched. Most developed countries have cultures characterised by religions with anthropoid gods, where the notion of Man as Master endures. Relatedly, under modern, internationalised, capitalism – now uncontestedly the dominant influence on world trade and national economies – we have conferred upon the market economy a life of its own, and, by defining our social purpose within this framework, we further distance ourselves from the rhythms of natural systems.

Those working in the health sciences, too, have been slow to perceive the significance of ecological disruption for population health. Some aspects seem clear enough – ozone depletion will enhance skin cancer rates and temperature rises will enlarge the habitat of malarial mosquitoes – but those are only the tip of a much bigger iceberg. Below the water-line loom wider-ranging hazards to human health. Meanwhile, despite the many uncertainties, the world's vital signs appear to be generally negative. The ozone layer is thinning faster than we expected. The 1980s was the hottest decade on record, sustained into the early 1990s. After three buoyant decades the world's per-person food output has recently faltered, and land degradation is occurring widely. The extinction of species and loss of genes, many directly useful to our future survival as food and pharmaceuticals, continues to accelerate. The arms race has yielded increasingly dreadful weapons that can destroy whole ecosystems. Underlying all of these, the burgeoning world population and the debilitating burden of Third World poverty and desperate subsistence agriculture weigh heavily on the environment.

These seemingly disparate problems arise from the sheer scale and intensity of human economic activity. If these problems continue, their impact will be geographically uneven. Land degradation, deforestation and climate change will occur mostly in poor countries at low latitude; direct exposure to increased ultraviolet radiation will increase most in rich countries at high latitude. Eventually, however, weather instability, climatic impairment of crop yield, rising seas and loss of genetic resources would affect the health of human populations everywhere. Of course, there is much that scientists do not yet understand about these ecological disturbances and their consequences. But we cannot ignore the probability that these global environmental changes will have various adverse effects upon the health and wellbeing of *Homo sapiens*.

Some of the predicted effects may not become serious for a generation or two. Much of the impact of today's environmental excesses will be to impoverish the environment in which future generations must live. This would be the first time, at a global level, that one generation has conferred a *negative* legacy upon future generations. That poses an unprecedented moral problem, since the usual expectation of human society (in particular, modern western society) has been that each generation will increase, or at least preserve, the store of scientific knowledge, technological skills and the material infrastructure of society for future generations.

Finally, a more personal comment. Writing about environment and health within an ecological framework has required ranging over a wide terrain – further widened by the need to consider political, social and ethical aspects. Although no-one can hope to be fully informed over so wide a terrain, I am reassured by the comment of an Australian philosopher, John Passmore, who, in *Man's Responsibility for Nature*, says: 'Everybody who writes about ecological problems is, in respect to certain of the topics he is discussing, an amateur.'[2] The import of

Passmore's remark is that these problems cannot be meaningfully addressed *unless* they are considered within a multidisciplinary context. Passmore goes on to say: 'So far as the Western tradition discourages communication between specialists, it presents an obstacle to the adequate examination of ecological problems. Inter-disciplinary investigations are in this area not a luxury, but a necessity.'

Accordingly, I have attempted a broad analysis which I hope will provide a useful synthesis, particularly for those who have not previously thought much about human population health within an ecological context. This should inform and strengthen our response to the challenge posed by global environmental overload. Many commentators judge that we may not have long to develop the far-reaching social responses required to solve these problems. If a clearer understanding of the risks to human health facilitates such responses, this book will have achieved something worthwhile.

Acknowledgements

I have sought advice from many colleagues in writing this book. Comments on early drafts were made by Matt Gaughwin, Tony Worsley, David Shearman, Stephen Boyden, Sara Parkin and John Powles. Brian MacDermott and Ken Dyer gave subsequent advice on the ordering of ideas. Comments on particular sections came from John Moss, John Hatch, Michael Manton, Philip McMichael, Richie Gun, Mary Beers, Alistair Woodward, Bruce Armstrong, Andrew Oates, Basil Hetzel, Harvey Marchant, Barrie Pittock, Ernesto Kahan, Philip Weinstein, John Young, Graeme Hugo and Tord Kjellstrom. I discussed the general issues with Andrew Haines and John Last, two health scientists who saw early the importance of this topic. Louise Stafford, in correcting various drafts, ably decoded my tortuous annotations. My wife, Judith, encumbered with writing her doctorate, put up with my elastic working-hours. The enthusiasm of my daughters, Anna and Celia, for the ideas and practice of environmentalism, gave me added impetus.
Adelaide, January, 1993   A. J. McM

References

1. World Bank. *World Development Report* 1992. *Development and Environment.* Oxford: Oxford University Press, 1992.
2. Passmore J. *Man's Responsibility for Nature.* London: Duckworth, 1974 (Second Edition, 1980).

Introduction

Homo sapiens has existed for less than one ten-thousandth of Earth's lifespan – and, indeed, for less than one-thousandth of the time since animal life ventured from the oceans onto the dry land. Humans are newcomers, with no special immunity against the usual fate of biological species on Earth: extinction. Indeed, it is just now becoming conceivable that within several generations the human species may face threats to its survival because of its disruption of Earth's life-supporting ecosystems.

Could humans, *really*, be an 'endangered species'? Isn't it more likely that any such threat to our survival would, at worst, be confined to certain hapless populations, and therefore would not threaten our species as a whole? Besides, we call other species 'endangered' when their population numbers *fall* below a critical level. Yet human numbers are assuredly not falling! Nor is human habitat shrinking. Indeed, we are commandeering more and more of the world's surface area and incoming solar energy for our own needs, and are now using (or preempting) an astonishing 40 % of Earth's most basic resource, the incoming solar energy stored by terrestrial plants.[1] As we take control of more of this 'net primary production', via agriculture, pastoralism, forestry, land-clearing and urbanisation, there is less available to sustain other species. It sounds as if *they*, not us, are endangered.

Yet that is very much part of the problem. We humans cannot live apart from nature, remote from the great web of life. The emerging risks to human population health do not arise from local environmental contamination with direct-acting toxic chemicals, nor from a Malthusian outstripping of Earth's available material resources (oil, metals, timber, etc.). Rather, the risk arises from the disruption of natural systems because we are exceeding the biosphere's carrying capacity – i.e. we are overloading the planet's 'metabolic' capacity to absorb, replenish and restore. Through

1

our aggregate impact, various natural balances are tipping in directions that, if sustained, would make the world less able to support life. Some aspects of this overload, particularly land degradation, have occurred before on a localised scale. Most aspects, however, are the product of aggregate human activity in recent decades: increased emissions of greenhouse gases, damage to stratospheric ozone, depletion of aquifers and large-scale destruction of rainforest. These do not act by direct toxicity nor by exhaustion of non-renewable materials. Instead, they impair the productivity (soil, forest, oceans, biodiversity) or stability (climate, sea-level, ultraviolet filtration) of Earth's natural systems. John Powles has recently said of them: 'These effects are intrinsically difficult to predict and to counter, but this is now the main challenge facing those concerned with the health of human populations.'[2]

Fossil fuel combustion illustrates the above distinctions. Our earlier concern was over the resultant local pollution by noxious gaseous emissions which, at 'smog' concentrations, caused extra deaths. Subsequently, we became concerned that our energy-intensive society would run out of fossil fuels – oil first, probably. Today we realise that, even with emission-controlled smokestacks and proven reserves of fossil fuels that will last centuries, the problem we face is a quite different one. We are loading the atmosphere with a heat-trapping gas, carbon dioxide, that, eventually, will disrupt various of the biosphere's natural cycles, processes and conditions upon which we fundamentally depend for life-support.

Compared with the hunter-gatherer era, which predominated until a short 10,000 years ago, human numbers have multiplied one thousand-fold (including a massive ten-fold increase in the past 250 years) and our average, daily, per-person energy use is also about one thousand times greater. Our aggregate impact upon the biosphere is therefore about one million times greater than in those pre-agrarian days. We are consequently overloading Earth's capacity to absorb otherwise *non*-toxic waste gases, to replenish slowly-renewable resources such as soil and groundwater, and to sustain genetic and ecological diversity. It is these disruptions that comprise an unprecedented threat to our life-support systems.

This is unfamiliar territory. Historically, population health crises have had some immediacy about them: wars, floods, the Black Death, air pollution, cholera, smoking-related diseases and the AIDS epidemic. During 10,000 years of human settlement, there have been two major categories of environmental health problem, one animate, the other inanimate. First, there has been the age-old problem of contagious infectious diseases, associated with increased population density. A myriad

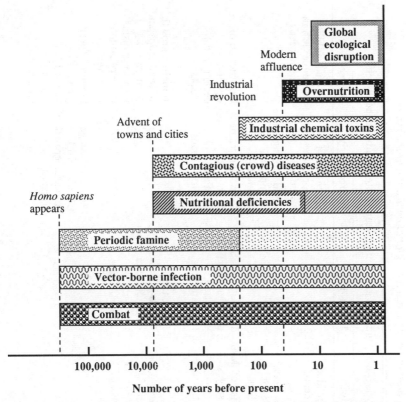

Figure. Emergence of the major categories of population health hazards during human cultural evolution.

of microbes survive by feeding on, and breeding in, humans, and some cause infectious diseases. We have found ways of countering many of those infections. Second, more recently we have been exposed to noxious chemical exposures resulting from industrial activity. We have learnt much about the toxicological effects of heavy metals, chemical pesticides, asbestos and assorted air pollutants, and we have begun to limit and control these exposures. Also, as shown in the figure, the agrarian and urbanised lifestyle has entailed a (sometimes precariously) reduced range of nutrients, followed, later, by the 'diseases of affluence'.

What now confronts us, however, is quite different. There may well be a convergence of population pressures, land degradation, climate change, depletion of groundwater and genetic impoverishment of breeding-stocks to cause significant food shortages next century. Human population health may also be adversely affected by such things as the increased spread of

Table. *Possible adverse effects upon human health caused by global environmental changes.*

Environmental change	Manifestation	Type (direct, indirect) and timing[a] (early, late) of adverse health effect			
		Direct, early	Direct, late	Indirect, early	Indirect, late
Enhanced greenhouse effect	*Global warming and other climate change*	Heatwave-related illness and death Natural disasters: cyclones, floods, landslides, fires		Extension of vector-borne infections Food shortages due to impaired agriculture	Altered viability of (edible) fish in warmed oceans
	Sea-level rise	Increased risk of flash floods and surges	Inundation → social disorder, impaired sanitation, farmland loss	Consequences of damage to foreshore facilities, roads, etc.	Destruction of wetlands → decline in fish stocks
Stratospheric ozone depletion	*Increased UV-B flux at Earth's surface*	Sunburn, conjunctivitis Suppression of immune system → increased risk of infection	Skin cancer Ocular effects: cataracts, pterygium		Impaired growth of food crops and of marine microorganisms (base of aquatic food web)
Acid aerosols (from burning of sulphurous fossil fuels)	*Acid rain*	Effects on respiratory system (?)		Aquatic damage (reduced fish) Impaired growth of crops	Impaired forest growth → reduced ecosystem productivity

Land degradation: intensive agriculture, overgrazing	*Erosion, sterility, nutrient loss, salinity, desertification*	Decline in agricultural productivity	Rural sector depression → migration to fringes of cities (see bottom row)	Exposure to pesticides and fertilisers (may also cause algal blooms)	Consequences of silting up of dams and rivers
	Depletion of underground aquifers	Lack of water for drinking and hygiene	Decline in agricultural productivity		
Loss of biodiversity	*Destruction of habitat*	Deforestation → disruption of local culture	Loss of potentially edible species		Deforestation → greenhouse enhancement
	Loss of genetic diversity; weakening of ecosystems			Loss of medicinals, and other health-supporting materials	Greater vulnerability of crops and livestock. Reduced vitality of ecosystems
Other effects of overpopulation (particularly in poor countries)	*Proliferation of crowded urban slums (due to migration and high fertility)*	Infections Malnutrition Homelessness Antisocial behaviours	Social disorder Chronic toxic effects of environmental pollutants		Consequences of overload of local ecosystems

[a] The designations 'early' and 'late' are notional, and indicate *relative* timing. (Based on McMichael, 1993.[3])

vector-borne diseases like malaria (because of climatic change), increased deaths among the frail (due to more frequent heatwaves), increases in skin cancer, blindness and infectious diseases along with reduced terrestrial and aquatic food yields (all due to greater exposure to ultraviolet radiation), death, injury and disease due to the disruptions of altered climate and sea-level, and resurgent infectious diseases (particularly from urban crowding, poverty and squalor, and from changes in patterns of travel and in sexual and other behaviours). A summary of the *possible* impacts on human population health – encompassing direct and indirect effects, immediate and delayed effects, and local and global effects – is shown in the accompanying table.

So, why has the prospect of damage to this planet's natural systems not prompted more discussion about the impact on human health? One reason for this veil of inattention is that, relative to the more obvious en-vironmental threats to economic productivity, aesthetic amenity and our favourite species of foliage, fur, feathers and fins, this category of threat to population health lacks tangibility and immediacy. As a public health issue it lacks the familiar, measurable and toxicological qualities of other environmental health problems. We have therefore only dimly perceived its implications for population health.

This book seeks to widen our field of vision. By considering the topic within an ecological setting, I hope to clarify the significance of global environmental change for human biology and health. Admittedly, there is much that is uncertain or speculative about the causes and consequences of global environmental changes. But the fact that there is much that we don't, or can't, yet know is not a reason to dismiss the possibility, nor to defer prudent social response. Scientists and policy-makers are going to have to learn to live with more uncertainty than in the past.

This point about uncertainty needs emphasis. Since it is not yet possible to make specific predictions and to forecast actual outcomes, scientists must deal with ranges of plausible scenarios. They may use *predictive* models to estimate outcomes for a given scenario, but that is not a *prediction* of what will necessarily happen. The UN's World Commission on Environment and Development (WCED) recognised this problem in its much-quoted report of 1987, *Our Common Future*. Noting that 'major, unintended changes are occurring in the atmosphere, in soils, in waters, among plants and animals, and in relationships among these', the Commission said:

The rate of change is outstripping the ability of scientific disciplines and our capabilities to assess and advise. It is frustrating the attempts of political and economic institutions, which evolved in a different, more fragmented world, to adapt and cope.[4]

Even though many of these global environmental changes may entail effects upon health not previously encountered, we cannot defer social action until we 'know' those end-effects. By the time the health consequences of ecosystem disruption are clearly evident in human populations it may be too late to reverse or repair the damage. The dynamics of ecosystems do not obey the linear orderliness of physical systems; instead, they are influenced by feedback loops and critical loads. Limits, once exceeded, may (rapidly) lead to decline or collapse.

The content of this book

If I had to reduce my argument to a simple 1–2–3, it would be this. First, the *one* underlying problem is the entrenched inequality between rich and poor countries, which predominantly reflects recent imperial history, power relationships and the global dominance of Western industrial technology and economic values. Second, the *two* central manifestations of this inequality are: (1) rapid, poverty-related, population growth and land degradation in poor countries, and (2) excessive consumption of energy and materials, with high production of wastes, in rich countries. Third, the *three* possible (perhaps coexistent) adverse outcomes of those manifestations are: (1) exhausting various non-renewable materials, (2) toxic contamination of localised environments, and (3) impairment of the stability and productivity of the biosphere's natural systems. Of those three possible outcomes, the exhaustion of non-renewables seems unlikely (although topsoil, aquifers and stratospheric ozone are only slowly renewable!), localised chemical pollution will only become a systemic problem if it disrupts ecosystems . . . but the third would, by definition, be a threat to human (and other) life. Some of today's global environmental changes seem to portend that third outcome.

Overall, the book has three parts. It explores, first, the evolutionary and ecological backdrop to human biology and human population health. It then examines the possible impact of each of the main incipient global environmental changes upon population health. Finally, it considers the implications for human society – particularly the impediments to, and opportunities for, effective response.

In a little more detail: In chapters 1–4, I discuss the nature of the problems posed by ecological disruption. These chapters offer an evolutionary perspective on life and the emergence of *Homo sapiens*, and on the consequent needs of human biology. The nature of ecosystems is explored, before considering how the sustained good health of a species is an expression of ecological balance. There is also some discussion of the concept of 'population health' and the difficulties in estimating the adverse health effects of ecological disruption. The ecological perspective is, in my view, crucial to our thinking about the future health of human populations in today's overloaded world. The long-term survival of every species depends on its continued access to energy, nutrients, water and respirable gases. For as long as we humans survive, we too are destined by our biological origins to be participants in an interdependent biosphere, taking in water and oxygen, and drawing a sustainable share of the net primary production from photosynthesising plants (i.e. biochemical energy) and of the nutrients cycling through the system.

Chapters 5–10 are the 'meat' of the book. In them I review, first, the underlying problem of excessive population growth, and then the five main categories of ecological disruption: climate change, stratospheric ozone depletion, land degradation and impairment of food production, loss of biodiversity and the burgeoning growth of cities. For each, there is an exploration of the origins of the problem, the social and political context, and a detailed discussion of the possible impacts upon human population health.

Chapters 11–13 step back from the ringside view of population health problems and discuss the underlying problems within a broader social context. Those chapters do not attempt to deal comprehensively with those complex questions, but they do seek to extend the dialogue in light of the evidence of the threats to human health. In chapter 11, some of the main impediments are explored. What do we mean by 'growth' and 'development'? Is the real world an embarrassment to orthodox economic theorists? What are the main impediments to understanding this potentially dramatic Public Health Problem? What are the intellectual assumptions or political values that obstruct societal response? Chapter 12 examines the impediments to international cooperation and ecological sustainability posed by the entrenched inequalities of wealth, trade and influence. It also examines the public health hazard of warfare, an ever more likely response to increased tensions over dwindling environmental resources. Military technology casts longer shadows over populations and ecosystems, while nascent intercountry tensions over supplies of water, fisheries and other 'commons' cast intersecting shadows.

Chapter 13 considers the prospects for moving forwards. Our scientific traditions are mechanistic and reductionist, and they militate against the integrated approach that is now needed. We need to attune our science to the pervasive uncertainties and, therefore, to the precautionary principle. We have applied our science and technology primarily to mastering and exploiting the environment, not to sustaining it. Ecologically sustainable solutions will require over-riding various deeply-ingrained aspects of human evolutionary inheritance and several thousand years of culture and technology that have pointed us in non-sustainable directions.

Finally, a word on the referencing of 'factual' matter. There is much information being published about global environmental problems. I have quoted many figures and made many factual assertions. Not all are referenced, since to give references for everything would have brought its own form of overload. Wherever possible I have corroborated figures and factual statements from several sources. Alongside the specialist scientific literature that I have referenced, there are various omnibus sources published by the World Health Organization (WHO), the United Nations (UN) and its many agencies, the World Bank, the Worldwatch Institute and others. Those are listed at the end of this chapter. I think that what is most important here, however, is not the factual *detail* (different sources give slightly different figures for many of the items cited) but the overall import of the analysis and argument.

Why this blind spot?

It is worth exploring further why we might think it implausible that the health, if not survival, of our own species could be jeopardised by global environmental change. For a start, the anthropocentric nature of the world's major cultures – particularly the dominant, technologically-oriented Western culture – obscures awareness of the ecological dimensions of human existence. Even so, the ominous *direction* of many of the global environmental trends is no longer difficult to see. The current trends in world population growth, in the widening wealth gap between rich and poor nations, and in the associated global ecological disruptions to atmosphere, soil and groundwater are all well documented. Are the spread of food shortages in sub-Saharan Africa and the recent downturn in the world's per-person cereal grain production early signals? (After all, no bells will be rung to signal the 'start' of abnormal trends; their existence is a matter of judgement, often in hindsight!)

It seems clear that we do not yet much understand the intimate relationship between ecological systems and the health of populations. Our day-to-day personal health appears to be determined by immediate circumstances: family history, behaviours, type of job, locally-circulating viruses, luck on the roads and so on. Indeed, the dominant view of health and medical care in developed societies focuses upon the individual; we do not usually apply a population perspective. The essential ecological underpinnings of population health are therefore well beyond our field of vision; the potential adverse health effects of global environmental change are usually distant in time and place. Biological evolution has 'programmed' us, like all other species, to react most decisively to current problems rather than to future possibilities; a 'biological predilection for short-term gain' has been built into human biology by the process of natural selection.[5]

It is the combination of this instinct for short-term individual gain (which is widespread in nature) with human brain-power that, through its consequences, jeopardises the longer-term needs of the human species. So, to overcome today's global environmental problems we will need both understanding and moral fortitude to compensate for this genetically-endowed 'predilection for short-term gain'. I hope that this book will help to increase that understanding. But do we have the collective will to take effective, equitable corrective action? Thomas Berry, a theologian and cultural historian, observes that there has been 'no sustained religious protest or moral judgement concerned with the industrial assault on the Earth, the degradation of its life systems, or the threatened extinction of its most elaborate modes of life expression.'[5] He is right to note the absence of institutional moral disapproval, but, if people do not understand the ecological predicament now confronting us, then it is ignorance rather than moral torpor that is the problem. (One hopes, somewhat against the evidence, that much of the prevarication and pursuit of national self-interest at the 1992 UN Earth Summit arose from a misunderstanding of the longer-term, ecological, implications of today's global environmental problems.)

Meanwhile, there are some signs that our understanding is increasing. In 1990, the UN's Intergovernmental Panel on Climate Change (IPCC) predicted that greenhouse-induced global warming would have various adverse consequences upon human health, human settlement and human social organisation. The Panel's report stated:

In coastal lowlands such as in Bangladesh, China and Egypt, as well as in small island nations, inundation due to sea-level rise and storm surges could lead to significant movements of people. Major health impacts are possible, especially in large urban areas, owing to changes in the availability of water and food and increased health problems due to heat stress and spreading of infections. Changes in precipitation and temperature could radically alter the patterns of vector-borne and viral diseases by shifting them to higher latitudes, thus putting large populations at risk. As similar events have in the past, these changes could initiate large migrations of people, leading over a number of years to severe disruptions of settlement patterns and social instability in some areas ... Global warming and increased ultraviolet radiation resulting from depletion of stratospheric ozone may produce adverse impacts on air quality such as increases in ground-level ozone in some polluted urban areas.[6]

These concerns were echoed in the subsequent work of WHO's Commission on Health and Environment.[7] Will this new awareness grow fast enough to avert serious ecological problems? The revolutionary ideas of civic hygiene and infectious disease control took over a half-century to enter public consciousness, and to cause significant social change and public health gain. In a world of accelerating environmental impact we may have precious little time to embrace the idea of a sustainable biosphere as the necessary, non-negotiable basis of long-term human population health.

Reevaluating priorities

What is the cultural and social significance of today's global environmental problems? Far from having reached what, in Fukuyama's ecologically myopic assessment, is 'the end of history' (marked by the triumph of liberal democracy),[8] we may well be entering a transitional stage in human history, moving 'between two worlds'.[9] Having created, through modern technology, a seeming capacity to override local ecological constraints, we must now contemplate a social reformation that reestablishes an ecologically sustainable way of life. This will require us to look to distant horizons for solutions that, in both time and content, extend well beyond the puny reach of conventional political decision-making.

Currently, we exalt the goal of 'development', framed essentially in economic terms and dominated by the pursuit of an expanding gross national product (GNP). Within this framework our national accounting systems ignore ('externalise') the costs of environmental damage and ecological disruption. As we will see in chapter 11, this is a form of disguised deficit budgeting that eventually must lead to ecological crisis. The prevailing values of economic growth and material consumption, in industrialised countries of all ideological persuasions, have led us into much of our current environmental predicament. Although we may protest that those values and behaviours are the same in kind as those of our ancient human predecessors, it is our particular generation, with its unprecedented population size and energy-intensive technologies, that is the first to overload the *global* environment.

Orthodox economists anticipate the future by linear extrapolation of the past. They presume that the system is 'open'; besides, limits can always be pushed back with new technology (innovation, substitution and discovery). Theirs is the dominant paradigm in political and financial circles. Ecologists, aware of the complex, changeable and non-linear aspects of natural systems, perceive 'development' as being subject to the confines of the biosphere. Where the economist views ecological concerns as a subset of economics, to be dealt with market-place corrections, the ecologist views the economy as a subset of the global ecosystem. (Meanwhile, an ecologically-attuned minority of economists argues that economic practice need not be inimical to the environment, and that good ecological practice would be good long-term economics.)

Some of today's environment-damaging trends will take decades to slow and reverse. Decisions to desist from use of ozone-damaging chemicals or to eliminate the emissions that cause acid rain are *relatively* easy. In contrast, strategies for capping the growth of world population or for curbing the release of greenhouse gases are unprecedentedly difficult; and they strike at the value base of our society. Lasting solutions will therefore not come from the fine-tuning of markets nor from ingenious technical 'fixes'. The problems go deeper than that. The realisation that much of the environmental health impact will, eventually, bear on *everyone* – not just on the world's poor – should stiffen our resolve to act. The developed world has more resources with which to adapt to climate change, and more power to ensure access to food, water and fuel, should their supplies dwindle. However, in the longer term, we share a common global future. The rich countries of the northern hemisphere became noticeably more concerned about the effects of ozone layer depletion in early 1990s as it transpired that it was no longer a problem confined to the southern hemisphere! (Similarly, legislative reforms accelerated in nineteenth century England when it became clear that infectious disease epidemics, such as cholera and typhoid, were no respecters of persons, and that the smell of the Thames was pervading the Houses of Parliament!)

Returning to the general theme of this book, empiricists might object that past predictions of environmental doom have generally not eventuated. For example, predictions made in the 1960s of widespread famines in the 1980s did not materialise. However, it is important to stress that those earlier predictions were based on anticipations of resource limitations (that is, supply-side constraints) whereas today's problems arise from exceeding the kinetic capacity of ecosystems to regenerate, transport, process and degrade. Globally, there are too many people doing too much too fast. That poses a new type of environmental problem.

The human species has the same ultimate dependence upon life-supporting ecosystems as do other biological species. However, we are a distinctive species with the latent capacity to understand, and therefore to aright, our own self-made macro-environmental problems. René Dubos, the microbiologist-cum-human-ecologist said: 'wherever human beings are concerned, trend is not destiny'.[10] *So far*, he seems to have been right. We have the wit, but do we have the wisdom? By a mix of luck and political management we have averted nuclear war between the two superpowers during the latter half of this century. We could therefore be tempted to hope somehow to scramble through these global ecological problems. But scrambling will not suffice – and, anyway, we are leaving our run a bit late. The global environmental changes that we are setting in train will throw increasingly long shadows over Earth's ecosystems and, therefore, over human population health. They present a formidable challenge to our collective rationality, morality and will.

A note on terminology

There are over two hundred countries in the world today. There is no single satisfactory way of categorising these countries, in order to describe and compare. The main options are: developed versus developing; First World versus Third World; industrialised versus non-industrialised; North versus South; and rich versus poor. All such broad categorisations obscure the finer-grained differences. Classification in terms of economic 'development' begs a basic question about the desirable form and direction of societal development. That aside, variants such as 'less developed' and 'more developed', or 'overdeveloped' and 'undeveloped', shed little extra light – although they may protect national sensibilities. Meanwhile, the bustling small-country economies of East Asia (the 'tigers') are often referred to as 'newly industrialising'. Notions of the First, Second and Third Worlds emerged during the 1950s, as international communism expanded and as European powers shed or lost colonies.[11] In the first flush of independence, many post-colonial countries proudly identified with the aspirations of strong nationalist leaders like Indonesia's President Soekarno, who foresaw solidarity and economic progress for the newly liberated Third World. Much of that dream has subsequently tarnished.

The simplest classification is to describe countries as either rich or poor. This highlights a fundamental difference between countries in today's world, and I have used this classification frequently since it is relevant to much of my argument. The closely related North–South classification has the merit of simple imagery; most of the poor countries are south of latitude 35 °N. (One could excuse an Australian's discomfort with that classification.) From a political economy perspective, the terms 'core' and 'non-core' may best describe the positions of these countries within the structure of the capitalist world-economy. In the upshot, I have used various of these terminologies, sometimes reflecting the source of information. To a large extent, they are interchangeable, and all are in wide use elsewhere.

The phrase 'human species' also needs clarification. Primates, as an 'order' of mammals, arose about 60 million years ago (see also Fig. 1.1). Those early prosimians branched into various 'families': monkeys, apes and hominids. The hominid and ape families diverged a brief 7 million years ago. The ape family is today represented by three branches ('genuses'): orangutans, gorillas and chimpanzees. The *Hominidae* family differentiated around 3–4 million years ago into two main branches – one culminating in *Australopithecus robustus* (which died out 2 million years ago) and the other, via another Australopithecine line, becoming the *Homo* genus. *Homo sapiens* is the most recent species within that genus, arising several hundred thousand years ago. The 'modern' (Cromagnon) strain of *Homo sapiens* eventually predominated over the Neanderthal strain a short 35,000 years ago. Recent molecular studies show that we humans share 98.4% of our genes with chimpanzees and around 97% with gorillas and orangutans, which raises an interesting question about the logic of the conventional classification of primates.[12] I have used the phrase 'human species' often, and it usually refers to contemporary humans. From the context, it should be clear when the phrase refers to the full complement of experience as *Homo sapiens*, or when it refers to the even longer span of experience across the full succession of human species within the *Homo* genus.

Finally, the various words and concepts relating to the biosphere and its ecosystems are defined and discussed later in the text (see especially chapter 2).

Key sources of factual material

Brown LR, Flavin C, Kane H (eds). *Vital Signs. The Trends That Are Shaping Our Future*. New York: Norton, 1992.

IPCC (Intergovernmental Panel on Climate Change). *Policymakers Summary of the Scientific Assessment of Climate Change. Report Prepared for the IPCC by Working Group I.* World Meteorological Organization and UN Environment Programme: Geneva and Nairobi, June 1990.

IPCC (Intergovernmental Panel on Climate Change). *Policymakers Summary of the Potential Impacts of Climate Change. Report Prepared for the IPCC by Working Group II.* World Meteorological Organization and UN Environment Programme: Geneva and Nairobi, June 1990.

IPCC (Intergovernmental Panel on Climate Change). *Climate Change. The IPCC Response Strategies.* World Meteorological Organization and UN Environment Programme: Geneva and Nairobi, 1990.

IPCC (Intergovernmental Panel on Climate Change). *Supplementary Report.* World Meteorological Organization and UN Environment Programme: Geneva and Nairobi, 1992.

UN Department of International Economic and Social Affairs. *Statistical Yearbook.* New York: UN, various years.

UN Department of International Economic and Social Affairs. *Long-Range World Population Projections.* New York: UN, 1992.

UNDP (United Nations Development Programme) *Human Development Report 1991.* Oxford: Oxford University Press, 1991.

UNEP (United Nations Environment Programme). *Environmental Data Report* (Third Edition). Oxford: Basil Blackwell, 1991.

WHO Commission on Health and Environment. *Our Planet, Our Health.* Geneva: WHO, 1992.

World Bank. *World Development Report 1992. Development and the Environment.* Oxford: Oxford University Press, 1992.

World Resources Institute. *World Resources 1990–91.* Oxford: Oxford University Press, 1990.

World Resources Institute. *World Resources 1992–93.* Oxford: Oxford University Press, 1992.

References

1. Vitousek PM, Ehrlich PR, Ehrlich AH, Matson PA. Human appropriation of the products of photosynthesis. *Bioscience* 1986; **36**: 368–73.
2. Powles J. Changes in disease patterns and related social trends. *Social Science and Medicine* 1992; **35**: 377–87.
3. McMichael AJ. Global environmental change and human population health: A conceptual and scientific challenge for epidemiology. *International Journal of Epidemiology* 1993; **22**: 1–8.
4. World Commission on Environment and Development. *Our Common Future.* Oxford: Oxford University Press, 1987.

5. Potter, van Rensselaer. Getting to the year 3000: Can global bioethics overcome evolution's fatal flaw? *Perspectives in Biology and Medicine* 1990; **34**: 89–98.

6. Houghton JT, Jenkins GJ, Ephraums JJ. *Climate Change. The IPCC Assessment.* Cambridge: Cambridge University Press, 1990.

7. WHO Commission on Health and Environment. *Our Planet, Our Health.* Geneva: WHO, 1992.

8. Fukuyama F. *The End of History and The Last Man.* New York: Free Press, 1992.

9. Caldwell LK. *Between Two Worlds. Science, the Environmental Movement and Policy Choice.* Cambridge: Cambridge University Press, 1990.

10. The ideas of the human biologist René Dubos are discussed further in chapter 3.

11. The term Fourth World is sometimes used to describe stateless peoples like the Palestinians and Kurds – and, by extension, relegated minorities like the Australian Aborigines and indigenous forest-dwellers of Brazil and Southeast Asia. It is also occasionally used to describe the very poor nations, such as Bangladesh.

12. Jared Diamond argues that, since the hominids are genetically closer to the chimpanzees than are the other apes, there is now a strong case for reclassifying humans as the 'third chimpanzee', sharing a genus with our closest relatives, the common chimp and the pygmy chimp. Orangutans and gorillas, who diverged earlier from the ape line, 10–15 million years ago, would then be in another genus. See: Diamond J. *The Rise and Fall of the Third Chimpanzee.* London: Radius, 1991.

1

First things

1.1 Introduction

In recent centuries, we humans have swarmed over most of the world's habitable and productive land, and with our rapidly evolving technologies we have trodden increasingly heavily. The aggregate impact of a burgeoning population, using more energy and materials and overworking the land, has brought us close to the limits of Earth's carrying capacity. This is a 'first' in human history.

Certain aspects of this overload problem have been foreseen often enough. A well-known bleak prediction for the human species was that of Thomas Malthus, an English clergyman well educated in the social sciences. He wrote, in 1798, in the oblique idiom of the time: 'there is no bound to the prolific nature in plants and animals but what is made by their crowding and interfering with each other's means of subsistence.'[1] In other words, it is in the nature of things that expanding populations will outstrip their food supply: population numbers tend to increase exponentially, while food supplies increase only arithmetically. (See also section 5.2.3.) Malthus concluded that society's poor and weak were likely to miss out in the struggle for a place at the dining table. His prediction that human numbers would tend persistently to outstrip food supplies, causing cyclical disasters (especially famines), has reverberated controversially through the subsequent two centuries. His ideas about the competitive struggle to subsist provided Charles Darwin, forty years later, with the critical – indeed (against the prevailing creationist orthodoxy) subversive – clue to explain the occurrence of biological *mutability* within and between species.

Today's main problem is a problem of a planetary 'metabolism' that is being overloaded and therefore disrupted. Putting it overly simply, some of the environmental problems are unforeseen consequences of industrial

17

technology, such as changes in atmospheric composition due to emissions of carbon dioxide and chlorofluorocarbons; while some result from demographic pressures and environmental mismanagement, such as deforestation, land degradation, depletion of groundwater, and the consequences of hugely over-crowded cities. The two major underlying influences are, in the Third World, an extraordinary population growth over the past half-century and, in rich countries, an ecologically distorted lifestyle that depends on very high consumption of energy and which creates substantial wastes. We will examine the connections between these environmental stresses in later chapters. However, to understand how *Homo sapiens* has reached this unfamiliar and dramatic cross-roads, we must first look back to our evolutionary origins and outwards to see ourselves in cosmological context.

Caution is needed here! Knowledge and theories in these scientific arenas are changing at an increasingly hectic pace, as astronomers, subatomic physicists, archaeologists and molecular biologists peel away successive layers of mystery. Currently, ideas about the timescale and sequencing of hominid evolution, the sources of life on Earth, the origins of the solar system, and the creation and composition of the universe itself seem to change about as often as one picks up a good science journal. Likewise, variants of modern Darwinian evolutionary theory jostle with one another – while, on the periphery, the Gaia hypothesis postulates Earth as a self-regulating 'superorganism' (see chapter 2).

Some readers may already have detailed knowledge about aspects of this chapter. However, it is not the details that matter here. What is important is that if the needs of human biology for the attainment of optimal biological functioning – i.e. good health – are to be understood *within an ecological context*, then we must have insight into the origins and nature of the interdependent relationships between species of organisms and their environment. Through understanding the general process of biological evolution, we can interpret human biology and its distinctive attributes and requirements within this ecological framework. Humans have evolved as versatile generalists, able to adapt to different environments. However, our particular specialisation is an extraordinarily large and complex brain that has allowed not just simple tool-using but an overlay of techno-logically-sophisticated culture (the ultimate tool) – and that culture, progressing ratchet-like, has now accelerated past the point of easy control.

Our scientific understanding of the origins, dynamic functioning and limitations of ecosystems is both recent and incomplete. Ecology and systems-based earth sciences are still young sciences. Nevertheless, we have increasing insight into the ultimately limited resilience of natural systems – i.e. their 'carrying capacity' (see glossary). We can also appreciate the provisional status of our species' existence in light of the following two precepts: that life on Earth, despite its own resilience, has been decimated on several occasions; and that extinction is the *overwhelmingly normal* outcome for each species.

1.2 Biological Origins

1.2.1 Why on Earth?

The odds against such a sequence of encounters leading to the first living entity are enormous. On the other hand, the number of random encounters between the component molecules of the Earth's primaeval substance must have been incalculable. Life was thus an almost utterly improbable event with almost infinite opportunities of happening. So it did.

James Lovelock[2]

We earthlings live in a rather average street in the backblocks of cosmic suburbia. Our solar system is now about 4.6 billion years old, and will eventually self-destruct. The Sun, with an estimated lifespan of about 10 billion years, is now in mellow middle-life. It is gradually becoming hotter and, in older age, as its nuclear-fusion fuel becomes heavier, it will flare up and char the inner planets to a crisp. Then, as a fading glowing ember, it will dwindle to a rather dull, cold, 'brown dwarf', too small to make the spectacular explosive farewell of a supernova. All life on Earth will die out before then – although some predict that future humans might gain a reprieve by migrating to another life-supporting planet (presumably taking ecosystem starter-kits with them).[3]

Biological life, which is characterised by intensive capture of energy and nutrients, growth, reproduction, information storage and (often) movement, seems to be a temporary and incidental phenomenon within the universe. Life on Earth has existed for about the last quarter of the universe's approximately 15 billion years. Biological life as we know it, or can imagine it, cannot materially influence the ultimate fate of this vast store of interconvertible energy and matter. Rather, life appears to be an optional extra, something that has arisen opportunistically within particular physico-chemical circumstances. We living species seem merely to be going along for part of the ride, so long as those circumstances stay within life-supporting bounds.

There are several arguments why life on Earth may not have been a 'sure thing'. First, concurrence of the circumstances conducive to 'life' anywhere in the universe is probably uncommon – unless one imagines fantastic forms of life able to live, for example, in liquid methane, or which are structurally based on other non-carbon elements. How have those circumstances arisen on Earth? Our planet is a tiny speck within a vast ether of mysterious 'cold dark matter' that makes up most of the mass of the universe. The universe's primordial event, the Big Bang, unleashed a rapidly-expanding fireball of superheated subatomic particles which, after 'cooling', formed an incandescent mist of low-weight elements: hydrogen, helium and lithium. Localised coalescence of these gases occurred, some of which, as extraordinarily remote quasars near the receding edge of the universe, we can still 'see' today in grand cosmic action-replay.[4] Several billion years later, trillions of galaxies of stars formed within those brilliant gaseous coalitions, and around some of those stars planets formed by gravitational clumping of gas and cosmic dust. We know that at least one such planet had characteristics within which organic 'life' could evolve. This was Earth, the third of nine (or more?) planets orbiting around one modest star in an otherwise unremarkable galaxy, the Milky Way.

The concurrence of basic life-supporting conditions – moderate and stable temperature, an adequate atmosphere, and liquid water – is unlikely enough. Even so, there are other auspicious circumstances on Earth. Because the local solar system formed long after the Big Bang, our sun was able to gather in, by gravitation, interstellar elemental debris that had been progressively enriched by nuclear fusion reactions within long-deceased stars. Hence, the solar system's planets, particularly the inner, solid, planets Mercury, Venus, Earth and Mars, contain various heavier elements. Many of these elements, such as iron, zinc and copper, are important in higher-order molecular biological events. Iron, for example, is the centre-piece of the vital oxygen-carrying haemoglobin molecule. Further, chemically-reactive elements like iron and sulphur in Earth's ancient watery environment enormously influenced the chronology of changes in atmospheric composition, and therefore the paths of biological evolution.

A second reason for wonderment over the existence of life is that self-replicating organic molecules – the foundation stones of (our type of) life – somehow arose within a lifeless cosmic laboratory. The origins of life remain unclear.[5,6] Since there is no relevant fossil evidence from those earliest times, we must settle for theory and simulation. The basic question is how did elemental atoms and simple molecules form complex molecules capable of replication and, later, cooperative teamwork?

Earth's prebiotic chemical mix, essentially devoid of free oxygen, is thought to have been 'reducing' and therefore rich in hydrogen and simple hydrogen-laden molecules – particularly water (H_2O), ammonia (NH_3) and some methane (CH_4) – along with garnishings of other elements such as iron.[5] Later, with volcanic outgassing and a loss of low-weight hydrogen from the upper atmosphere, carbon monoxide (CO) and then carbon dioxide (CO_2) were added to the atmosphere. Meanwhile, there was a turbulent welter of extraneous energy entering the primordial soup: cosmic rays, lightning flashes, unfiltered ultraviolet radiation, and violent geo-thermal events. Assisted by this energy, atoms and molecules were able to consummate their chemical affinities for one another and form bigger molecules, such as hydrogen cyanide (HCN) and formaldehyde (H_2CO). These, in turn, provided a starting point for the formation of more complex organic molecules, comprising a linked core of carbon atoms to which hydrogen, oxygen and, often, nitrogen atoms are attached. From this profusion of chemical reactions arose molecules that subsequently became the building blocks of 'life': amino acids, purines and simple sugars such as ribose. Indeed, in laboratory simulations of prebiotic Earth's chemical chaos, beginning with the famous Urey-and-Miller experiment of 1953, scientists have succeeded in creating such building-block molecules.[6] These, linked together in macromolecular polymers, were presumably the forerunners of proteins and the gene-bearing nucleic acids. There may have been further input to this prebiotic chemistry from the hyper-bombardment of Earth by remnants of unregimented inter-planetary debris – of the kind that pock-marked the face of the Moon – bringing a rich assortment of preformed complex molecules that became part of the assembly kit of 'life'.[7]

At some time during those first billion years the jackpot was achieved when, by chance, organic molecules formed that were able to replicate themselves, using local chemical energy sources and the supply of molecular building-blocks. This 'self-replication' model has shaped much recent thinking about the origins and evolution of life.[8,5] Molecules that reproduce themselves by template-based organisation of smaller molecules are the first step towards a self-sustaining, repetitive, chemical process ('biochemistry') that uses local environmental sources of energy and building materials. That's life. Further – and this is a point of extreme importance – among the accidentally varying forms of self-replicating molecules the most efficient replicators soon outnumber, and eventually displace, the rest.

Somehow, through the play of chance, some of these primitive self-replicating nucleic acid molecules then teamed up with other protein-like molecules. That is, the replication of nucleic acids – perhaps RNA (ribonucleic acid) specifically – was coupled with a capacity to shape complementary protein molecules as their metabolic accomplices.[9] Whenever such a combination conferred functional benefit, by the differentiation of molecular labour, those particular molecular assemblies and their replicated offspring would have prevailed in the basic competition for energy and materials. As further advances in molecular function and self-replication occurred, more complex life-forms emerged.

A third reason for wonderment at the existence of life is that, in a violent and ever-changing universe, the continuation of life on Earth has been a ceaseless struggle. Many mass extinctions have occurred, and several have left few survivors. However, life has so far hung on, occasionally against steep odds. When pressed, the intrinsic genetic diversity of life-forms has lowered the odds enough to ensure the survival of at least some life-forms and, therefore, of their radiative speciation following each such disaster. The longer-term unfolding of the evolutionary story, and the emergence of the human species, is taken up in section 1.3. The following section examines the essential features of the biological process underlying that story. The account is simplified – but not, I hope, in any misleading way.

1.2.2 Biological evolution

The diversity of 'life' on Earth is the result of biological evolution, responding to the challenges and vicissitudes of environmental variation. Charles Darwin, who in 1859 published the first major, scientifically-grounded, theory of the origins of biological diversity, assumed a generally stable environment in which gradual adaptation improved the fit between a species and its local environment. He envisaged 'the accumulation of slight modifications of structure or instinct, each profitable to the individual under its conditions of life'.[10] Recently, we have gained new insights into the complexities and discontinuities of the relationship between genes and an erratically changing environment. In contrast to Darwin's gradualism, Gould has proposed that 'punctuated equilibrium' is the evolutionary norm, with periods of relative evolutionary quiescence rudely interrupted by occasional adaptational turbulence in response to major environmental change.[11] Details aside, Darwin's theory remains the bedrock for any understanding of biological evolution. Although his notion of random genetic variation in populations as the menu from which natural selection chooses its sustenance is generally known, some widespread misconceptions about biological evolution remain. These include assumptions about 'progress', 'purpose' and the 'culmination' of evolution in the human species. Since biological evolution both shapes and is shaped by the dynamic processes of ecological systems, I will briefly review the essentials – even though this will be superfluous for some readers. This should also help us understand the serendipity, conditionality and changeability of the destiny of *Homo sapiens*.

Evolutionary change is based on nature's exceptions, and not on those individual organisms that follow the high-fidelity rules of self-replication. How is this high fidelity achieved? How did early life-forms code, package and transmit the genetic information – from parent to identical offspring? From those primitive self-replicating molecular assemblies through to microbes and mammals, the genetic information has been carried in long-chained nucleic acids. These nucleic acids, called DNA (deoxyribonucleic acid) and RNA (ribonucleic acid), are polymers of the four different nucleotide molecules – guanine, adenine, cytosine and thymine (or uracil in RNA) – which originally formed from purines and pyrimidines in the prebiotic soup. Each sequence of three nucleotides within the chain is called a 'codon' because it codes for a specific amino acid molecule, which it attracts alongside by a type of molecular docking. Via this template-based action, a particular sequence of codons directs the assembly of a particular protein molecule, comprising a unique sequence of amino acids. These proteins then act as structural materials or as enzymes that facilitate biochemical reactions within the cell. We call the blueprint-like sequence of codons a 'gene' – the unit of hereditary information. It is thus the genes that control the structure and function of the living cell. In every known life-form on Earth the genetic information is carried by these 'smart' nucleic acid molecules.

Now, the fidelity of gene replication from parent to next-generation cell is remarkably high – but not perfect. Damage can occur to the genes, either during normal cell activities or during replication. By bureaucratic analogy, the instruction sheet can be damaged while on the desktop or while being photocopied. If the former, the damage may cause one or more nucleotides in the genetic sequence to be 'misread' while assembling a protein molecule, thus causing an error in its composition – and perhaps, therefore, in its function. If the damaged gene is not detected and repaired before the cell next divides, this molecular damage is passed on as legitimate inheritance to the next generation of cells, and it thus becomes an unrecognisable heritable mutation in which the coded message has been permanently altered.[12] In single-celled organisms such mutations are passed invariably to successive generations. In multicelled organisms this only happens if the mutation occurs in the germ cells – i.e. in either the male sperm-producing cells or the female ova. Mutations occurring in non-germ cells may cause abnormal function (including cancer) or cell death, but they will not be inherited by the organism's offspring.

Inherited mutations, if not seriously detrimental to the offspring, increase the amount of genetic variation within oncoming generations. Genetic variability thus keeps Nature's options open; chance may favour the accidentally prepared gene. Some mutations may be immediately beneficial to the offspring's prospects for successful breeding; others may become beneficial in future. Whenever a mutation confers some competitive edge in reproduction it will become more prevalent within the population gene pool. This, unknown to Charles Darwin, is the molecular mechanism that drives the ongoing trial-and-error 'experiment' of Nature – the process of 'biological evolution' that leads to the formation and modification of distinct species. With ruthless impartiality, natural selection sifts through the chance offerings, discarding most, favouring a few.

The occurrence of genetic mutations is essentially random (although there is speculation that the evolutionary process *itself* may have been subject to selection pressures in favour of evolution-facilitating molecular mechanisms[13]). Many mutations are caused by external factors that impinge on the cell, such as ionising radiation or various DNA-binding chemicals. Alternatively, spontaneous errors can occur during the intricate process of gene replication. Indeed, it is quite remarkable that the whole process does not end in the sort of tangle that besets the fishing lines of inexpert anglers. In the humans species, the long molecular strings of genes are arranged as forty-six coiled chromosomes packed into a microscopic cell nucleus; yet, extraordinarily, the total DNA of just one cell, if fully unravelled and laid end-to-end, would extend over 1–2 metres!

Changes in Earth's environment, which are usually gradual but sometimes cataclysmic, have thus evoked a continuing biological evolution of species. But it has not been a one-way street. Life itself has hugely changed the global environment. For example, 2–3 billion years ago blue-green cyanobacteria (algae) began to release oxygen, a waste product of their photosynthesis, into their local aquatic environment. In time it spilt over into the atmosphere. The advent of oxygen into a world that had previously been a predominantly chemically 'reducing' world had enormous consequences for the course of subsequent biological evolution; not only did oxidative (aerobic) metabolism become possible, but the oxidising of rocks released minerals into the environment and, later, atmospheric ozone began to form. Further, because of the photosynthetic conversion of atmospheric carbon dioxide to organic carbon, much of this carbon in decaying plant material became locked away in marine-floor sedimentary deposits of limestone. This would have contributed to lowering the atmosphere's natural greenhouse effect and may thus have helped to cool the young planet, from temperatures ten or more degrees higher than today's.[14]

In this unpredictable and sometimes violent world what are the prospects for species survival? If the environment changes gradually, then survival depends upon fortuitous genetic evolution. If it changes rapidly – as has happened many times – then the pace of biological evolution is overwhelmed by the speed of change and many species perish. An estimated 99% of all species that have ever evolved on Earth are now extinct. The spectacular discoveries of countless, fantastic, discontinued lines of evolution among the Cambrian fossils in the famed Burgess Shale, in Canada, bear long-silent testimony to this, the usual fate of species.[15] Generally, the 'loser' species were probably the finely-adapted specialised majority who found themselves up ecological blind-alleys when the environment changed.

Although the several primary branches of the tree of biological evolution have never been severed, various large branches have been lopped off, along with countless smaller branches and twigs. While many of those extinctions have been caused by slow changes in the world's environment, others have been caused by cataclysmic disruptions. Biological evolution, for ever testing the environmental boundaries of survival, is actually more like a seasonal and outwardly-growing bush, subject to violent occasional pruning, than a perennial and upwardly-growing tree. One vigorous but young shoot on a quite recent branch is the hominid (i.e. the erect-walking 'humanoid') family.

To appreciate better the dependency of *Homo sapiens* upon the world's ecosystems, we should now explore the evolutionary saga further. Our current concerns over the world's deteriorating environment entail fretting over one or two contemporary frames in a long-running movie film. However, if we don't know the background plot that has unfolded over thousands of earlier frames in the film, and if we don't have insight into the formative aspects of the character and role of the players, then we will be less able to predict how the film might progress – and end.

1.3 Life on a restless planet

1.3.1 From algae to apes

For almost three billion years, beginning around 3.6 billion years ago, life was confined to the oceans, waterways and sludge. Early single-celled organisms extracted chemical energy from soluble compounds, such as hydrogen sulphide, in the watery environment. With time, those early organisms 'learnt' to enlist the help of solar energy to break down such compounds. In this oxygen-free environment life depended on anaerobic metabolism and, like today's swamp-dwelling and sediment-dwelling anaerobes, it produced methane (CH_4) as its major waste. Small and soft, this early life left few fossil remains. Sometime between 2 and 3 billion years ago (see Fig. 1.1) 'modern' photosynthesis evolved, enabling cells to use the Sun's energy to break the relatively strong bonds between oxygen and carbon (CO_2) or hydrogen (H_2O) in order to build energy-rich organic molecules. The consequent release of thermodynamically reactive oxygen, as a waste gas accumulating in the environment, enabled the evolution of energy-efficient *aerobic* metabolism.

Then, a billion or so years ago, two other major developments occurred. Cell nuclei appeared, enclosing the genetic material in specialised metabolic command centres. These more highly organised 'eukaryotic' cells enabled the development of complex multicellular organisms. Sexual reproduction also appeared. This meant that single-parent cells no longer had to reproduce by the solitary, ancient act of mitosis. This 'primitive' process entails a replication of the full set of genetic material (packaged as chromosomes) and then, through cell division, its separation into two identical halves. Mitosis thus maintains a dynastic succession of genetically identical offspring cells from a common ancestor. Sexual reproduction, however, entails not mitosis, but meiosis; this entails scrambling of ancestral genes in each new generation. This is possible because the cells of sexually reproducing species have two versions (alleles) of every gene, packaged on paired chromosomes (humans have 23 pairs of chromosomes). In meiosis, each of two would-be consenting adult cells – from two individual organisms, A and B – subdivides its genetic material in a two-step process. The first step mixes up the cell's original 'maternal' and 'paternal' genes, and the second step allocates just one 'recombinant' set of unpaired genes to each of the resultant 'haploid' cells. Then two haploid cells – one from each of the A and B lines – unite to produce a unique diploid (paired) recombination of their haploid sets of genes. The triumph of sexual reproduction, then, was that this two-step genetic recombination

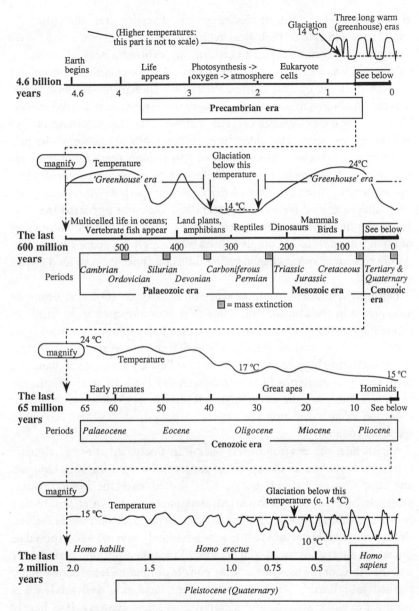

Figure 1.1. Geological eras, evolution of life, and fluctuations in global temperature. The more recent time periods are shown at successive magnifications. (Multiple sources)

greatly increased the intra-species genetic diversity and, therefore, the complexity and rate of biological evolution.

In another evolutionary corner-turn, life evolved a way of acquiring energy that was more efficient than extracting it from inanimate molecules or from decaying organic matter. Predatory 'animal' life-forms emerged that fed on photosynthesising plants, with their stored Sun-derived energy. That is, 'consumer' species evolved that were capable of dining on live 'producer' species.[16] The oldest animal-like fossils are those of the pre-Cambrian soft-bodied chitinozoans of 750 million years ago. They are presumed to have ingested phytoplankton, a primitive microscopic plant. Subsequently, after the onset of the Cambrian period 570 million years ago, aquatic animal forms with mineralised skeletons appeared (see Fig. 1.1). Later, the development of teeth and jaws enabled the herbivorous diet of marine animals to be complemented increasingly by a carnivorous diet. As big consumer species began eating little consumer species, food chains emerged.

During the preceding billion years or so, oxygen (O_2) had begun to accumulate in the atmosphere. Some of it was converted to a mantle of ozone (O_3) by the sun's ultraviolet rays. Gradually, plant life spread from water onto land, and eventually, about 400 million years ago, sufficient ultraviolet-absorbing ozone had formed to allow a wide range of plant and plant-eating amphibious life to survive on dry land. ('Come on out, the weather's fine', say the cartoonists.) Radiative explosions of terrestrial evolution occurred during the Devonian period, as this new ecological space was filled.

Meanwhile, the environmental backdrop fluctuated through climatic extremes. However, as the ancestral supercontinents (Pangaea, Laurasia and then Gondwanaland) periodically drifted over the polar regions, protracted global ice ages occurred, particularly in the later Ordovician period and during the Carboniferous-Permian period. Continental shelves, already battered by tectonic drift, were alternately covered and exposed as glacial ice sheets advanced and receded. Volcanic eruptions periodically rent the fabric of the biosphere. Stray comets or huge meteors occasionally crashed into Earth's surface, sending dust, acid rain and tidal waves around the globe and causing precipitous climatic changes. It is hardly surprising, then, that the fossil record shows that five mass extinctions, each entailing the loss of over three-quarters of all species, have occurred in the last 500 million years (see Fig. 1.1), along with many smaller ones.[17]

When the Ordovician period ended about 440 million years ago, two-thirds of all marine species were destroyed. Later, towards the end of the

Permian, 250 million years ago, an estimated 96% of all land and sea species were wiped out in the greatest mass extinction to have yet occurred. The dramatic fall of sea level during the Permian glaciation enabled massive oxidation of exposed sedimentary deposits, thus removing atmospheric oxygen and raising carbon dioxide levels. This probably contributed to the prolonged increase in global temperatures during the Mesozoic Era (Fig. 1.1). More recently, at the end of the Cretaceous period, 65 million years ago, three-quarters of species disappeared – including all of the dinosaurs and most of the early mammals. This mass extinction may have occurred in a geological 'moment' as brief as several thousand years, in which case a catastrophic explanation is most likely. Alternatively, since there had been exaggerated tectonic plate activity over the preceding several million years, any such catastrophe may have been merely a *coup de grace* following prolonged climatic havoc.[18] There are two favoured catastrophic explanations for this extinction: a massive impact by an asteroid or a spectacular volcanic eruption. Transient aberrations in geological strata, dated precisely at 65 million years ago, and comprising a worldwide paper-thin layer of iridium and microspherules of rapidly melted and refrozen basaltic rock, seem to be compatible with either explanation.[19,20,21] Scientists are increasingly confident that the massive Chicxulub crater in the Gulf of Mexico has many of the geological fingerprints of a massive impact at that exact time. Either type of catastrophe could have filled the upper atmosphere with dust, ash, sulphates and carbon dioxide, causing initial planetary cooling and strongly acidic rain, followed by prolonged greenhouse warming due to the excess carbon dioxide.

The dramatic disappearance of the dinosaurs and many other cohabiting species opened up broad swathes of ecological space. This was a moment of evolutionary opportunity for the mammals, whose ancestors had originated early in the dinosaur dynasty, around 200 million years ago. For millions of years, the early mammals (predominantly the ancestral marsupials and monotremes) had been small, non-flamboyant, nocturnal bit-players on the fringe of the dinosaur-dominated stage. They would, however, be offered leading parts in later productions. Meanwhile, they scurried around in the undergrowth and shallows, avoiding being trodden on or swallowed. For one hundred million years mammalian brain size, relative to body size, remained constant. Subsequently, the remnants of the early mammals that survived the Cretaceous-Tertiary extinction successfully took over the vacated niches. Within ten million years, new families of *placental* mammals – with increasingly large brain:body ratios

– had spread profusely on land (rodents and prosimians), as well as in the sea (cetaceans, the ancestral whales and dolphins) and air (bats). Meanwhile, in the oceans, the fishes recovered and proliferated.[22]

During much of the 500 million years of animal life that preceded the Cretaceous-Tertiary extinction, Earth had been a warm and humid place. With the exception of the ice ages, the global temperature was on average, about 5 °C warmer than today's temperature (15 °C). At the height of the Mesozoic Era, the era of reptiles and dinosaurs, around 100 million years ago, global average temperatures were 7–8 °C higher than today. Tropical breadfruit trees thrived in Greenland, and dinosaurs roamed a warm and iceless Antarctica. In contrast, the current Cenozoic Era, beginning 65 million years ago, has entailed a generalised and prolonged global cooling. Warm-blooded mammals, just one of Nature's ever-ready options, turned out to be well suited to this cooler environment. After an initial gradual cooling, global temperatures dropped by around 5 °C about 40 million years ago. Then, after some resurgent warmth, the temperature began falling again about 15–20 million years ago, during the Miocene period. Hence, at about the time that the ancestral hominoids (i.e. the prehuman Miocene apes) were evolving, 20 million years ago, Earth was approaching another protracted period of ice ages during which the polar ice advanced and retreated rhythmically.

During the past several million years, this ebb and flow of the ice, hundreds of metres thick, repeatedly drew water away from the equatorial regions to the polar regions, causing the forests and oceans to recede. Grassy plains replaced some of the forest in Africa. The early hominids – first the australopithecines and then the early *Homo* species – struggled to survive amidst this rhythmic contraction and expansion of equatorial forests. Walking on hind limbs probably evolved as a better, safer, means of moving between dwindling patches of forest. Some hapless hominids may have eventually been forced onto the surrounding plains to eke out a living in the less familiar open environment.[23] This intense selection pressure on the displaced big-brained, upright-lumbering, hand-using fringe-dwellers is thought to have brought about a burst of biological innovation, resulting in a bigger-brained and more fleet-footed hominid with a richer repertoire of communicative grunts, a more finely-controlled manipulative hand and primitive tool-making capacity. Enter the humans – *Homo habilis*, then *Homo erectus* and, most recently, *Homo sapiens*.

Over the last 2–3 million years, coincident with the *Homo* genus, an unusually cold and glaciated world has been the norm, punctuated by brief interglacial periods. Over the past million years there have been an estimated eight glaciations, each of 50,000–100,000 years duration, interspersed with shorter interglacial global warmings of around 10,000–20,000 years. At the height of the most recent glaciation, ice sheets covered most of what, today, are England, northern Europe, Scandinavia, Canada and north-east USA, and the relatively shallow Mediterranean Sea was reduced to a few puddles. The last glaciation began to recede rapidly around 13,000 years ago, and the world is now well into another interglacial period.

1.3.2 The getting of wisdom?

Judged by the usual criteria, the ape family has not been a particularly successful branch of evolution; many apes have become extinct. The one *apparent* exception is *Homo sapiens*, whose hominid forerunners diverged around seven million years ago from the ancestral chimpanzee (our closest genetic relative). *Homo sapiens* is still a short-lived species, and is estimated by fossil records, language trees and increasingly precise molecular genetics to have originated from *Homo erectus* approximately 200,000 years ago, most probably in eastern Africa – although the scientific jury is still out on this.[24,25] Although phenomenally successful by the primary biological criterion of population growth, as well as by the secondary (human-proclaimed) criterion of control of environment, *Homo sapiens* is still an evolutionary newcomer of unproven staying power. Tenured or not, with a brain like ours we have been able to exercise naming rights of all species, including our own. We call ourselves 'sapiens' – which means wise. Wisdom, however, comes with experience and age. It includes a capacity to learn from past mistakes and to foresee and plan for the future. We humans are clever, inventive, adaptable – but, it is clear from our current global predicaments that we are not yet very wise.

Over the past 15,000 years, a sixth mass extinction has been gathering momentum. This began in Eurasia and North America with the demise of most of the large mammals, such as the woolly mammoth, woolly rhinoceros, ancestral elk, ox and bison, and the giant sloth of North America. Around 80% of large animals became extinct in North and South America. Accelerating climatic change and shifting vegetation, as the ice age waned, may have been important, but most scientists now think that humans were at least accomplices in these late-Pleistocene mega-faunal extinctions.[26] In temperate North America, much of the extinction was due to the arrival of humans 11,000–12,000 years ago.[27] These hunter-gatherers, who had previously migrated east from Siberia across the land-bridged Bering Straits, finally managed to make a north-south traverse of

Canada at the end of the last ice age. In temperate America they met a food bonanza – huge animals, unacquainted with humans, and therefore easy prey. As local meat supplies consequently became depleted, the well-fed, proliferating human predators spread south. Within about a thousand years, most of the remaining large animals in the Americas had become extinct. Similarly in Australia, where climatic change was much less severe, the early Aboriginals, who had arrived across narrow seas and temporarily-exposed land-bridges from south-east Asia around 50,000 years ago (perhaps even earlier), probably contributed to similar widespread extinctions of the giant kangaroo and many other large marsupials such as the 'marsupial lion'. Meanwhile, the mega-fauna of Europe were decimated as ice receded, climates changed and roving bands of human hunters advanced.

More recently, human groups arriving on previously uninhabited islands have decimated the native populations of large flightless birds and other animals. Within the last two thousand years, the Maoris hunted the Moas and other huge flightless birds of New Zealand to extinction, Polynesians arriving in Hawaii wiped out the flightless geese and dozens of smaller birds, and immigrants from Indonesia and Africa overhunted the giant lemurs and elephant birds of Madagascar.[27,28] Similar extinctions occurred on the islands of Crete and Cyprus. While some controversy persists about the cause of continent-based extinctions in, for example, America and Australia, there is little doubt about these more recent island-based events, where vast piles of animal bones are closely associated with the archaeological remains of human tools, fires and ovens.

This sixth mass extinction, then, has not been caused by catastrophic disruptions of the physical geosphere of the kind that, in earlier eras, wrought sporadic global havoc. Instead it may reflect a mix of factors, including climatic changes at the end of the last ice age and increasing human depredations as early stone-age hunters spread around a warming world. (As we shall see in chapter 9, modern humans have continued the current mass extinction of animals on land not by slaughter but, increasingly, by destruction of habitat. The spread and dominance of European industrial technology and values over the past 200 years has ruptured the equilibrium widely attained with local species by less intrusive older civilisations. At sea, we are now rapidly destroying marine life by overfishing and wanton slaughter.) This post-glacial radiation of humans became, eventually, the first near-global spread of a relatively new species which, until around 50,000 years ago had been largely cooped up in the warmer climes of Africa, southern Europe and parts of Asia. In place after place, this uniquely migratory, adaptationist species arrived to find that an ample and often submissive meat dinner was near at hand.

Perhaps that was when humans first lost their 'ecological innocence'. The overkilling of edible or otherwise useful species was a departure from a basic principle of sustainable ecosystems – namely, that life can only be supported in the long haul by living off the ecosystem's interest (surplus productivity), and not by squandering its capital.

1.3.3 The hominid 'wild card'

Each species is an experiment of Nature. Only one such experiment, *Homo sapiens*, has evolved in a way that has enabled its biological adaptation to be complemented by a capacity for cumulative cultural adaptation. This unprecedented combination of the usual biologically-based drive for short-term gain (food, territory and sexual consummation) with an intellectual capacity to satisfy that drive via increasingly complex cultural practices is what distinguishes the human 'experiment'.

There is little argument about the immediate survival benefits of expanded brain-power; the dominance of mammals over birds, reptiles and fish bears ample witness. Scientists postulate a sequence of selection pressures on mammalian brain development: the need, post-dinosaurs, to combine daytime sensation (sight) with older nocturnal senses (smell and hearing); the omnivore's need for conceptual, behavioural and motor versatility; and, later, the need for communication and social coherence in larger primate groups. Apes have twice the brainpower (brain:body ratio) of monkeys, and humans have twice the brainpower of apes. We can therefore imagine that somewhere along the evolutionary corridor down which the heavy-browed hominids were lurching several million years ago, an opening labelled 'C' was found, leading to a previously untried passage way. There the cerebrating hominids discovered consciousness, culture and control over environment. As their brain size continued to increase, conferring a succession of marginal survival-enhancing gains, early humans acquired foresight, imagination and abstract thought. The unusual functional assymetry ('lop-sidedness') of the human brain, with the dominant left half conferring the ability to walk upright, to be right-handed (usually) and to use language, further amplified the biological evolutionary distance between humans and their primate forbears.[29]

That biological gain, invested in the development of human culture, has cumulated over the ensuing million years – and has done so spectacularly in the last few thousand years. It is that acquired cultural magnificence and

technological mastery that, in our own partisan view, 'sets us apart' from
the rest of the animal kingdom. We can therefore easily be tempted to
imagine that we represent a pinnacle, a culmination, of biological
evolution. Our modern religions resonate to, and affirm, that theme. In the
succinct phrase of Jacob Bronowski we fondly contemplate 'The Ascent of
Man'.[30]

In other higher animals, knowledge and skills are transmitted between
generations by parental instruction. Lion cubs have to be shown how to
kill game; the older female elephants in the matriarchal family carry the
knowledge of where the emergency water sources are for use in times of
drought. The same teaching is recycled between successive generations.
However, the advent of cumulative culture is an unprecedented occurrence
in nature. It acts like compound interest, allowing successive generations to
start progressively further along the road of cultural and technological
development. By travelling that road, the human species has, in general,
become increasingly distanced from its ecological roots. The transmission
of knowledge, ideas and technique between generations has given humans
an extra, and completely unprecedented, capacity for surviving in
unfamiliar environments and for creating new environments that meet
immediate needs and wants.

As with all steps forward or sideways in biological evolution, the longer-
term consequences of the human brain – even had they been in some sense
'knowable' – were not relevant; the logic and process of natural selection
proceeds strictly in the present tense. The occurrence of a species able to
acquire progressive control over the environment is therefore a 'wild card'
in the evolutionary game, a card that has not been dealt before. Nor has it
been on the table for very long. *Homo sapiens* has appeared in the last ten-
thousandth of the planet's history (i.e. the equivalent of the last ten seconds
on the 24-hour clock). Now that the hominid wild-card has been dealt, the
range of possible outcomes for life on Earth has multiplied.

1.4 Future prospects

From the saga of biological evolution, it is clear that, while the biologically
fittest survive best in the short-term, a good bit of luck is also needed to
secure an ecological niche and to survive in the long-term. The extra
requirement for the long-term survival of *Homo sapiens* will be a greater
understanding of the ecological context within which we live – to offset our
unique destructive capacities. Without that understanding, we modern

humans will continue to think of ourselves as not only special – which I think we are, by any usual criterion – but as being apart from, or above, the world's ecosystems. This we cannot be.

To argue, as has the US National Academy of Sciences,[31] that humans are uniquely adaptable, that they already live in disparate environments, and could therefore cope with a hotter, drier or more ultraviolet-irradiated world, is to miss the point. Such adaptation has usually taken a long time, much trial-and-error, and has been backstopped by the chance to move on and try again elsewhere. Over the centuries, geographically disparate human settlements have achieved adaptation to stable local ecosystems, with supplies of water, healthy soils and waterways, stable weather patterns and an interdependent fabric of plant and animal life. In each such setting, local human culture has incorporated its hard-won adaptations to local environmental circumstances. Where such adaptation has become inadequate, in the face of environmental change, then human settlements have failed. The long-established Nordic settlements in southwest Greenland died out nearly five hundred years ago when global temperatures fell 1–2 °C during three centuries and their food supplies failed. If significant and rapid global warming of 2–3 °C occurs within the next century, then the important point is that the *ecological framework* that supports life will become unstable and, almost certainly, less productive.

The main global environmental changes now being caused by humans entail reversal of features of the biosphere that have accrued over billions of years. Since the ancient advent of photosynthesis in plants, three great changes have occurred. First, most of the carbon dioxide in the atmosphere has been removed and laid down as limestone sediment and fossil 'fuels'. Second, the stratospheric ozone layer has formed. Third, fertile soil has formed from the oxidative, plant-assisted, weathering of rocks, the decay of organic plant material and the evolution of microbes and other soil-dwellers that aid nutrient redistribution. More recently, vast amounts of freshwater have been laid down underground during the ice ages. Meanwhile, the winnowing and balancing processes of natural selection have caused stable ecosystems to form. Today we are reversing each of those great changes. Extra carbon dioxide is being added to the atmosphere, stratospheric ozone is being destroyed, topsoil is being degraded and lost, aquifers are being depleted and species are being destroyed.

One way or the other, the human species is destined for extinction. The same shadow hangs over us as has hung over all other species that have ever lived on Earth – our contribution to the saga of evolution could be cut short by global or cosmic catastrophe over which we have no control. Meanwhile, we are beginning to see that we are now at risk of prematurely pulling down the pillars of our life-support system. Echoing in the background are the words of Theodosius Dobzhansky, the evolutionist, who said that 'selection promotes what is immediately useful, even if the change may be fatal in the long run.'[32]

1.5 Summary

For approximately the first 1 billion years of Earth's 4.6 billion years of existence as a globule of cosmic condensate, the elements and simple molecules remained in lifeless form. Then, by chance, simple organic molecules with the power of self-replication were formed within the primordial sludge. So life began. Through the logically irresistible process of biological evolution, driven by the natural selection of organisms best suited to an ever-changing environment, life-forms became more complex. Countless species have come and gone as the planet's environment underwent continual, sometimes violent, change. Very recently, the hominid line has evolved as a more specialised branch of the earlier primate line. Equipped with an unprecedented biology-supplementing capacity for cultural acquisition, the human line is the wild-card in the evolutionary pack.

The modern human species, *Homo sapiens*, was puny in its environmental impact at first. More recently, the human portfolio of burgeoning population size, overworked land, energy-intensive technology and waste-generating consumerism has resulted in accelerated and massive changes to the environment. As a consequence, the world's natural systems are today coming under increasing overload from one of its own resident species. This is a 'first' in Earth's history, and it has widespread implications for the health and survival of all species.

References

1. Malthus TR. *An Essay on the Principle of Population*. (Originally published 1798) London: Macmillan and Company, 1926.
2. Lovelock J. *Gaia: A New Look at Life on Earth*. Oxford: Oxford University Press, 1979.
3. Debate continues over whether extra-terrestrial life would require planetary conditions similar to Earth's. Scientists think there may be many medium-sized stars – perhaps 5% of the thousand billion stars in our galaxy – whose dust-and-gas nebulae have formed into planets. A minority of such planets may resemble Earth, being of solid rock-and-metal core, temperate and wet,

and could already be inhabited! Some scientists have envisaged the 'terraformation' of Mars by human intervention to create a custom-built atmosphere amenable to human life. (See: McKay CP, Toon OB, Kasting JF. Making Mars habitable. *Nature* 1991; **352**: 489–96.)

4. Radiation from some of these superdense high-energy quasars, near the receding edge of the observable universe, is still reaching us today after a journey of 12–14 billion years across time. In them we can 'see' the state of the universe as it stood just 2–3 billion years after its beginning. In 1992, we 'saw', with the COBE satellite, to within 300,000 years of the Big Bang by detecting irregularities in the very ancient microwave radiation just now reaching Earth. (These irregularities, if confirmed, may indicate the onset of clumping of matter within an evenly-spread new-born universe.)

5. Joyce GF. RNA evolution and the origins of life. *Nature* 1989; **338**: 217–24.

6. Horgan J. In the Beginning. *Scientific American* 1991; **264(2)**: 100–9.

7. Chyba C, Sagan C. Endogenous production, exogenous delivery and impact-shock synthesis of organic molecules: an inventory for the origins of life. *Nature* 1992; **355**: 125–32.

8. Dawkins R. *The Selfish Gene*. Oxford: Oxford University Press, 1976.

9. There is debate about the evolution of the nucleic acids. Were these first 'naked' genes, unenclosed by cells and directing the production of protein helpers, made of deoxyribonucleic acid (DNA) or ribonucleic acid (RNA)? DNA is the dominant genetic material in modern cells. However, some recent evidence suggests that RNA provided the breakthrough, because of its special combination of informational and functional properties – enabling it to both code for and catalyse the formation of proteins. (See Joyce GJ. 1989, above; also: Moore PB. RNA catalysis. The universe expands. *Nature* 1992; **357**: 439.)

10. Darwin C. *The Origin of Species by Means of Natural Selection*. London: Penguin, 1968. (Originally published by John Murray, London, 1859.)

11. Gould SJ. Punctuated equilibrium – A different way of seeing. *New Scientist* 1982; **94**: 137–41.

12. The p53 'tumour suppressor gene', on human chromosome 17, is currently exciting enormous interest among cancer biologists. Its more general function appears to be to arrest cell division whenever damage to DNA is detected, thus preventing transmission of the genetic damage. The p53 gene is thus a 'genome policeman' – and its deletion or mutation enables a damaged cell to elude 'arrest' and become the progenitor cell for what may turn out to be a colony of cancer-prone cells.

13. There is some evidence for 'directed mutation', wherein the type of mutation that occurs (or is retained) maximises the adaptation to a particular environmental change. (See Cairns J, Overbaugh J, Miller S. The Origin of Mutants. *Nature* 1988; **335**: 142–5.) Other mutation-influencing genes have also been tentatively identified – e.g. 'jumping genes' that relocate within the DNA molecule in order to activate or suppress other genes. (Wills C. *The Wisdom of the Genes*. Oxford: Oxford University Press, 1991.)

14. Schwartzman D, Volk T. When soil cooled the world. *New Scientist* 1991; **1777**: 27–30.

15. Gould SJ. *Wonderful Life*. London: Penguin, 1991.

16. During the earlier period, anaerobic scavengers (bacteria) had dined on dead producer organisms. Today, animal species, from termites to humans,

carry anaerobes in their intestines to aid the digestion of enzyme-resistant
plant materials. Hence those host animals continue the age-old tradition of
excreting methane as a waste gas.

17. Raup D. *Extinction: Bad Genes or Bad Luck?* New York: Norton, 1991.
    (Also: Jablonski D. Extinction: A paleontological perspective. *Science*
    1991; **253**: 754–57.)
18. Eldredge N. *The Miner's Canary: Unravelling the Mysteries of Extinction.*
    New York: Simon and Schuster, 1991.
19. Sigurdsson H *et al.* Geochemical constraints on source region of
    Cretaceous/Tertiary impact glasses. *Nature* 1991; **353**: 839–42.
20. Alvarez LW, Alvarez W, Asaro F, Michel HV. Extra-terrestrial cause for
    the Cretaceous–Tertiary extinction. *Science* 1980; **208**: 1095–108. (See also:
    Alvarez W, Asaro F. An extraterrestrial impact. *Scientific American* 1990;
    263(4): 44–52.)
21. Courtillot VE. A volcanic eruption. *Scientific American* 1990; **263(4)**:
    53–60.
22. Novacek MJ. Mammalian phylogeny: shaking the tree. *Nature* 1992; **356**:
    121–5.
23. Gribbin J, Gribbin M. *Children of the Ice.* Oxford: Basil Blackwell, 1990.
24. Vigilant L *et al.* African populations and the evolution of human
    mitochondrial DNA. *Science* 1991; **253**: 1503–7. And: Cavalli-Sforza L.
    Genes, peoples and languages. *Scientific American* 1991; **265**: 72–8.
25. Thorne AG, Wolpoff MH. The multiregional evolution of humans.
    Scientific American 1992; **266**: 28–33.
26. Donovan SK. *Mass extinctions: Processes and Evidence.* New York:
    Columbia University Press, 1989.
27. Diamond J. *The Rise and Fall of the Third Chimpanzee.* London: Radius,
    1991.
28. Adams D, Cawardine M. *Last Chance to See.* London: Heinemann, 1990.
29. Corballis MC. *The Lopsided Ape. Evolution of the Generative Mind.* Oxford:
    Oxford University Press, 1992.
30. Bronowski J. *The Ascent of Man.* Boston: Little Brown & Co., 1973.
31. National Academy of Sciences (US). Policy Implications of Greenhouse
    Warming: Report of the Adaptation Panel. Washington DC: National
    Academy Press, 1991.
32. Dobzhansky T. Evolution at work. *Science* 1958; **127**: 1091–8.

2

The ecological framework

2.1 Environmental and ecological perspectives

2.1.1 Introduction: ecosystems

We have scanned, in chapter 1, the origins of our species within the turbulent saga of life. As a product of biological evolution, humans have been honed by selection pressures to be able to survive temporary hardship and to thrive in times of plenty. So it is for every species. But each species is not just a separate bit-player in an uncoordinated cast of thousands; each is also a thread in the close-woven fabric of an ecosystem. To see more clearly how humans fit within, and depend on, the fabric of Earth's natural systems we must first explore some basic concepts.

The environment is that part of the world that surrounds us. We citizens of modern societies typically refer to the environment as if it were a resource, a platform, for human activity – but not actually the place where we *live*. This is not surprising, since, through culture, we humans have progressively distanced ourselves from the reality and rhythms of nature. Consequently, we seem to stand outside the framework of nature, unconstrained by the underlying ecological dimensions of our existence, and free to act according to the dictates and values of our fabricated culture. Yet, the essence of that ecological framework is an obligatory interdependence between living organisms and their physical and biological environment. It is that web of interdependence, of mutuality, that makes up an 'ecosystem'.

Before discussing ecosystems, we should briefly examine the distance that has opened up between the natural environment that shaped human biological evolution and the environment in which we now predominantly live. In the physical environment of the Pleistocene savannah, over the past several million years, our *Homo* predecessors existed as small extended-family groups, gathering food, hunting, mating, raising children and attempting to avoid the daily dangers of predators and rival groups.[1] Over the past few thousand years, our unique capacity to supplement our biological attributes with cultural adaptation, in an increasingly fast-track fashion, has led us to reshape our living environment by the domestication of plants and animals, the creation of cities and the industrial, technological and informational revolutions. In terms of our biological health, these cultural developments have made for more physically secure lives, surer food supplies, and the control of some infectious diseases. But they have also tipped some ecological scales in the other direction, fostering crowd-dependent infections, potentiating the spread of infectious agents through accelerated long-distance travel and transport, increasing the risks of nutritional deficiencies, creating health-damaging dietary excesses (chapter 4), and causing the various environmental and psychological strains of overpopulation and urbanisation (chapter 10). More broadly, these cultural-technical developments also underlie the major incipient disruptions to global ecosystems discussed in chapters 5–9.

The term 'ecosystem' was coined in the 1930s, and refers to a bounded system of dynamic interdependent relationships between living organisms and their physical and biological environment. While some ecosystems have a clear and separate identity (as on small remote islands), generally there is continuity between adjoining ecosystems. In aggregate, the world's ecosystems occupy the biosphere. The 'biosphere' notion was developed early this century by a Russian scientist, Vernadsky, and refers to the domain on and near Earth's surface where the prevailing conditions enable solar energy to produce the geochemical changes necessary for life. The biosphere comprises all living things along with the lower atmosphere, the hydrosphere (predominantly the oceans) and the upper lithosphere.

Ecosystems provide the framework within which incoming solar energy is captured and channelled through a hierarchy of life-forms (Fig. 2.1). Plant life, on land and at sea, provides nature's solar panels that convert sunlight to storable – and edible – chemical energy. Animals feed on these plants and on other animals. The quest for 'food' is thus the central organising principle of life within every ecosystem. These systems embody mechanisms by which nutrients are retained and recycled and by which water and respirable gases are disseminated and replenished. (As we shall see later, one of humankind's greatest ecological disruptions is the accelerating loss of nutrients from land to sea, as sewage, wastewater, agricultural runoff and eroded topsoil.) Life does not create new *matter*. Instead it recycles nutrients, using solar energy to build and maintain the

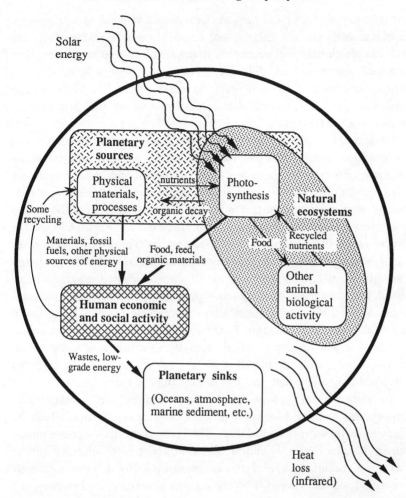

Figure 2.1. The biosphere is essentially a closed system; only radiant energy enters and leaves. Natural ecosystems need nutrients, physical resources, energy-capture by photosynthesis and the recycling of nutrients. Humans – dependent, like all species, on ecosystem support – deplete some planetary sources (e.g. soils and aquifers) and generate unusual wastes which may overload planetary sinks. (Based on multiple sources.)

temporary biological structures. Each of the atoms in our body is on short-term loan from nature's storehouse.

Through the coevolution of species, ecosystems acquire self-stabilising mechanisms and a dynamic internal balance. Within a stable ecosystem, one species usually does not decimate another; otherwise the food supply of the predatory 'victor' species would disappear. This is not to say that the panoply of species and the form of their interdependent relationships is immutable, nor that there is not competition and change. Over the millenia the global environment has changed hugely, and ecosystems have changed continually as different evolutionary players have strutted, flapped or crawled their hour upon the world's stage. But, in the shorter term, ecological systems tend to maintain their own internal balance at all links along the food chain.

Ecological relationships have, until this century, been poorly understood by biologists. Not surprisingly, therefore, there is still little understanding of these concepts within the world's dominant cultures. Within the Christian culture, humans have usually been viewed, in biblical terms, as having 'dominion over the fish of the sea and over the birds of the air and over every living thing that moves upon the earth' (Genesis, Chapter 1). Within Islam, there is a similar anthropocentrism. The notion that humans ultimately depend upon complex ecosystems seems to be better understood, intuitively, within stable primitive cultures living close to the natural world, where the limits of ecosystem support have been found by trial-and-error. We will consider culture-bound perceptions of the natural world more fully later in this chapter. The important point here is that the notion of the 'carrying capacity' of an ecosystem, with its subtexts of interdependence and sustainability, is critical to an ecologically-based understanding of the determinants of human population health.

To illustrate this ecological perspective, consider the human health implications of coal-burning and of killing malarial mosquitoes by spraying with the pesticide DDT. The main concerns of conventional 'environmental health' would be with the direct toxicological hazards of noxious air pollutants and of the contamination of food, water and air by pesticide residues. Viewed from the ecological perspective, however, our concerns about coal-burning embrace the myriad of health consequences due to greenhouse warming and climate change and the damage done to productive ecosystems by acid rain. Similarly, ecological concerns over DDT spraying include the accelerated genetic evolution of DDT-resistant mosquitoes, the adverse effects of the pesticide upon other species (and consequent imbalances between crop-damaging insects and the birds that eat them) and, perhaps, the depletion of mosquito larvae that are critical for wetlands fish breeding. Further, there will be demographic consequences of averting childhood malarial deaths in populations in which compensatory reductions in human fertility are unlikely to occur. The ecological perspective thus includes the disruption of dynamic systems.

2.1.2 Balance within ecosystems

Introduction

The internal balance within ecosystems entails a web-like interplay between many species and their inanimate environment. This balance between species results either from the coevolution of mutually supportive species or from what Richard Dawkins calls the 'biological arms race' between competing species.[2] Consider a simplified example of the latter. Since the quickest owls catch the slowest mice, and the quickest mice elude the slowest owls, so the quick owls and the quick mice are therefore more likely to survive, and to transmit their quickness genes to the next generation. Thus, the biological evolution of owls and of mice proceed in parallel. Balance is maintained and both species survive. However, if – extraordinarily – a mutant strain of superfertile owls with infrared telescopic-vision eyes suddenly evolved, then, during several burgeoning generations of well-fed owls all the mice would be caught and eaten and owls would then go hungry. In fact, such extreme scenarios do not happen in nature, where change is slower and systems-feedback occurs. However, that sort of wholesale eradication *can* occur when an introduced predator or competitor disturbs an otherwise stable ecosystem.

In Australia, for example, predatory foxes and feral cats along with voracious competition from rabbits and goats – i.e. four recently introduced species – are progressively wiping out the indigenous small marsupials. Indeed, in just 200 years Australia has accounted for one-third of all of the world's mammal extinctions over the past 500 years. On the other hand, a striking co-evolutionary accommodation has occurred between rabbits and the mosquito-transmitted myxoma virus. This once virulent virus, released by scientists in the 1950s as a targeted biological missile, wiped out all but a minuscule fraction of Australia's rabbits. With rabbits succumbing rapidly, both species were under intense selection pressure: the rabbits to 'find' a virus-resistant strain that could survive the onslaught, the virus to 'find' a less rapidly lethal, and therefore food-preserving, strain. Nature turned up trumps on both counts, and so rabbit numbers have rebounded while the less virulent virus circulates freely within a secure food supply.

Evolution occurs in response to current ecological opportunity, not in relation to future possibilities. Ecological balance is therefore not a 'goal' of evolution; but it is a logical consequence of the fact that those ecosystems with well-buffered stability and balance will survive best.

Nature's continual probing and exploring of new or vacated ecological niches is not purposeful exploration; the occupancy of new or altered niches happens, unplanned, as a result of continual genetic variation. Thus, what might look like 'purpose' is merely the end-result of selective processes. The Devonian lung-fish, faced with a drying up of its watery habitat several hundred million years ago, underwent genetic adaptation in favour of stronger fins able to assist the fish to retreat to the receding waters. Those same strong fins subsequently enabled some of the stranded fish to eke out a survival as early amphibians on land.[3] Thus, one way or the other, niches get filled, the ecosystem gets topped up and internal balance is achieved.

Insights from the HIV/AIDS virus

The human immunodeficiency virus illustrates vividly the interplay between species, as (hopefully) it iterates towards an ecological equilibrium with its hapless human host. This thus-far successful experiment of nature most probably originated, like some other viruses, from African monkeys.[4] Although we do not know when the virus first entered the human species, it seems that an auspicious new ecological niche arose among *Homo sapiens* during the 1960s and 70s. There was a new permissiveness in Western society's recurrent cycle of social attitudes towards sexual behaviour, combined with the rise of commercialised multiple-partner homosexuality, long-distance air travel and intravenous drug use. We do not know how many other viruses probed this particular ecological opportunity; but we do know that at least two viruses have benefited from it: the AIDS virus (HIV) and the hepatitis B virus. (There are actually at least two AIDS viruses, HIV1 and HIV2, which are genetically and clinically distinct.)

Both the AIDS virus and hepatitis B virus are spread by direct contagion and neither causes serious illness quickly.[5] HIV is less infectious than the hepatitis B virus, but it is ultimately more likely to kill the infected individual. This accords with its being a recently-arrived microbe that has not yet established the sort of equilibrium between its needs and those of its host that would enhance the virus' prospects for long-term species survival. The hepatitis B virus, longer-established and wider-spread in many developing countries, is more infectious but it is unlikely to kill – other than in a delayed fashion by sometimes causing liver cirrhosis or cancer. Both viruses thus allow plenty of time and opportunity to be spread from one individual to another. (Any of their mutant strains or close viral relatives that might have killed their hosts quickly will have already perished, through natural selection, because of the premature elimination of their medium of transmission.)

The outlook for the newly-arrived AIDS virus is still uncertain. Being of relatively low infectivity, the virus would disappear if high-risk human behaviours were either stopped or rendered 'safe'. Alternatively, if its infectious spread within the heterosexual community became more efficient, then it could decimate human populations. Unlike for hepatitis B, there is no long-term effective human biological (antibody) response to HIV infection. The virus, like other 'retroviruses', is highly genetically mutable and, in this case, the genes controlling the outer antigenic coat of the virus mutate readily. This quick-change capacity of HIV ensures that there are always some surviving genetic strains that can elude the body's immune defences – and frustrate our attempts to develop a vaccine.

The biology and the dynamics of HIV infection of humans will inevitably change. Many infectious agents eventually attain, through evolution, a stable and well-adjusted relationship between themselves and their infected host. Often, virulence is reduced in the parasite's self-interest. But it need be reduced no further than is compatible with ensuring its propagation and dissemination.[6] (Hence, the chronic or recurrent debility associated with the protozoal infections that cause African sleeping sickness or malaria.) We should remember, too, that the underlying principle in coevolution is self-interest. If changes in human behaviour reduce viral transmission, then selection pressures are heightened in favour of an alternative mode of transmission. For example, it has been speculated that such pressure may have caused the skin-transmitted spirochaete that causes yaws (which used to be common in pre-Renaissance Europe when inadequate clothing required winter-time huddling) to evolve to the closely-related sexually-transmitted syphilis spirochaete.

Overall, the induction of clinical disease is the exception rather than the rule in parasite-host relationships. Disease is usually incidental to the infecting parasite's needs, although disease manifestations may provide the means of spread – by sneezing, coughing or disseminating diarrhoeal material. The number of bacteria involved in an occasional infected sore throat is trivial compared to the vast numbers of unintrusive bacteria we carry within our alimentary tract. The fermentation of dietary fibre within our large bowel, with the release of metabolically useful (to us) fatty acids, would not occur without these legions of methane-producing anaerobic bacteria feeding on otherwise indigestible fibre. Thus do both the host and the colonic bacteria benefit from a symbiotic relationship. Many other intestinal bacteria quietly feed on other undigested material, intestinal secretions, and mucosal cells shed by the host. Indeed, about half the weight of human faeces is bacterial mass. Whether we perceive it or not, we humans are inextricably enmeshed in ecological relationships.

Humans – tipping the scales

As we have seen in the previous chapter, for the first 99 % of their existence, humans were an undoubted success. They fed, they bred and they spread. They achieved increasing dominance of their local environment and brought hunting pressures upon various animal species. They began to change their environment in many ways. Much of the Earth's grassland is the result of fires lit by humans and of the subsequent grazing of domesticated livestock. The clearing of forests by agricultural humans over the past several millenia is thought to have contributed substantially to regional climate change.[7] William McNeill, an historian who has interpreted the tides of human history in relation to infectious diseases, argues that the very recent imposition of language-based culture upon the age-old processes of biological evolution has upset the older balances of nature:

Time and again, a temporary approach to stabilization of new relationships occurred as natural limits to the ravages of humankind upon other life forms manifested themselves. Yet, sooner or later, and always within a span of time that remained minuscule in comparison with the standards of biological evolution, humanity discovered new techniques allowing fresh exploitation of hitherto inaccessible resources, thereby renewing or intensifying damage to other forms of life.[8]

It is only within the past 200 years, however, that the ecological balance between humans and their biological and geophysical environment has begun to be widely disturbed. The scale of this disruption was becoming evident over 50 years ago, when Sauer wrote: 'In the late 18th century the progressively and rapidly cumulative destructive effects of European exploitation became marked ... In the space of a century and a half – only two full lifetimes – more damage has been done to the productive capacity of the world than in all of human history preceding.'[9] In the ancient world, localised civilisations such as Mesopotamia surged, receded and collapsed as their ecological fortunes waxed and waned. Today, however, ecological imbalance is emerging for the first time at a global level as we put unprecedented strains on the carrying capacity of this small planet.

2.2 Non-linear systems

2.2.1 Exceeding limits

It is axiomatic that an ecosystem cannot yield substantial and sustained increases in materials and energy to the human species without, in the first instance, depriving other species and, eventually, causing damage to the productivity and viability of the ecosystem. The problem then is not just one of overdrawing against nature's capital assets, but of disrupting the productivity and workability of nature's systems. Earlier arguments, from Malthus to Meadows ('The Limits to Growth'[10]), were based on simple models of resource limitation, predicting crisis after the intersection of the graphs of falling supply and rising demand. However, neither argument foresaw the technological innovations that have repeatedly rolled back those limits, thereby buying time. In 1987, the UN's World Commission on Environment and Development (WCED) took a longer, and more ecologically oriented, view of the notion of limits, arguing that the exploitation of natural resources must be constrained if it is to be indefinitely sustainable 'for future generations'.[11] Where Malthus had asked about the imminent limits to the rate of food production, the WCED asked about the long-term viability of the agricultural production base – i.e. the world's soil and freshwater. On this view, the critical limits apply not to the supply of materials but to the planet's capacity to absorb wastes such as carbon dioxide and to replenish soil and freshwater. (In Fig. 2.1, above, this represents the overloading of planetary sinks and the depletion of certain of the sources.)

Underlying much of this debate about when the graphed lines will cross, or when the limits will be met, is an assumption that we are dealing with simple systems that conform to a mechanistic Newtonian world in which relationships are obligingly linear. But ecosystems are not like that. At some point an external load will overwhelm the system's resistance and rapid change or collapse then ensues. For our early nineteenth-century predecessors, including legions of British naturalists and clergymen enthralled by recreational bug-hunting, the study of natural history assumed 'creationist' orderliness and constancy. By contrast, modern twentieth-century ecology (as with aspects of the contemporary theories of 'chaos' and 'complexity') recognises uncertainty, complexity and changeability.

Viewed within an ecological framework, it becomes plausible that overloading the biosphere will, at some point, disrupt life-support systems. However, we cannot easily foresee when we will overstep the mark, since thresholds in non-linear systems are difficult to anticipate. Today's problems of ozone depletion and incipient climate change have occurred because self-correcting mechanisms within the biosphere have become overloaded. Once such a threshold is passed the momentum of induced change continues, and feedback mechanisms that amplify or constrain the change may be activated. If regenerative powers or absorptive capacities are long exceeded, the system may collapse, causing what the catastrophe theorists call (with nice understatement) a 'discontinuity'.

Non-linearity has been well illustrated by the sudden appearance and then rapid spread of forest dieback in Europe and North America after decades of air pollution with acid-forming gases; and by the recent algal blooms ('red tides') along the eastern Italian coast caused by the increasing pollution of coastal waters with industrial, agricultural and domestic chemical wastes that have eventually exceeded the aquatic system's self-cleansing capacities. Because of these sorts of non-linearities we may have misestimated the impending warming effect of anthropogenic greenhouse gases. While the increase in consumption of fossil fuels may be linear, the environmental response is not. As is discussed further in chapter 6, the many biogeochemical feedbacks that influence greenhouse warming ensure that the climate change system is non-linear.[12]

2.2.2 A homeostatic biosphere?

Claude Bernard, the great nineteenth-century French physiologist, recognised the body's capacity to act in a coordinated way to ensure the constancy of the 'milieu interieur'. His physiological experiments, conducted on a range of animal species (against the shrill protests of his devoutly religious wife), established the concept of homeostasis. Homeostatic mechanisms enable complex interrelated systems to achieve a constancy of environment and function despite the perturbations posed by external stressors. As we have seen, similar mechanisms ensure resilient and dynamic equilibrium in local ecosystems.

More radically, a similar phenomenon has been proposed for the whole biosphere by James Lovelock, an eminent British atmospheric scientist.[13] Lovelock's Gaia hypothesis posits that the components of the Earth's biosphere – comprising all living organisms, the geosphere and the cycles of water, gases, nutrients and energy – make up a global homeostatic mechanism that ensures constancy of the environment. Gaia (from the Greek goddess 'Mother Earth') thus acts as a coordinated, self-regulating superorganism, using feedback mechanisms to counter externally imposed disturbances.

Lovelock's formulation of the concept of Gaia rested on his asking two fundamental questions. First, how is it that the configuration of chemical, physical and thermal circumstances of Earth's biosphere are so different from those observable on the surface of other planets? Second, how are these conditions maintained over long periods of time in the face of external perturbation? For example, over several billion years, the heat output of the Sun has increased by around 30%, and yet the Earth has maintained a relatively constant temperature in contrast to the effect that an increase in external temperature would have had on an inert object. Lovelock reasons that the increase in solar energy would have stimulated increased photosynthesis, which would have removed carbon dioxide from the atmosphere (and deposited much of it in land and ocean sediments), thereby reducing the natural greenhouse capacity of the atmosphere and facilitating compensatory global cooling.

Other planets do not have free oxygen in their atmosphere, and neither did Earth before the appearance of photosynthesising plants. Atmospheric oxygen and its derivative ozone shield are the *product* of life on Earth, and, subsequently, they have enhanced the evolution and survival of life on Earth. According to Lovelock's Gaia hypothesis, oxygen has accumulated in the atmosphere to a level that is optimal for biological life, thus reflecting the balance of positive and negative feedbacks from dependent living organisms when the oxygen level is below or above the 'optimal' level, respectively. (The fact that the once-dominant anaerobic organisms were decimated by the advent of poisonous oxygen is, of course, something of an embarrassment to the hypothesis.)

The Gaia hypothesis is controversial for several reasons.[14] First, it is not possible to test the hypothesis according to the canons of conventional science. (Lovelock has subsequently published several simplified cybernetic models of Gaian processes to show how it *might* work.) Second, the hypothesis has been taken by some to imply a purpose and altruistic farsightedness in assembled Nature that is not evident from our observations of unthinking plants and instinct-driven animals. Nor does it fit comfortably with the mechanistic view of the world that has long prevailed in contemporary Western culture. Finally, how on Earth could such a planet-wide homeostatic mechanism have originated? Could there have been some higher-order selective process at work? Or is it an as yet poorly understood consequence of the aggregate adaptations of individual components of the biosphere?

The Gaia hypothesis stretches contemporary credulity. Nevertheless, its heuristic qualities have provoked wide debate and stimulated awareness of the interdependencies within ecosystems and of the homeostatic mechanisms that, within limits, act to stabilise and sustain this planet's life-support systems.

2.3 Ecological ideas in human culture

We are beginning to think of the world as a biosphere, and not just as a geosphere.[15] For most of the past 500 years of European-led exploration and colonisation, the interest of those countries has been in discovering, mapping, measuring, acquiring and controlling territory. The world has been viewed by politicians and merchants as real estate, to be owned and developed. This view peaked several decades ago in the ethos of the Space Race, at about the same time that China's beholden masses were proclaiming the 'winning of many victories over nature for the Great Helmsman, Chairman Mao'. A different, potentially subversive, view of the Earth was fostered by the remarkable image, seen from space in 1968, of a small and fragile planet suspended in a vast and disinterested black void. The world was visibly a biosphere of finite size. Ironically, as a new ecological awareness is seeping into the rich countries, via the increasingly populist 'green movement', their economic values and free-market ideology are enveloping the poor and middle-income countries of the developing world, often distorting traditional cultures that have their own mature ecological wisdoms.

The historian, Lynn White, in his seminal essay of 1967, *The Historical Roots of Our Ecologic Crisis*,[16] examined the sedimentary values imparted to European culture by Western Christianity (and, before it, by the Judaic ethic). These values include an explicit emphasis on ownership and domination of the environment, the virtues of hard work, utilitarianism and the role of science as the means of finding out how God's creation works and of technology as the means of applying that knowledge. Western Christianity ultimately prevailed over its more contemplative sibling, Eastern Christianity, which construed the world as a set of symbols through which God's will was revealed. White's influential essay drew attention to the anthropocentric conceits we have inherited from the ethical and religious traditions of European culture. He argued that the exploitative, destructive and unseeing attitude towards the biosphere that characterised the world's dominant culture arose irresistibly from the Judaeo-Christian ethic. (Others, however, draw a distinction between the Judaic and Christian ethics.[17]) Further, he argued, as the Age of

Enlightenment unfolded during the eighteenth century and a mechanistic view of the universe emerged, responsibility to God lost its meaning – and no further codified basis for responsible behaviour remained.

During that same century the radical idea of 'progress' emerged. Previously, most societies had lamented a lost Golden Age, and many had imagined the world to be inexorably declining towards a day of reckoning. In Europe, the steady advances of science and technology began to impress upon people the idea that history might be a chronicle of progress rather than of decay.[18] Later that century, the free-market doctrine proclaimed that what was good for the individual entrepreneur must be good, in aggregate, for society. The view that nature was there for humans to transform and use entered the mainstream ideology of modern industrialised societies. This century it has been a central assumption of capitalist and communist societies.

Several years before White's essay appeared, a seedbed of new awareness had been nurtured by Rachel Carson. In *Silent Spring* she argued that these same environmentally insensitive values were decimating the wild-life that is an integral part of the biosphere within which we humans have evolved.[19] The disruption of ecosystems, by profligate use of modern agricultural chemicals such as DDT, would do more than silence bird-song in Spring; it would fracture interspecies dependencies and would, ultimately, impair our food chain and water supply. Carson revealed and explained *ecological* relationships in a way that a popular audience had not previously encountered.

Despite these recent stirrings, Western intellectual traditions militate against our thinking within an ecological framework. During the seventeenth century, certain European philosophers such as Spinoza and Locke challenged this dominant human-centred view of the world.[20] In nineteenth-century Britain, various scientists began expressing concerns about the impact of deforestation and agricultural mismanagement on local ecological systems in the colonies, on the wider climate, and, indeed, on the long-term survival of humans.[21] In America, John Muir propounded the view that humans were the stewards of nature, within which everything had intrinsic value.[22] In recent years there has been some 'greening' of Western philosophy, embracing notions of the intrinsic 'rights' of nature and the more radical ideas of 'deep ecology' which reject the primacy of the human species.[23] But these have been tiny cross-currents within the predominantly materialist and utilitarian mainstream of Western philosophy.

Many traditional 'primitive' cultures, of negligible impact on today's world, have an animistic religion in which humans are part of the cosmos. The historian, John Young, has said of these cultures, both ancient and current: 'Man sought to manipulate nature to his advantage, to utilise its forces, using the spiritual technology which defined his culture, but he did not think of mastering it.'[24] The Australian Aborigines have an elaborate set of personal totemic links with the spirits of other animals and of physical sites, in what is a part-real part-fantasy ecosystem. This spiritual bonding of tribal aboriginal groups to particular land sites has created chronic tensions in modern Australia; mining company executives cannot comprehend a non-utilitarian, non-propertied view of land. More influential in the world, including in floral-powered California in the 1960s, has been one of the world's great and ancient religions, Hinduism, which has a pantheon of animistic, multi-limbed deities. The idea of active management of nature by humans forms no part of Eastern religion, be it Hindu or Buddhist. All creatures, including humans, are part of a suffering world. For Buddhists, active stewardship of the Earth is not compatible with the goal of spiritual salvation through freeing oneself from all Earthly bondage.

Now, formal philosophies aside, how compelling is the argument that we must rethink our social values in order to protect and sustain our life-supporting ecosystems? Most of us who read books such as this can look around and see a portion of the world that *appears* to be intact. We see trees; we have plenty of food; and we can find pleasant holiday locations. However, those appearances can be deceptive. We see the blue of the sky – but not the gradual increases in ultraviolet radiation. We have little personal memory of how many trees were there half a century ago or of how much louder was the croaking of the frogs in nearby streams. Many other forms of environmental degradation and ecosystem disruption proceed either invisibly (e.g. water table depletion; toxic waste accumulation), slowly (e.g. global warming; rising oceans) or – for the book-readers – remotely (e.g. soil erosion; the projected spread of infectious-disease vectors; increased food shortages; and Third World urban congestion, squalor and social breakdown). Yet these environmental changes are global in scope and, increasingly, they endanger the health of human populations. They also challenge our social and economic value system.

Under current economic orthodoxy, the neoclassicist's free-market model 'externalises' the costs incurred by resource depletion, pollution, ecological disruption and health impairment. The result of this fantasy is that increased economic productivity is seen as 'yield', as the creation of wealth, when in fact much of it depends on the expenditure of environmental 'capital'. The World Resources Institute has recently estimated, for example, that the 4% increase in Indonesia's cereal productivity over the past decade has been almost exactly offset by a 4% reduction in soil fertility.[25] Yet, the UN-approved system of national accounts reports the former gain, but not the latter loss. Historically, all settled human cultures have squandered their environmental capital to some extent – i.e. their non-renewable resources such as fertile soils, freshwater, forests and fuels. Usually, however, there have been additional resources and new frontiers nearby. That option is rapidly being closed off in today's overcrowded world. Economics has evolved as the science of rationally allocating competitively sought-after resources; it must now address itself to the unfamiliar prospect of finite, exhaustible resources.

Finally, there is a major ethical dilemma here also. Environmental degradation and its adverse societal impact is not equally shared between countries. Rich countries generally do not directly bear most of the consequences of their actions, whereas poor countries do. Most of the greenhouse gases and the ozone-destroying gases have, so far, come from the rich countries; yet their impacts, via the overloading of natural sinks, will tend to be global. In contrast, in populous poor countries the consequences of the escalating overuse of renewables, particularly soil, forests and fuelwood, tend to press directly on the local population. Internationally, our sense of fairness is about to be tested mightily.

2.4 Summary

Biological evolution, in filling the available niches in a changeable environment, has established dynamically balanced ecosystems. Ecosystems, by definition, support the biological functioning, health and survival of their member species. They provide each of those interdependent species with energy, nutrients and water, along with environmental cleansing processes. The human species, like all other living species, has evolved within the planet's ecosystems and, in the final analysis, depends upon them for food, water, air and other life-supporting resources.

Humans, unique within evolution's hall-of-fame for their powers of cultural adaptation, have assumed increasing control of their environment. The resultant pressures on the environment raise the question of how accommodating and resilient are the world's ecosystems. Thresholds, once exceeded, may trigger a chain of responses that impoverishes and impairs the functioning of ecosystems. We are now beginning to see that the disruption of natural systems will reduce the carrying capacity of the biosphere – for our own species and for many others.

Western philosophy has dealt little with ecological perspectives. Its rationalist and materialist themes have obscured understanding of the nature and needs of ecosystems. However, as we begin to understand the profound problems caused by ecological disruption so we can also sense the need for a reordering of social values. One little-considered problem is the risk to the long-term health of human populations as the capacity of natural systems is eroded. The next chapter lays the foundations for considering that problem in detail.

References

1. Eaton SB, Shostak M, Konner M. *The Paleolithic Prescription*: *A Program of Diet, Exercise and a Design for Living*. New York: Harper and Row, 1988.
2. Dawkins R. *The Selfish Gene*. Oxford: Oxford University Press, 1976.
3. A similar process has been postulated to explain the emergence of bipedalism in hominoids. Stranded away from the receding forests during recurrent ice ages, this conservative trait was initially an improved means of retreating to the old habitat. It subsequently enabled the successful occupancy of a new, open, non-forest habitat.
4. The predecessors to this virus may have long been unobtrusively circulating in other species or ecosystems – probably in Africa. The simian (monkey) immunodeficiency virus (SIV) is a likely precursor, although it is not as closely related genetically as is popularly thought. Other viruses have made the inter-species leap from monkeys to humans – for example, the O'nyong-nyong virus in Uganda in 1959 and the lethal Marburg virus, transmitted to laboratory workers, in 1967. (See: Ehrlich PR, Ehrlich A. *The Population Explosion*. New York: Simon and Schuster, 1990.)
5. The immediate symptom-producing virulence of a parasite seems to be closely related to its mode of transmission. Those spread by skin contact (e.g. leprosy) or mucous membrane contact (e.g. HIV) 'need' less virulence than do those spread by water (e.g. cholera) or air (e.g. influenza) – which require copious diarrhoea or sneezing, respectively, to counter environmental dilution. Those spread by vector (e.g. malaria) may debilitate the human host while generating enough parasite progeny to ensure passage to the next itinerant vector (e.g. malarial mosquito). (See: Ewald PW. Culture, transmission modes, and the evolution of virulence with special reference to cholera, influenza, and AIDS. *Human Nature*. 1991; **2**: 1–30.)
6. Massad E. Transmission rates and the evolution of pathogenicity. *Evolution* 1987; **41**: 1127–30.
7. Talbot LM. Man's role in managing the global environment. In: Botkin DB, Caswell MF, Estes JE, Orio AA. *Changing the Global Environment. Perspectives on Human Involvement*. London: Academic Press, 1989, pp 17–33.
8. McNeill WH. *Plagues and Peoples*. Middlesex, England: Penguin, 1976.

9. Sauer CO. Theme of plant and animal destruction in economic history. *Journal of Farm Economics* 1938; **20**: 756–75.
10. Meadows DH *et al*. *The Limits to Growth*. New York: Universe Books, 1972.
11. World Commission on Environment and Development. *Our Common Future*. Oxford: Oxford University Press, 1987.
12. Schimel D. Biogeochemical feedbacks in the Earth System. In: Legget J (ed). *Global Warming. The Greenpeace Report*. Oxford: Oxford University Press, 1990, pp 68–82.
13. Lovelock J. *The Ages of Gaia. A Biography of our Living Earth*. London: Norton, 1988.
14. Kirchner J. The Gaia hypothesis: Can it be tested? *Reviews of Geophysics* 1989; **27**: 223–35. Also: Resnik DB. Gaia: From fanciful notion to research program. *Perspectives in Biology and Medicine*. 1992; **35**: 572–82.
15. Caldwell L. *Between Two Worlds. Science, the Environmental Movement and Policy Choice*. Cambridge: Cambridge University Press, 1990.
16. White L. The historical roots of our ecologic crisis. *Science* 1967; **155**: 1203–7.
17. Passmore argues that White has misunderstood the Old Testament (Jewish) view of nature as the work, and glorification, of God, available for man's use – but not intended for man's domination and exploitation. Much of the early writing coincided with nomadic pastoralism, where humans shared the world with other living things not directly under man's control. Later, after the Greek Enlightenment, the view arose that the world exists purely and simply for man's sake. It is that intellectual tradition that Christianity has inherited, and which led, in the seventeenth century, to the Baconian–Cartesian view that nature was there for man to transform as he pleases. See: Passmore J. *Man's Responsibility for Nature*. London: Duckworth, 1974.
18. Ponting C. *A Green History of The World*. London: Penguin Books, 1991.
19. Carson R. *Silent Spring*. Boston: Houghton Mifflin, 1962.
20. Hargrove EC. *Foundations of Environmental Ethics*. London: Prentice Hall, 1989.
21. Grove RH. Origins of Western environmentalism. *Scientific American* 1992; **267(1)**: 22–7.
22. Eckersley R. The road to Ecotopia? Socialism versus Environmentalism. *The Ecologist* 1988; **18**: 142–7.
23. Nash RF. *The Rights of Nature: A History of Environmental Ethics*. Wisconsin, USA: The University of Wisconsin Press, 1989.
24. Young J. *Post Environmentalism*. London: Belhaven Press, 1990.
25. Repetto R. Accounting for environmental assets. *Scientific American*. 1992; **266(6)**: 64–70.

3

The health of populations

3.1 Persons, populations and public health

3.1.1 Introduction

For thousands of years, human populations have been confronted by famines, floods, plagues and war. Crowded urban populations have provided a rich culture medium, an ecological bonanza of human-borne nutrients and energy, for infectious disease microbes. More recently, in industrialised countries, the gains in health and longevity associated with increased wealth and education have been accompanied by two other population health hazards: the increase in chronic diseases attributable to the spread of affluent lifestyles (especially diet) and the toxic effects of environmental chemicals. Today, as mentioned in earlier chapters, the texture of environmental hazards is changing; the disruption of the stability and productivity of the biosphere's natural systems poses a qualitatively different type of health hazard. While toxic environmental pollutants typically affect small groups or local communities, global environmental changes may cause serious or irreversible changes in ecosystems that will affect the health of populations in the immediate and more distant future.

Let us consider two simple examples of how the disruption of Earth's natural systems might adversely affect the health of whole populations. First, while many Australians now reduce their personal cancer risk by wearing protective clothes outdoors, depletion of stratospheric ozone is beginning to change the population-wide skin cancer risks posed by that outdoors environment. Second, the use of bed-nets influences which particular African villagers get malaria, but, in a warming world, it will be the climate-related expansion of the malarial mosquito's habitat that primarily determines which populations have endemic malaria. A more

56

general example is that ozone depletion and greenhouse-induced climate change will each have various, and mostly adverse, effects on the productivity of the world's agriculture and fisheries, thus affecting food supplies. The important general point – variations in personal behaviours and local culture aside – is that the average levels of health risk within populations are largely determined by wider environmental factors, including the stability of ecosystems.

On a day-to-day basis, however, we mostly think about health at the personal level, often as a series of events. We evaluate our personal health in comparison to our past experience or to the health of other people. Local culture affects our approach towards matters of personal health. Funerals are festive affairs in some cultures, sombre affairs in others. Rich countries spend half of their total health care budget on the last six months of their citizens' lives, whereas poor countries smooth the dying pillow and nomadic peoples abandon the elderly infirm. (Some would say that, within certain rich countries, we have found euphemistic and more expensive ways of doing much the same.) The unique human capacities to experience many emotions, to empathise at a distance with others' suffering, to understand notions of fairness and equity, and to be aware of the inevitability of death lead us irresistibly to set great store by good personal health.

Meanwhile, nature uses a different scorecard. The poor health of any one person is of little account within the dispassionate, morally neutral framework of ecological systems – and of biological evolution. The success of a species requires that sufficient of its members remain healthy enough to breed. Natural genetic variation within the species ensures that, in any particular environment, there are at least some individuals who will thrive. Nature's 'interest' begins and ends there.

This chapter examines, first, the notion of population health and its fundamental dependence on natural systems. There follows a brief review of the historical development of the population perspective and the strategies of modern public health. Subsequently, there is a discussion of the complex relationships between economic development and population health.

3.1.2 Population health: Utopia versus ecological realism

The World Health Organization has long defined 'health' as being not just the absence of disease or infirmity, but a state of complete physical, mental and social wellbeing. This is a fine ideal for each individual, but it is not an appropriate way to think about the health of populations. Populations never are free of disease, disability and premature death. Indeed, interindividual differences in health and fitness are a continuing expression of the process of genetic variation and biological evolution. For example, persons born with the gene(s) for 'thrifty' metabolism are thought to store surplus dietary energy more efficiently than those without. Hence, in circumstances of abundant food, they will become obese and may develop diabetes mellitus, while in times of famine they are the survivors.[1] This interindividual genetic variation is one of nature's most precious resources in the shaping and sustaining of life in the face of environmental change.

A more realistic goal, therefore, is to maximise the reductions in rates of infant mortality, maternal mortality and other measures of disease and disability, thus minimising the number of people with poor health and increasing the average life expectancy. Such goals are framed in relation to 'population health', and refer to the average state of health in a whole population. They can be measured by such things as life expectancy or the average level of some index such as blood pressure. These measures summarise the health status of the population, and they allow comparisons of health status between populations or within a population over time. For example, we can observe that life expectancy has increased rapidly in Japan since 1940 or that childhood death rates are ten times higher in poor countries than in rich countries. A basic comparison of demographic and population health measures, for rich and poor countries, is shown in Table 3.1.[2]

Nevertheless, the unblemished health of the entire population has been part of every Utopian vision. Many social philosophers of the Age of Enlightenment, in the eighteenth century, believed in the perfectibility of the human species. Likewise, Benjamin Franklin spoke fondly of a future when 'all diseases may by sure means be prevented or cured'. Today – the Human Genome Project notwithstanding – we can recognise that this is a naive view, ignorant of the ecological context within which the human species exists. This ideal of perfect health for the human species is simply not attainable; it is a mirage. For a start, we share the biosphere (and much of our food production) with countless other species. Many of them derive their sustenance from us, using human tissues as a culture medium for their own propagation (i.e. 'parasites'). The infecting parasite bears its host no ill-will but merely seeks to ensure its own propagation, true to the inbuilt mission of all genetic material on Earth. Indeed, the tempo of genetic evolution of the tens of millions of species of microbes and insects operates on a timescale which makes that of the human species look glacial. Where

Table 3.1. *Demographic profile of major groupings of countries, 1990.*

Countries	Population in millions	GNP per person ($US)	Adult literacy (%)[a]	Crude birth rate per 1000	Crude death rate per 1000	Population growth rate (%) 1980–90	Total fertility rate	Infant death rate per 1000	Life expectancy at birth (years)
Low and middle income	4,146	840	64	30	9	2.0	3.8	63	63
Sub-Saharan Africa	495	340	50	46	16	3.1	6.5	107	51
East Asia (inc. China) & Pacific	1,577	600	76	23	7	1.6	2.7	34	68
South Asia (inc. India, Pakistan, Bangladesh)	1,148	330	47	32	11	2.2	4.2	93	58
Europe (eastern)	200	2,400	85	19	9	0.1	2.0	30	70
Middle East & North Africa	256	1,790	53	40	10	3.1	5.7	79	61
Latin America & Caribbean	433	2,180	84	27	7	2.1	3.3	48	68
High income	816	19,590	96	13	9	0.6	1.7	8	77
OECD members	777	20,170	96	13	9	0.6	1.7	8	77
Others	40	not reported	73	17	5	1.8	2.2	13	75
World	5,284	4,200	65	26	9	1.7	3.4	52	66

[a]'Literacy' is defined as the ability to read and write simple statements about daily life.
Source: World Bank (1992)[2]

the generation of new genetic strains takes the HIV/AIDS virus several hours and the malarial parasite days, it takes the larger mammals decades. Against these odds, it is inconceivable that we will ever be entirely free of pests or parasites. New microbes will continue to arise and will always tend to cause widespread disease in human populations that are weak, poor or socially disorganised.

The ideal of perfect health for everyone reflects an erroneous view of the human condition. It overlooks our inextricable involvement with the ecological systems within which human life exists. On such a view, in a business-as-usual scenario, increased wealth leads to improved health care systems and preventive strategies which yield progressively better health.[3] Well, it is true that we humans (at least in the world's more privileged communities) have managed to distance ourselves substantially from many of the age-old scourges of human health. However, the world in which we live is also inhabited by rapidly evolving microbes, its agricultural productivity is hugely influenced by the vagaries of the weather, cataclysmic physical events will continue to occur, and the current trends in human fertility and environmental overload will cause further instability. Not only do we live in a closed-system world where the economists' linear graphs do not apply indefinitely but we must take our chances in the natural world to which we belong and from which we draw our sustenance.

This distinction between the more mechanistic, social-behavioural perspective on human health and the ecological perspective is fundamental to the central argument in this book, and warrants further exploration. At the turn of the eighteenth century in Western society, there was renewed interest in the classical Hippocratic doctrine of the importance of clean air, water and food to individual health. These ideas allowed the further development, via philosophers such as Rousseau and the practical sanitarians, of the 'environmental' concept of health – a concept which shifted the focus from the individual to the population. The nineteenth century's enormously important sanitary ideal embodied this same population focus. It developed initially without any supporting evidence from laboratory science – indeed, various of its early proponents discounted (erroneously) the 'contagion' theory since they believed that collective cleanliness, by eliminating pervasive 'miasmas', was the source of good health. In like vein, in 1875 Benjamin Ward-Richardson published his ideas for 'Hygiea', an urban health-oriented Utopia with clean air, public transport, small community-based hospitals, community homes for the aged, and no tobacco or alcohol. His vision led to Hygiean city planning, house design and household products on both sides of the Atlantic.

Then, in the 1880s, came the Germ Theory, based on experimental confirmation of earlier contagionist ideas. On this view, so long as microbes were controlled by disinfection, vaccination and 'magic bullet' drugs then exposure to waste and filth did not matter. Germs were the main source of diseases. The age of microbiology had arrived; a new biomedical view of health replaced the previous emphasis upon the need for a sanitised and safe environment. Those two views are, nevertheless, distant relatives. Both were shaped, historically, within a paradigm that attributed disease to localised, remediable factors, whether contagious or otherwise. Neither view imagined that humans did not have the ultimate capacity to improve and control their environment such that diseases would be eliminated. Not surprisingly, therefore, during this century, as we have preempted more and more of the world's biophysical resources for our ever-expanding economic activities, we have continued to imagine that the achievement of universal better health could be achieved by technical control of disease-causing factors and by being fairer in our social dealings. This view, preoccupied with the local and tangible, and viewing human society as a self-contained entity, overlooks our ecological dependence upon the web of natural systems.

The difference between mechanistic and ecological thinking is well illustrated by two of the giants of microbiology, Robert Koch and René Dubos. Koch, in 1892, propounded the criteria by which it could be established that Microbe X is the cause of Disease Y. His view was that, for each infectious disease, there was a microbe that was the specific 'sufficient cause'. Dubos, over half a century later, stressed the influence of the environment upon both the behaviour of the microbe and the susceptibility of the host. He contended that a microbe is a necessary, but not sufficient, cause of disease. The bacterium that causes tuberculosis, for example, does not automatically destroy lung tissue. It can persist in a dormant state within the body for long periods, liable to flare up when the equilibrium is disturbed – as indeed has now happened in approximately five million HIV-infected persons, particularly in Africa and India. On the other hand, normally benign and widespread microbes can, under certain conditions, cause human disease. An example is the widespread and usually-quiescent herpes simplex virus, which causes facial 'cold sores' in response to stress and sunlight.

Dubos also argued that human populations can never be disease-free because they live in, and indeed cause, a changing environment to which they must continuously adapt. He wrote: 'To be healthy does not mean that you are free of disease; it means that you can function, do what you want to do and become what you want to become.'⁴ This pragmatic definition recognises the different needs for 'healthy' functioning between individuals and cultures. For example, during the sixteenth and seventeenth centuries, in southern Europe, women with fashionably enlarged thyroid glands (due to iodine deficiency) were preferentially selected as wives. Pursuing the argument that humans are an integral and balancing factor in the health of ecosystems, subject to the biological stresses and challenges of a changeable environment, Dubos wrote:

The concept of perfect and positive health is a utopian creation of the human mind. It cannot become reality because man will never be so perfectly adapted to his environment that his life will not involve struggles, failures, and sufferings ... The less pleasant reality is that in an ever-changing world each period and each type of civilization will continue to have its burden of diseases created by the unavoidable failures of adaptation to the new environment.⁵

He also argues that the high priests of modern medicine 'assume that the achievement and maintenance of health depend not upon living in accordance with the laws of nature, but upon medical care based on scientific biotechnology.'⁵ This view, he says, overlooks the social–ecological dimension of human existence, the fact that living in an ancient agrarian society required very different adaptations from those required during the industrial revolution or in the impending Automation Age. René Dubos recognised the intimate connection between the ecology of human living and the health of human populations earlier than most.

Historically, the major influences upon the health of communities have been those that derive from cultural values and practices, material standard of living, social organisation and characteristics of the physical environment. These influences set the range within which health lies; thus, population health reflects population circumstances.⁶ Within that range, the health of individuals is largely determined by personal behaviours and circumstances. But these variations in personal health are like corks bobbing on the ocean surface, while deeper unseen currents determine the average level of the population's health. For example, societies with a high intake of saturated animal fats (such as North America, Scotland, Finland and Eastern Europe) have much higher rates of coronary heart disease than do those with low intakes (such as Japan, China, Italy and Greece). Likewise, ambient environmental exposures – the quality of air, water and food, and the presence of infectious diseases – have been fundamental determinants of the level of population health throughout history.

Global environmental changes belong par excellence in the deep-current category. They are large-scale processes that affect the health of populations, reflecting reductions in the carrying capacity of life-supporting ecosystems. Sustaining the good health of *Homo sapiens* over generations will require maintaining those things we most take for granted: clean air, safe water, adequate food, tolerable temperature, stable climate, biodiversity and protection from solar ultraviolet radiation. We can adapt to varied environments, we can control aspects of the environment and increase the productivity (for humans) of coopted ecosystems – but we cannot live without this ecological support system.

3.1.3 The rise of the modern public health movement

Public health is the organised attempt by society to understand the determinants of poor health and to improve the level, and the equality, of health within a population. A society's particular public health goals reflect the prevailing values and perceptions about the preventability of health problems. The First Public Health Revolution unfolded in nineteenth-century Europe as that society sought to ameliorate the adverse health effects of squalid living conditions: poor sanitation, poor housing, dangerous work environments and air pollution. These health-directed interventions were embedded in a broader matrix of social progress and economic development that contributed other improvements in population health. Some further exploration of this topic will clarify the historical foundations upon which much of contemporary public health practice is built, and will elucidate the relationship between social-economic development and population health.

In contrast to individual-oriented clinical medicine, public health is a social enterprise that studies, improves and manages the health of populations. While clinical medicine seeks to 'cure', public health pursues 'prevention'. These two complementary enterprises began to take formal shape last century, supported by educational, professional and legislative underpinnings. While, historically, some European countries had sought various improvements in the public's health during the Renaissance, the sustained and explicit public health movement began in England last century in response to recognition of the wider environmental influences upon the health of the people. Indeed, even more than parliamentary democracy, railway systems and cricket, England's greatest contribution to modern civilisation may have been civic hygiene. As England's larger cities underwent an explosive expansion in population early last century, a new underclass of crowded, impoverished, poorly fed and powerless factory workers and their families was created. Most were from the countryside where the mechanisation of agriculture, the closure of the commons and the aggregation of landholdings had displaced many rural workers, for whom the lure of possible employment in the urban factories gave some faint hope of survival if not betterment. However, conditions in the cities were unprecedentedly filthy and unsanitary, malnutrition was widespread, infectious diseases rife, and in most cities death rates exceeded birth rates.

The British Parliament finally commissioned Edwin Chadwick, an engineer, to review the 'sanitary conditions of the labouring poor'. Chadwick's report (1842)[7] became a best-seller and, despite some initial societal resistance, stimulated far-reaching social reforms in sanitation, personal hygiene and housing design. It led to the first Public Health Act, of 1848, which enshrined the notion of State responsibility for the public's health. Society's motivation for these early reforms was complex. Lorenzo Tomatis, exploring the historical relationship between poverty and cancer, has argued: 'Disease and destitution may have been considered to be part of the inscrutable plan of the Almighty, but when by injuring and killing the workers they interfered with industrial production and put profit in jeopardy, and when infectious diseases spread from poor to rich districts, then it was time to take action.'[8] Further gains in occupational safety, food quality, personal hygiene and maternal education followed in the late nineteenth century and contributed to the long-term and momentous decline in death rates already well underway. From 1840 to 1940, the overall death rate in England declined by nearly three-quarters, including a dramatic reduction in deaths from infectious diseases.

While much of this gain came from reforms that were directly and deliberately health-related, social change and economic growth on a broad front contributed substantially to improved health. The historian Thomas McKeown has argued that steady improvements in nutrition, starting in the eighteenth century, greatly increased resistance to infectious diseases. This, he says, along with reduced exposure to infectious agents in water and food, laid the foundations of the decline in mortality that ushered in the 'demographic transition' in Europe.[9] That transition entailed a marked decline in mortality, especially infant and child mortality, followed within a generation by a reactive decline in the birth rate (see also chapter 5). Other commentators, while not disputing the strong influence of the invisible hand of rising living standards and social modernisation, argue that targeted public health interventions, including sanitation, vaccination, personal hygiene and improved child health-care, were equally import-ant.[10,11,12]

Here, then, in the nineteenth century, was the first major triumph of modern public health. Similar developments occurred elsewhere in Europe and in the USA. States in the USA began to assume responsibility for quarantine and sanitation from the 1850s onwards, and State Boards of Health were created. Following the national disasters of the Civil War in the 1860s (when 70% of all deaths among troops were from infectious diseases) and an epidemic of yellow fever that swept up from New Orleans in the late 1870s, a National Board of Health was created. The proclamation of the Germ Theory in the 1880s vindicated and consolidated the sanitation strategies, which, by then, were well established in Europe and North America. The beneficial effects of these public health and social measures preceded and exceeded the later gains conferred by clinical medicine. In developed countries this century, immunisation has contri-buted further to population resistance to infectious disease. By the time that systemic antibiotics were developed, many age-old infectious disease scourges such as tuberculosis and diphtheria had already been greatly diminished. Clinical medicine, while easing pain, achieving or hastening recovery from infectious diseases, assisting long-term management of certain chronic non-infectious disorders, and providing personal reassur-ance, has had a relatively small effect on average life expectancy.

To date, few major public health problems have had a supranational or a global dimension. The nineteenth-century developments in sanitation and pollution control arose, and were tackled, at a municipal level. Natural disasters (floods, cyclones, famines) are usually localised. Most outbreaks of infectious disease have been within circumscribed populations, although some infectious disease outbreaks have occurred on a wider front – particularly the bubonic plague (initially known as the 'Black Death') in Europe between the fourteenth and seventeenth centuries, cholera and tuberculosis in nineteenth-century Europe, and the global pandemics of influenza in 1919–20 and AIDS and cholera in recent decades. Health problems caused by direct-acting environmental toxins have usually occurred in a localised context. The more dramatic examples include the London smogs, Japan's Minamata Bay methylmercury toxicity, photo-chemical smog in Los Angeles and the chemical disaster at Seveso (Italy). In the background are other lower-key localised environmental health problems arising from lead in air, soil and water, from traditional urban air pollutants due to fossil fuel combustion, and from countless industrial and agricultural chemicals.

Today, however, the potentially most serious environmental health problems exist on the much larger canvas that describes the biosphere's natural systems. Before moving on to explore their implications for human population health, we should examine the relationship between economic development and population health. It has great relevance to the main topic of this book since the conventional argument that increased population wealth is a prerequisite to improved population health must now contend with the argument that consumption-based economic growth is the source of much of the world's incipient ecological disruption.

3.2 Health, wealth and environment

3.2.1 Introduction: Inequality and health

Health hazards and poor health are distributed unevenly both between and within populations.[12,13] Inequalities in power and wealth between rich and poor countries, amplified by the inegalitarian world trading system, continue to create disparities in the material standard of living, the quality of the local physical environment and the productiveness of local ecosystems. Major disparities in population health are the result.

Within a population, the health of individuals and subgroups is profoundly influenced by local opportunities, resources and social support; social inequalities cause health inequalities. If society's prime commitment is to economic 'growth', unleavened by policies of social justice, then the less powerful sections of society usually bear the brunt of local environmental degradation. This often includes localised exposure to industrial pollutants, reflecting the physical location of lower-class housing. The 1984 Bhopal chemical disaster in central India, killing thousands and maiming many tens of thousands, underscored vividly how the urban poor are at heightened risk because of where they live, often in the shadow of industry. Similar inequalities can be expected in the distribution of the adverse health effects of climate change since, for example, concrete-bound unventilated inner-city housing will be most affected by heatwaves, and coastal populations of peasants and slum-dwellers will be most exposed to rising seas.

Social inequalities in health status may be increased further by associated differences in susceptibility to the effects of health hazards. For example, malnourished children are more vulnerable to infection, particularly respiratory infection. In many countries, both rich and poor, a majority of those living in poverty are women. Typically, this reflects the plight of single-parent women (with low employability) and that of elderly widows. In many countries it reflects the lowly status accorded to women, often associated with their bearing a disproportionately heavy work-load, paid and unpaid. The problems of social class thus become entwined with those of gender. Not only do impoverished women have few resources, less education and little political power with which to protect themselves from adverse environmental exposures (e.g. temperature extremes, industrial pollution, diarrhoeal disease and respiratory infections in their young children) but they often have less time and opportunity to attend to their own health needs.

Good health, however unevenly it is shared, is widely valued. Among society's most underprivileged and socially alienated members there may be a devaluing of personal health, associated with self-destructive behaviours. For most people, however, good health is a desirable end in itself and a means to greater fulfilment and enjoyment. For society as a whole, good health can be a means to many political ends: a healthy labourforce, a healthy army and lower health-care costs. The health of the population thus can be viewed both as an input and as an outcome.

3.2.2 Health as 'input' to economic development

A healthy population can be viewed as an economic resource. The WCED Report (1987) argued that a healthy and well-nourished population facilitates economic development in poor countries.[14] There are several documented examples of how this can happen, although many of the earlier studies were too coarse-grained and insensitive to clearly identify such relationships. By reducing illness and debility, the malaria eradication program in Sri Lanka in the 1940s and 1950s caused a rise in incomes. In Sierra Leone, increases in food and improved nutrition for farmworkers have resulted in greater agricultural productivity.[15] These appear to have been socially beneficial experiences.

However, it is important to examine the sense in which health is an 'input' to economic development. Is good health to be used? – or to be used up? Historically, in Europe and elsewhere, industrial development has not depended on the sustained good health of individuals, but, rather, has depended on a sustained supply of initially-healthy young workers, expendable in the longer term. If the labour supply is seemingly endless and people are desperate for work, as in early industrialising England, in the gold mines of South Africa, or in many migrant or refugee groups today, then an indifferent or callous employer will use up health. In such situations, the source of surplus value in the product is not just the worker's labour, but draws also upon the damaged health of the worker – and, of course, the damaged environment or depleted resource.

Further, if good health is regarded merely as a resource for economic development, then we should ask what is society's purpose in pursuing economic development? There would, I think, be widespread agreement that human society has not evolved simply to promote ever-higher levels of consumption of goods and services, irrespective of people's fulfilment and happiness. Indeed, the population's sustained good health should be part of the overall 'development' of society, a development that encompasses not just capital accumulation but ecologically sustainable economic activity and broad social progress.

Some commentators argue, however, that good health, while socially desirable, must be a deferred outcome – i.e. a cost which can be met by society once sufficient development (whatever that means) has been attained. Indeed, that has been part of the general orthodoxy of 'structural adjustment programmes' recently imposed as a form of economic discipline by the IMF and World Bank upon loan-dependent poor countries. On that view, social progress, greater economic equality and improved population health will occur later as 'trickle-down' consequences of export-earning economic development, and are therefore not made integral to a broader process of development.[16]

The trickle-down view has not been confined to the problems of poor countries. Thomas McKeown's views, despite the radical nature of his critique of the historical irrelevance of organised clinical medicine to last century's decline in British death rates, have given some sustenance to this view. Neoclassical economists perceive in McKeown's thesis support for the notion that gains in health and welfare are best achieved via economic growth.[9] In 1981, the recently-elected Conservative Government in Britain claimed that to pursue equality in health – as had been recently advocated by a Labour-appointed committee of enquiry – would impede economic growth. The Conservatives' preferred strategy was to endure the inequalities, thus permitting faster economic growth that would eventually give more real benefit to the poor than would a policy of instant equality for all.[17] In the upshot, the Thatcher decade achieved no closure of the gap, no incipient trickle of redistributed wealth. How then might better health be achieved as a planned outcome of economic development?

3.2.3 Health as 'outcome' of economic development

The general historical experience of developed countries has been that, over the long haul, increases in economic wellbeing have foreshadowed improved population health. However, linking economic development with better health and lower fertility is not that simple. For a start, unlike Europe last century, today's poor countries do not have access to cheap commodities from exploited colonies. Further, there is the ominous possibility that the selective introduction of medical and sanitary techniques that reduce child mortality, within an otherwise unchanged setting of persistent poverty, population growth and environmental degradation, will amplify the mounting problem of overpopulation. This controversial issue is discussed further in chapter 5.

Much of the health deficit in poor Third World countries reflects the rank poverty, the welfare constraints of export-oriented economic development and the social and environmental adversity that flows from exploitation. With a world-economy that disadvantages poor countries, exacerbated land degradation, rural unemployment, food shortages and urban crowding all contribute to health deficits for the rural dispossessed, the underfed and the slum-dweller. Within many poor countries, the net flow of economic value is from the politically weak rural majority engaged in traditional work-forms to the urban-based market sector. Gradients between rich and poor are as great within countries as between countries. Consequently, in many African, Asian and Latin American countries, the average life expectancy is 20–30 years less than for rich Western countries, and is lower still for the socially and economically disadvantaged in those countries. For example, in Kenya, children living in Central Province are about three times more likely to survive to age five years than those living in nearby, poorer, Nyanza Province, where water supply and sanitation are much worse.

The link between population wealth and health, however, is not rigid. This is illustrated by considering one of the most basic measures of population health – life expectancy at birth.[18] During 1985–90, life-expectancy figures around the world were: Africa, 52 years; Asia, 61 years; South America, 66 years; USSR and Eastern Europe, 72 years; Western Europe, North America, Australia and New Zealand, 75 years. Fig. 3.1 shows the considerable variation in life expectancy among the world's poorest countries. However, when the figures for all countries are looked at in more detail, the graph of life expectancy against per-person income (i.e. GNP divided by population) contains many surprises. Some countries

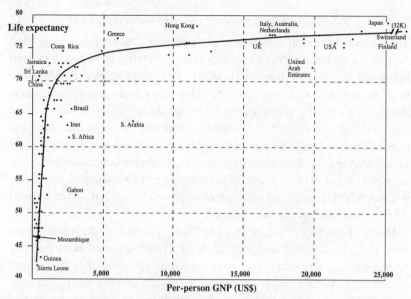

Figure 3.1. Relationship between life expectancy (at birth), and per-person GNP, 1990 figures. (Data source: World Bank, 1992. Graph is drawn freehand)

have much higher life expectancies than their economic indices might predict.

Countries with commitments to social justice, lesser gradients in income, mass education and primary health care tend to get more health for their per-person wealth. Sri Lanka, where life expectancy is expected to reach 74 by 2000 (having been 60 in 1950), Costa Rica, Kerala State (southwest India) and China are four well-recognised examples.[19,20] Their relatively good health has been attributed to the deliberate strengthening of primary health care systems, to the integration of traditional with modern health care and to aspects of social reform. Costa Rica has boldly decided to spend no money on armaments and defence, and, instead, invests its national wealth in social assets.[15] In China, much of the advance must have been due to better nutrition. Between 1950 and 1985, per-person food production in China increased by 75% (despite massive famines due to ideological extremism in the late 1950s), and there were corresponding increases in birth weight, childhood growth rates and a big reduction in infant mortality. In Sri Lanka, Kerala and Costa Rica, the provision of primary health care throughout the population and the education of women are the two strongest predictors of gains in population health.[19] Kerala's matrilineal system confers higher status, better education and more employment on women. Cuba, Sweden and Finland also have higher life expectancy and lower infant and child mortality relative to per-person GNP. This, too, appears to be due to long-term governmental commitments to education and health. Likewise, there have been unexpectedly large health gains in the 'newly-industrialising countries' of eastern Asia (Japan and the four little 'tigers': South Korea, Taiwan, Hong Kong and Singapore) – who must therefore also have been doing something more than just earning money. These examples indicate that it is the configuration of social and economic development rather than economic 'growth' (i.e. increased GNP) *per se* that is the determinant of population health. Indeed, it has been argued that the rapid rise of Japan to the top of the world's life expectancy ladder reflects not just its accumulation of wealth but the relatively even distribution of this increased income throughout the population.[21]

By contrast, life expectancy figures in Hungary, Romania and certain of the oil-rich OPEC nations are below their GNP-related expectations – and, in the cases of Hungary and Romania, have recently been falling.[15] Whereas elsewhere the gains in life expectancy in poor countries have been mainly due to reductions in infant and child mortality, the health deficits in some of the more recently industrialised countries have been due to recent increases in adult mortality. The decreases in male life expectancy in Hungary, Romania and the USSR have been predominantly due to increased cardiovascular disease death rates.[22] Some of these Eastern European experiences illustrate that the headlong pursuit of industrialised material growth can be damaging to both environment and health. Cubatao, in Brazil, and some of the other such 'industrial export zones' created to generate desperately-needed national income have become environmental hazards to the health of those who live and work there.[23]

In developed countries that are already well advanced in material standard of living, the critical determinant of the population's average health status (e.g. life expectancy) may not be the average personal income but the extent of disparity in wealth between rich and poor subgroups.[21] In those countries the prime determinant of personal health seems not to be a person's absolute income but his/her relative status and assets within that society. That is, a sense of deprivation, of being socially and materially excluded from the normal life of society, is a key determinant of personal health. Blacks living in New York's Harlem have had a deteriorating life expectancy over recent decades, and, in the early 1980s, their adult male survival had become worse than for men in Bangladesh.[24] One important consequence of this effect of inequality on population health indicators is that an increase in national wealth, in the presence of continued income

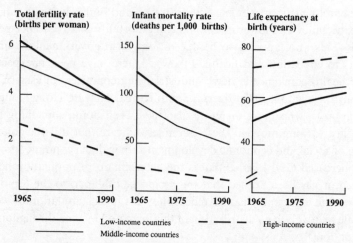

Figure 3.2. Changes in key demographic indices in selected countries, 1965–90, by national income. (Source: World Bank, 1992)

inequality, will not reduce intrapopulation disparities in health. Nor will it lead automatically to commensurate additional gains in population health.

Overall, the broad sweep of human history suggests that economic development – at least up to a certain level – is associated with gains in population health (see Fig. 3.2). The initial, poorly-controlled phases of industrialisation and agrochemical usage entail widespread occupational and residential health hazards to those of lower socioeconomic status. In time, however, health gains have generally flowed on to most or all sections of the population. Nevertheless, the relationship between economic development and improved population health is neither automatic nor linear, and social priorities in the expenditure of national wealth may produce very different health outcomes. Further, population health indicators gloss over inequalities within populations and thus side-step questions about whose health is improving and whose health is paying for the economic development? Most important, in the long run the relationship between economic development and improved population health can only hold so long as the ecological infrastructure is sustained.

3.3 Human health within an ecological framework

Ecology is concerned with the healthy interaction of living creatures in a closed system. Humans interact with each other as well as with other living creatures, and these interactions can have important effects on the health of all partners in the complex closed ecosystem of our planet. We ignore this reality at our peril.

Last, 1987[25]

3.3.1 *Introduction*

In the long-term, the health and survival of a population cannot be sustained if the carrying capacity of its ecosystem is exceeded. For the human species (whether considered as component populations or as a whole) this criterion of living within the carrying capacity of ecosystems is less straightforward than for other species. We have a unique ability to adapt to environmental adversity, to extract additional energy and materials beyond those that flow naturally through ecosystems, and to offset a shortfall in ecological productivity in one location by trading with humans in another location. We can thus live at one remove from direct accountability to local ecosystems – as, for example, do rich urban populations eating year-round imported tropical fruits or dry-land farmers irrigating their crops. For the human species as a whole, however, we cannot sever the connection. Ecological debits may be displaced or deferred, but cannot be indefinitely ducked. For example, food production pressures may lead initially to energy-supplemented, soil-eroding, mechanised agriculture, or the industrialised production of basic goods and services may require increased combustion of fossil fuel. From these activities short-term economic gains will accrue, but the longer-term consequences will be widespread land degradation and global greenhouse warming, respectively. It is this ecological underpinning of population health that is now threatened by the burgeoning size and economic activity of *Homo sapiens*.

Much of the health experience of human populations over the ages, particularly plague, pestilence and famine, has been a reflection of local ecological disruption. The collapse of the agriculture-based civilisation of Mesopotamia 5,000 years ago, the drought-assisted epidemics of plague in densely settled parts of Egypt, Italy and Africa in the second and third centuries AD, the Black Death of fourteenth-century Europe and the decimation of remote aboriginal populations by infectious diseases from European settlers and invaders – all these testify to the potentially disastrous impact upon human health of the disturbance of ecosystems. During this century, the spread of the cholera pandemic to Africa and Latin America, health problems in industrially-blighted Eastern Europe and the former USSR and the famines of Sahelian Africa all remind us of the critical ecological balance between humans, microbes, climate, land, water and air.

3.3.2 Health as an ecological indicator

Various biologists have wrestled with definitions of human health within an ecological framework. Half a century ago, the Russian environmental scientist, Vernadsky, stated: 'No living organism exists on Earth in a state of freedom. All organisms are connected indissolubly and uninterruptedly, first of all through nutrition and respiration, with the circumambient material and energetic medium. Outside it they cannot exist in a natural condition.'[26] In 1957, WHO supplemented its basic definition of health with the statement that health is a condition of the human organism which expresses adequate functioning under given genetic and environmental conditions. Dubos, echoing Darwin, argued that health is 'an expression of fitness to the environment.'[5] These definitions underline the notion that, for any species, optimal biological functioning requires sustaining of the ecosystem within which, and for which, the species evolved.

The health of a population reflects recent and current experiences. Recent gains in population health, however, do not foreshadow further gains. (As the wise managers of investment funds say: 'Past performance is no guide to future profits'.) A plentiful supply of insects this summer improves the prospects for insect-eating birds, and so their fledglings thrive. But if the climate falters and the insects don't breed, then the bird population will exceed the ecosystem's carrying capacity and many underfed offspring will die next year. Similarly, as we begin to disrupt the atmosphere, and, potentially, the climate and weather patterns, average life expectancy is rising in most countries. Those rises reflect recent improvements in the material and social aspects of the human living environment, but, of course, they neither predict nor guarantee the future health of the human species.

We should note here a paradox that is often raised by the advocates of business-as-usual economic growth. If there are hazards to human population health in environmental degradation and ecosystem overload, they argue, then how is it that life expectancy is higher in the rich countries than in the poor countries, and that in most developing countries around the world life expectancy is increasing? As we have already seen, many of the changes in society that accompany economic development are beneficial to health. However, much of this material gain has come from our using up of natural resources, and from our reliance on the capacity of the planetary sinks (see Fig. 2.1) to absorb our material wastes. Meanwhile, the human population has been expanding ever faster. Good health can be attained in the shorter term, whether in Mesopotamia or Massachussets, by living off environmental 'capital'; human culture, via changes in social organisation and technology, can find and mobilise that capital. But, ultimately, a business-as-usual approach will exhaust that capital. Our cultural cleverness cannot avert the long-term need to live within the biosphere's carrying capacity.

This need to sustain the stability of the world's natural systems is a simple, but fundamentally important, notion. Globally, when the load of the world's human population exceeds the carrying capacity of the biosphere – as has recently happened with the stratospheric ozone layer – then biological health is threatened. Regionally, it is clear that population size is already excessive in many parts of the Third World, particularly in sub-Saharan Africa, where local food production (even if it were effectively distributed) cannot match local demand. Those populations are, to some extent, supported by the cultural adaptations of international trade and aid, which, for the moment, disguise the breaching of this basic ecological principle.

3.4 Estimating the effects upon population health

3.4.1 The difficulties

Toxic environmental pollutants act directly on human health. Therefore, their effects can usually be measured – and the aggregate health risk to an exposed population can be estimated. However, for global environmental changes there are two basic obstacles to estimating risks. First, many effects are likely to occur via indirect mechanisms and not by direct toxicity. Second, various of the predicted problems are without precedent, and so we lack empirical (i.e. proof-of-the-pudding) evidence of their effects. Thus, in estimating the impacts of some of the environmental changes, scientists may have to 'fly blind', or at least with partial vision, since we do not know the full range of ecologically-mediated consequences. Nor can we afford to wait for the body-counts. The early multilateral decision to curb the use of chlorofluorocarbons was taken before their ozone-damaging capacity was scientifically confirmed – and well before there was any evidence of adverse effects upon biological systems. That policy change was based entirely on theoretical predictions, although in this case the process and consequences of the damage are easier to assess than are those of climate change or loss of biodiversity.

Environmental health scientists will have to work more by extrapolation, prediction and analogy, and less by observation-based calculation. They will need to communicate more with other scientific disciplines, to understand better the relevant theories and ecological processes. This will not be easy since scientists, in general, like tidy answers. Indeed, the compartmentalised specialisation of most modern science tempts its practitioners to believe that they are working ever closer to indivisible, objective 'truth'. However, there are certain scientific problems which Weinberg referred to as 'trans-science'. These are problems which can be stated in the language of science but which are actually unanswerable via the canons of conventional science.[27] He gave as an example the estimation of the probability of catastrophic nuclear reactor accidents. With current scientific knowledge and methods, predicting the number of extra deaths worldwide that would result from more frequent heatwaves in a warmer world may seem no easier. At best, there will be a wide penumbra of uncertainty surrounding such predictions.

Before considering briefly some of the problems in estimation, we should clarify the concept of 'risk'. Risk refers to the probability of occurrence of a particular event, and it can be expressed as a time-based rate (e.g. 15 heart attack deaths per 1,000 persons per year) or as a cumulative lifetime probability (e.g. one in every ten women develop breast cancer). Such measures enable description of the average risk within a specified population or group and allow the comparison of risks between different groups of people. Now, it is one thing to understand the notion of risk, but, as we shall see, estimating the risk to population health due to ecological disruption poses qualitatively new challenges. Such 'estimates' will nevertheless be crucial inputs to social decision-making in the near future.

3.4.2 Examples of estimations

To estimate the consequences of most global environmental changes, we necessarily depend on extrapolation from analogous exposure circumstances, or else we must do simulation experiments or computer modelling. Laboratory experiments are unlikely to be of much use since they typically entail simplified systems to test the effect of varying just one factor while all other factors are held constant. The real world of ecosystems is not like that; everything else does not stay constant.

Consider ozone layer depletion. We could estimate the increase in the incidence of skin cancer caused by, say, a 5% increase in ground-level exposure to ultraviolet radiation (specifically the cancer-causing band, UV-B) by extrapolation from the already-documented variations in skin cancer rates between geographic locations that naturally experience that same difference (5 %) in UV-B. But if we want to estimate the impact of increased UV-B irradiation upon crop growth, then we may need to carry out a simulation experiment in the field laboratory. The results of that experiment might subsequently be factored into a computerised model of the overall effects of complex climate change upon agricultural production. Those results, in turn, could be entered into a computerised model of the effects of altered food supply upon population health. So the plot thickens with each step!

Alternatively, we could wait and see what actually happens in the fullness of real time! But if we sit back and wait for an unequivocal upturn in the incidence of ultraviolet-induced skin cancers we will lose valuable decision-making time. Given the decades-long 'latency' period for skin cancer, an upturn could take at least 20–30 years to become evident, by which time ozone layer depletion might have increased substantially. Besides, as public awareness increases over coming decades, behaviours related to personal exposure to sunlight will change. People will wear more protective clothing; they will avoid going outside at high-exposure times; and they will use more sunscreen. Those behaviour changes would therefore obscure the essential environmental problem – i.e. a change in the skin cancer-inducing potential of our ground-level environment.

So, what else can we do? We monitor the thickness of the ozone layer; we monitor the UV-B flux at the Earth's surface. Yet how will we know if there are any adverse biological consequences? Perhaps we could monitor early biological effects in exposed human skin – e.g. molecular damage to the genetic material of epithelial cells. But how do we allow for changes in exposure behaviour over time? Are there some sentinel populations of humans whose sunlight-exposure behaviour will stay constant over time? Probably not – although traditional groups like the Inuit eskimos may change little. Perhaps we will need to identify groups of animals, either free-range or experimental, that can provide standardised measures of biological response over time. Hairless pink mice, in cages, could be placed daily in the midday sun for two high-season weeks each year. If we are interested not in skin cancer but in effects on the eye, such as cataracts, then perhaps we should study time-trends in cloudiness of the lens in the eyes of grazing animals (e.g. Australian sheep) or of penguins in Antarctica.

Clearly, we face a generalised difficulty in seeking to estimate the health impact of global environmental change. Stephen Schneider, an American climatologist, argues that while we often lack the information needed to make actual predictions, we can estimate the outcomes of particular scenarios. In relation to global warming, he says: 'it is first necessary to invoke behavioural assumptions about future population, economic, and technological trends ... Although these may be impossible to forecast with confidence, a set of plausible scenarios can be derived. The differential consequences for the climatic forecast of each of these scenarios can be evaluated.'[28] Thus, scientists may use predictive models to estimate outcomes for a given scenario. While that is not a prediction of what will necessarily happen, we may well be able to attach estimates of probability to each scenario.

There are many other sources of uncertainty in our computer modelling of these scenarios. Consider the assessment of the impact of climate change on food production and human health. Four stages of modelling could be envisaged, dealing with: first, the effects of greenhouse gas accumulation upon heat retention, temperature and climate; second, the interactive effects of weather with other variable factors such as increased ultraviolet radiation, groundwater depletion and microclimatic change due to deforestation; third, the patterns of political and social response by humans; and fourth, the impact of altered food supplies upon human health. To date, the simplified computer models used have been confined to the first stage. Uncertainties would dominate the second and third stages. The fourth stage, if and when reached, should be feasible.

Such difficulties are not peculiar to the task of estimating health impact; ecologists, climatologists, agronomists, economists and many others must grapple with similar complexities in relation to the effects of global environmental changes. Many of the problems are of global scope, and we therefore do not generally have the traditional environmental health option of observing before acting. In the more general words of the WCED Report: 'the speed with which changes in resource use are taking place gives little time in which to anticipate and prevent unexpected effects'.[14]

3.5 Summary

The health of populations has always been determined by fluctuating environmental and political fortunes, along with an overlay of cultural practices and social values. As the needs of personal health have been the concern of clinical medicine, so the fortunes of population health have been the concern of public health. The rise of public health ideas was grounded in the alleviation of insanitary living, local environmental pollution, occupational hazards and substandard child-rearing. Nutritional deficiencies, infectious diseases, social deprivation and the effects of exposure to noxious environmental pollutants have been the grist upon which the wheel of public health research and policy has traditionally turned.

Today, the disruption and erosion of the world's natural systems is posing a potentially much greater and much more complex public health problem. If atmospheric pollution increases, if UV flux through the ozone-depleted stratosphere increases, if infectious disease vectors extend their domain because of climatic change, if food supplies become endangered by loss of arable land and by acidification of waterways and if social organisation breaks down under demographic strain, economic depri-vation or competition for dwindling resources, then the health and survival of human populations is endangered. This is a new problem area that will require a new generation of insights, research methods and far-reaching policy responses.

Our burgeoning impact upon the planet will oblige us to take increasing account of the relationship between ecological systems and the optimal biological functioning of member species. That optimal functioning is what we think of as 'good health'. Across many successive generations, population health can only be maintained if the demands made on ecosystems are within their carrying capacity.

References

1. Neel JV. Diabetes mellitus: a 'thrifty' genotype rendered detrimental by 'progress'. *American Journal of Human Genetics* 1962; **14**: 353–62. (Recently this thesis has been questioned. The great increase in diabetes in Nauruans, for example, within the first generation exposed to Western dietary affluence, is less in later generations – and may therefore reflect a heightened susceptibility in persons who begin life mildly under-nourished. Their children may be more 'attuned' to energy-rich diets. See: Hales CN, Barker DJP. Type 2 (non-insulin-dependent) diabetes mellitus: the thrifty genotype hypothesis. *Diabetologia* 1992; **35**: 595–601.)
2. World Bank. *World Development Report 1992. Development and the Environment*. Oxford: Oxford University Press, 1992.
3. See, for example: Peto R. Statistics of chronic disease control. *Nature* 1992; **356**: 557–8; and: Lohman PHM, Sankaranarayanan K, Ashby J. Changing the limits to life. *Nature* 1992; **357**: 185–6. (The arguments offered are fine – as far as they go – but they address a world free of the major health hazards that ecological disruptions portend.)
4. Dubos R. *Mirage of Health*. New York: Harper, 1959.
5. Dubos R. *Man Adapting*. New Haven: Yale University Press, 1965.
6. Rose G. Sick individuals and sick populations. *International Journal of Epidemiology* 1985; **14**: 32–8.
7. Chadwick E. *The Sanitary Condition of the Labouring Population of Great*

80 The health of populations

Britain. Facsimile edition. Edinburgh: Edinburgh University Press, 1965
(first published 1842).

8. Tomatis L. Poverty and cancer. *Cancer Epidemiology, Biomarkers and
   Prevention.* 1991; **1**: 167–75.

9. McKeown T. *The Modern Rise of Population.* New York: Academic Press,
   1976.

10. Szreter S. The importance of social intervention in Britain's mortality
    decline 1850–1914: a re-interpretation of the role of public health. *The
    Society for the Social History of Medicine* 1988; **1**: 1–37.

11. Preston SH, Haines MR. *Fatal Years: Child Mortality in Late Nineteenth
    Century America.* Princeton: Princeton University Press, 1991.

12. Mercer A. *Disease Mortality and Population in Transition.* Leicester:
    Leicester University Press, 1990.

13. Powles J. Changes in disease patterns and related social trends. *Social
    Science and Medicine* 1992; **35**: 377–87.

14. World Commission on Environment and Development. *Our Common
    Future.* Oxford: Oxford University Press, 1987.

15. World Bank. *World Development Report* 1991. *The Challenge of
    Development.* Oxford: Oxford University Press, 1991.

16. Blomstrom M, Hettne B. *Development Theory in Transition. The
    Dependency Debate and Beyond: Third World Responses.* London: Zed
    Books, 1984. But see also discussion of the current variability in the
    application of adjustment programmes, in Ribe H *et al*, 1990 (ref. 10,
    chapter 12).

17. Strong PM. Black on class and mortality: Theory, method and history.
    Journal of Public Health Medicine 1990; **12**: 168–80.

18. The life expectancy of the *average* new-born baby is an estimate. The
    estimate is based on combining the risks, calculated from *within the current
    population*, of dying at each successive age. That is, it estimates for how
    long the average baby would survive if its whole-of-life experience were that
    of today's population.

19. Caldwell JC. Routes to low mortality in poor countries. *Population and
    Development Review* 1986; **12**: 171–220.

20. Behm H, Soto AR. Costa Rica. In: UN Department of International
    Economic and Social Affairs. *Child Mortality in Developing Countries:
    Socioeconomic Differentials, Trends and Implications.* New York: UN, 1991,
    pp 38–52.

21. Wilkinson RG. Income distribution and life expectancy. *British Medical
    Journal* 1992; **304**: 165–8.

22. Rich V. Russia: Pollution takes its toll. *Lancet* 1992; **339**: 295–96.

23. Cooper Weil DE, Alicbusan AP, Wilson JF, Reich MR, Bradley DJ. *The
    Impact of Development Policies on Health.* Geneva: WHO, 1990. A further
    example comes from Indonesia – ostensibly one of the late-developing
    economic 'success stories' of Asia. Recent census data indicate that the
    decline in infant mortality in Indonesia levelled off during the late 1980s
    and the health differentials between rich and poor have widened (Hugo G,
    personal communication).

24. McCord C, Freeman HP. Excess mortality in Harlem. *New England Journal
    of Medicine* 1990; **322**: 173–7.

25. Last JM. *Public Health and Human Ecology.* Norwalk, Connecticut:
    Appleton and Lange, 1987.

26. Vernadsky VI. The biosphere and noosphere. *American Scientist* 1945; **33**: 1–12.
27. Weinberg A. Science and trans-science. *Minerva* 1962; **9**: 220–32.
28. Schneider S. The science of climate-modelling. In: Legget J (ed). *Global Warming. The Greenpeace Report*. Oxford: Oxford University Press, 1990, pp 53–67.

4

System overload: ancient and modern

4.1 Introduction

The previous two chapters have examined the nature of ecosystems and how human population health is affected by economic circumstance and ecological supports. I have argued that the carrying capacity of natural biophysical systems necessarily sets limits upon the growth of population and, for humans, the growth of material aspiration. This chapter explores aspects of these relationships between human biological health, the viability of society and the encompassing natural systems. The examples are both historical and contemporary.

Human population growth has been an episodic process. The historical pattern has been for human population growth to press against the limits imposed by technologically exploitable resources. All other living species exhibit similar population dynamics. However, as we have seen in chapter 1, the one (huge) distinction is that humans have extended their biologically-endowed powers of resource exploitation with an exponentiating cultural-technological capacity. This cultural overlay has radically altered the limits of exploitable resources. It has allowed humans to move to a much higher plane of resource use and, therefore, to maintain a much-expanded population. However, expansion of the environment's carrying capacity (see glossary) for the human species is usually achieved at the cost of reductions in biodiversity or ecological 'services'. This cooption and simplification of nature entails some unavoidable longer-term risks.

The archaeological landscape is littered with the remains of earlier human civilisations that flowered and withered, often reflecting the changing fortunes of the human–environment relationship. Looking back to the dawn of *Homo sapiens*, the American historian William McNeill contrasts the level of competition from pests and pathogens faced by early

82

humans in tropical eastern Africa with that in temperate Europe into which they subsequently migrated.[1] The domination of the simpler, non-humid, European ecosystems was much easier for humans – who were an introduced 'exotic' species. This, he argues, opened up a new power relationship for humans *vis-a-vis* their environment, a relationship that differed radically from the intrinsically limited power of humans in tropical Africa where ecosystems were (and are) much denser and richer in parasitism. Hence, over subsequent millenia, human numbers increased more rapidly in Europe than in Africa.

Human trade, conquest and colonisation have, with increasing intensity, crossed ecosystem boundaries, and have often transported infectious disease organisms between geographic regions. The introduction into fourteenth-century Europe of the bubonic plague (Black Death) organism, *Pasteurella pestis*, depended on its hitching a ride in rat-borne fleas aboard trade caravans returning from China. The bacterium, transmitted by flea-bite, infects the lymphatic glands (the 'bubos') and causes an acute, painful and florid disease that is almost always fatal. Expansion of the Mongol Empire in the thirteenth and fourteenth centuries had opened up trade routes across the Steppes of Asia, where burrowing wild rodent populations are naturally (and benignly) infected with the plague organism. Trade-borne black rats first carried the bacterium to China, where an outbreak of plague occurred in 1331. Subsequently, in 1346, the bacterium arrived at the Crimea, the trade-shipping gateway to Europe. The Black Death swept through Europe in the next three years, killing about one-third of the population. Waves of recurrent infection wiped out another 10–15% of the population over the next half-century. Simultaneously, there was a dramatic population decline in the Islamic Middle East which may also have been due to the trade-transmitted bubonic plague and its spread around the eastern Mediterranean.[2]

The subsequent decline of the plague may have reflected social and behavioural change in Europe as the Little Ice Age began and temperatures fell by 1–2 °C over several centuries (see Fig. 4.1).[3] After London's Great Plague, in 1665, the plague began to vanish in northwest Europe in the late seventeenth century. Earlier that century, the Nordic settlements in Greenland (the 'Westvikings') were extinguished by the unmanageably cold weather.[4] These once-thriving colonies had been established centuries earlier during the approximately 600 years of mediaeval warmth that peaked in the 'little climatic optimum' in the eleventh and twelfth centuries. This congenial climate had facilitated the spread of the Nordic empire throughout coastal and riverine Europe and up into the Arctic Circle. In

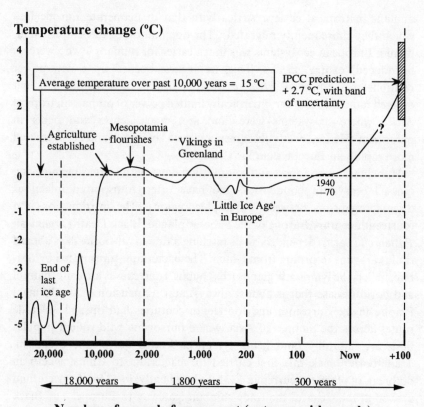

Figure 4.1. Variations in average global temperature over the past 20,000 years. The prediction of the International Panel on Climate Change (see chapter 6) for the coming century is shown. (Multiple sources)

similar vein, other historians have argued that the decline and collapse of the ancient Mycenean civilisation was exacerbated by climatic change. Shifts in the path followed by winter storms probably deprived the cities of Mycenae of water, much of which had already been committed to agricultural irrigation.[5]

4.2 The decline of ancient civilisations

Early civilisations depended critically upon the sustained productivity of local agriculture and supplies of freshwater. Societal stresses occurred often in response to climatic fluctuations and the overloading of agriculture by expanding populations. The demise of various earlier civilisations illustrates the process of collapse brought about by non-sustainable use of local ecological systems.

The ancient civilisation of Mesopotamia, located in today's Iraq along the valley of the twin rivers, Tigris and Euphrates, provides a classic example of the human-environment interaction question. This area was one of the first foci of widespread settled agriculture, with hundreds of small farming communities forming from around 5000 BC. Because the area lacked rainfall, farming depended on irrigation. Over the next two millenia, towns and then cities formed, as centralised political control, religious authority, social stratification and local warfare emerged. The early cities of southern Mesopotamia flourished as independent city states during the fourth millenium BC, and, by around 3000 BC, became the kingdom of Sumer. Around 2400 BC, outsiders conquered Sumer to form the kingdom of Akkad. Then after an increasingly unstable period of 600 years, the Babylonians of northern Mesopotamia conquered the rapidly declining Sumerian civilisation. This story is of interest to us here because it is widely argued that the fragility and abuse of Sumeria's artificial agricultural system eventually brought about that civilisation's downfall.

There is an unusual wealth of written documentation about human settlement in the Tigris–Euphrates Valley, including records on clay tablets reaching back to 3000 BC (Sumer was the first literate society in the world). Altogether, the record of settlement has been reconstructed over six millenia,[6] during which the population fluctuated dramatically between peaks of 600,000 around 2000 BC, then of 1.5 million around 800 AD, and today approximately two million. The early growth of the Mesopotamian population reflects the success of their irrigation system in exploiting the fertile alluvial soil. Eventually this large-scale irrigation caused increasing soil salinity and, during the third millenium BC, wheat began to be replaced by barley, a more salt-tolerant crop.[7,8] Between 3500 and 2500 BC, wheat declined from contributing one-half to one-seventh of the total crop yield, and after 2000 BC no more wheat was grown in southern Mesopotamia. (This substitution of barley for wheat is an example of what Stephen Boyden calls *antidotal*, or curative, cultural adaptation as opposed to *corrective*, or preventive, adaptation, in response to environmental deterioration.[9]) Even more ominously, records show that between 2500 and 1700 BC the total crop yield fell by two-thirds. Accompanying records, referring to the earth having 'turned white', indicate massive salinisation.[8]

Similar collapses due to the overloading of agricultural systems subsequently occurred in central and northern Mesopotamia. Adverse climatic changes may also have contributed. Progressive delays in the time of harvesting during the last millenium BC suggest the emergence of warmer drier conditions (see also Fig. 4.1).[10] These would have necessitated further increases in irrigation, exacerbating soil salinity. Meanwhile, demographic and social pressures mounted to boost crop production – and therefore to forego the long-established fallowing routine which allowed the seasonally-raised salty water table to recede again. Computer simulation techniques have been used by historians to explore the impact of climatic variation on the viability of the Mesopotamian civilisation.[6] They conclude that, with a fixed agricultural resource base and allowing for food storage and other social adaptations, prolonged climate change would have caused collapse. In these computerised models, if climatic variability is compounded with population overload, internal societal stresses and agricultural malpractice, then a cascade effect ensues and ecological collapse occurs. Other possible contributory explanations in an overloaded, decaying, society include increased bureaucratic inefficiency, excessive taxes on farmers (causing them to overirrigate) and armed conflicts with nomadic societies.[6]

Equally dramatic surges and declines in population occurred in other ancient agricultural centres – Egypt, the Indus Valley, China, and, later, Mesoamerica and Peru. In each case, the appropriation of agricultural surplus enabled the development of urban social-political superstructures. The Indus Valley, in northwest India (now Pakistan), flourished during the third millenium BC and then collapsed from salinisation and deforestation. The subcontinent's focus of agriculture then shifted to the watery, malarious, Ganges Valley. Egyptian civilisation has been more stable and durable, with a lesser reliance on artificial irrigation. While the vicissitudes of plague, drought, famine and armed invasion have made for a roller-coaster population profile, Egypt has been perennially nourished by the seasonal flooding of the Nile. As the mountain snows of Ethiopia melt, the resultant surge of water bears a rich mix of silt and nutrients down the Nile Valley. Recently, however, this life-sustaining annual cycle has been partially obstructed by the Aswan Dams. Further, as Ethiopia's population surges and its rain-fed highland topsoils erode, the increasingly desperate need for irrigation water threatens Egypt's seasonal supply of river water.

On the other side of the world, the collapse of the Mayan civilisation in Mesoamerica, during the ninth century AD, may have resulted from excessive deforestation and soil erosion.[11] This powerful civilisation, centred in Teotihuacan (southern Mexico), formed long after those of Eurasia because of the lack of animals suitable for domestication, the naturally low yield of maize (the staple crop) and the difficulties of rain-forest agriculture.[8] Subsequent selective breeding of maize enabled the production of agricultural surplus, the main stimulus for urban-based civilisation. Researchers who have studied hieroglyphic records think that the population increased to precariously high levels, at the limit of their agricultural technology, and that ecological disasters and starvation probably ensued. Likewise, a thousand years ago, the Anasazi of Chaco Canyon (today in arid New Mexico) appear to have thrived in a well-wooded and well-watered environment, only to succumb to a self-inflicted ecological disaster centuries later. Again, deforestation, soil erosion and falling water tables probably caused the collapse of this civilisation.[12]

Having seen how these early agricultural practices sometimes caused ecological disaster and social collapse, we will explore next the history of human food and nutrition. Animals eat to obtain the energy and materials needed for bodily structure and function. Hominid and, more recently, human biological evolution has necessarily been intimately related to the naturally available food sources. Hence, the consequences of human dietary variations, over time and between cultures, illustrate well the ecological determinants of our health. Subsequently, we will see how the drive to produce more food for an ever-expanding world population, with changing food preferences, is ushering in new ecological pressures and problems.

4.3 The human diet: 'evodeviation' and health

4.3.1 The three historical dietary eras

Food is a fundamental determinant of human health. Human biology, forged by evolution over several million years of relatively stable hunting-and-gathering by hominids, is attuned to a certain range of each of the various macronutrients and micronutrients. Diets that are deficient – or excessive – in these nutrients may cause biological malfunction. In the late twentieth century, the 'affluent diet' of rich countries, with its surplus energy intake and a proportional increase in energy-dense foods, is a major cause of many chronic health disorders.[13]

Human societies have moved through three major historical eras: hunter-gatherer, peasant-agricultural, and industrialised society.[9,14] Each transition entailed profound changes in food supply; the latter two have encouraged population surges. Anthropological studies of the diets of today's surviving hunter-gatherer societies, such as the Bushmen of

Table 4.1. *Comparison of nutrient profiles of the typical diets of the three main eras. Estimates have been averaged from several sources. The three major nutrients are expressed as a percentage of total energy intake.*

Nutrient	Hunter-gatherer 1,000,000 years ago	Peasant-agriculturalist Beginning around: 10,000 years ago	Industrialised 200 years ago
Protein (%)	30	10–15	15
Fat (%)	15	15	40
Carbohydrate (%)	55	70–75	45
Simple sugars[a]	5	5–10	20
Complex CHO (starch)	50	65–70	25
Dietary fibre (g/day)	40	80–100	15–20
Salt (g/day)	1–3	3–5	10
P:S ratio[b]	1.4	1.6	0.5

[a] Includes refined sugar and naturally-occurring simple sugars
[b] Polyunsaturated:Saturated fats. (The degree of 'saturation' of a fat refers to the extent to which its central chain of carbon atoms has a full complement of attached hydrogen atoms.)

southern Africa's Kalahari Desert, indicate that, historically, the fat intake in temperate climates was 15–20% of total calories – about half the current intake in developed countries – and had a much higher ratio of unsaturated to saturated fat. Fibre intake appears to have been about 40 grams per day, compared with 15 grams in developed countries, and ascorbic acid (vitamin C) intake was four times higher. Therefore, assuming that modern human beings (*Homo sapiens sapiens*) first appeared around 100,000 years ago, the modern species has subsisted for well over 99% of its history on low-fat, high-fibre diets rich in ascorbic acid and calcium. There is therefore a potential dissonance between human biology, as shaped by the palaeolithic diet, and the contemporary 'affluent' diet.

Long-hidden by the mists of ancient time, the profile of the early human diet is now emerging from the work of palaeontologists, anthropologists and comparative biologists.[15] Two American scientists, Eaton and Konner, have reconstructed the nutrient profile of the palaeolithic diet that prevailed from around 400,000 BC to 20,000 BC and have compared it with the contemporary Western diet.[16] An outline of the three main dietary eras is shown in Table 4.1. Relative to the diet eaten by our predecessors, the 'affluent' diet entails a high consumption of energy-dense foods, with a high intake of fat from domesticated animal sources and of foods prepared with added fat, sugar and salt. Saturated (storage) fats have burgeoned in the affluent diet while unsaturated (structural) fats have diminished. Complex carbohydrate and dietary fibre intake have decreased, as has that of micronutrients (vitamins and trace minerals, including potassium and calcium).

Boyden refers to this dissonance as 'evodeviation'. He argues that, for any species, if the conditions of life deviate from those of its evolutionary formative natural habitat, then its members will be less well suited, physiologically and behaviourally, to the changed conditions.[9] Human cultural evolution, determining the sources and methods of food production, has occurred at an ever-increasing pace, whereas human biology has simply not had time to adapt evolutionarily. With a few known minor exceptions (for example, the retention into adulthood of the intestinal enzyme, lactase, normally present in childhood for digesting mother's milk) probably no diet-related genetic evolution has occurred in humans since the agricultural revolution began 10,000 years ago, a mere 400 generations ago.

4.3.2 From palaeolithic palate to agricultural revolution

Humans, by evolutionary origin, are non-specialised omnivores, capable of surviving on a varied mix of natural foods. Recent evidence, from elemental analysis of fossil hominid bones, suggests that our ancestors of two million years ago were omnivores who ate some meat. This biological versatility would have enhanced human survival when food supplies were erratic. Although the supply of certain foods was often precarious, the mix of available nutrients (carbohydrates, fats and protein) was relatively constant. Our hominid ancestors subsisted for much of their approximately four million years of evolutionary history on a diet high in vegetables and low in animal content. The 'technology' of hunting wild game, by an essentially non-predatory species, was primitive – and usually unsuccessful.[12] Scavenging from half-eaten carcasses occurred, judging by the detritus of long (marrow-rich) bones associated with early human archaeological remains.

The economy of the early human hunter-gatherer could only support small roving bands. In the several millenia before the agricultural revolution, it is likely that humans, bolstered by a widening armamentarium of tools and knowledge, adopted a broader spectrum of subsistence foods.There was an increase in the hunting of large animals; the fossil records also feature fish, shellfish and small game, along with the tools used for processing plant foods.[15] The typical ratio of animal to plant foods was about 1:2. Dietary fibre intake, much higher than in contemporary developed society, was predominantly from fruit and vegetable, rather than from grains.The escalating success of humans as hunters, in the late palaeolithic period, is thought to have caused a mix of ecological pressures and rapid population growth that helped impel humans towards the cultivation of cereal grains and the herding of animals. Rapid and fluctuating climatic changes after the ice age contributed to this pressure, particularly in the populous eastern Mediterranean Levant where the Natufians were struggling to sustain a food production base, and in the Zagros mountains (near today's Iran–Iraq border). Over three millenia, from around 10,000 BC, there was a slow but steady transition to a combination of hunting with herding of goats and sheep and with the more systematic reaping of various wild plants. Gradually, emmer and einkorn wheats, barley and certain legumes were singled out for more intensive cultivation, leading inexorably to the preferential gathering and re-sowing of the larger-grained plant strains.[8,9]

With the gradual onset of agriculture, food supplies became simpler and less seasonal. A dramatic substitution of plant foods for meat occurred. There is human skeletal evidence that this qualitative change in diet, with lesser animal protein, was reflected in a 5–10% smaller body size in agrarian humans relative to their immediate hunter-gatherer predecessors. Nevertheless, since the game meat eaten by pre-agrarian humans had been very low in fat, particularly saturated fat, the agricultural revolution entailed only limited changes in dietary nutrient profile. These changes included small decreases in the intake of fat, protein and certain vitamins and a larger increase in the intake of carbohydrate foods. It also entailed a reduction in the diversity of foods eaten.[14,15]

Improved food security and the capacity to produce more food (in return for much more work) led to a surge in population. However, farming is intrinsically more vulnerable to climatic and other disasters than is opportunistic food-gathering, and bad seasons often meant hunger. As populations grew, the continuing struggle to produce sufficient food to feed everyone was exacerbated by a reduction in the proportion of actual food-producers in urbanised society. Undernutrition would thus have been a chronic problem within at least part of the population. Since densely settled agricultural communities are at greater risk of infection, any additional vulnerability because of undernutrition would have acted to slow the rise in population. Further, the reduction in the range of foods eaten would have often resulted in micronutrient deficiencies.[17] There is evidence, for example, of vitamin deficiencies such as night-blindness and beri-beri from the second millenium BC.

Because of the need to feed urban populations, problems of contamination of stored foods would have sometimes caused widespread illnesses. Ergot (wheat rust) appears to have caused outbreaks of poisoning in Babylonian and Assyrian cities and, later, in the cities of mediaeval Europe. Ergot scarring has been found in foods in the preserved stomach contents of Grauballe Man, an Iron Age individual from a Danish peat bog. Meanwhile, a further impact on human health, reflecting the radical change in ecological relationships caused by the domestication of animals, was the ready transfer to humans of new types of infections – either directly or after minor genetic adaptation. These included smallpox, brucellosis, anthrax and tuberculosis from cattle, influenza from hogs, leprosy from water buffalo, the common cold from horses, and measles, rabies and hydatid cysts from dogs.

As agriculturally-based civilisations rose and fell, Europe – for long a backwater of history[18] – gradually developed a momentum that reflected its fertile soils, benign climate and the incessant fertilisation of new techniques and resources from invading and migrating populations. The advent of crop rotation and the sod-turning plough, along with the social liberalisation in the wake of the fourteenth-century plague epidemic, led to improved agriculture. Nevertheless, the usual situation in Europe (as elsewhere) was of dietary monotony, chronic malnutrition, and periodic and sometimes massive famines. During the unusually cold seventeenth and eighteenth centuries, severe famines occurred widely. Throughout, susceptibility to infection and high death rates were the norm.

A second 'agricultural revolution', dating from the late eighteenth century in Europe, led to increased supplies of food. After a succession of debilitating wars of religion, a new system of law and order was established with the rise of nation-states. Improvements in agriculture occurred, assisted by the introduction of the energy-dense potato from the Americas. After being long treated with disdain, the potato rapidly became a dietary staple in Europe from around 1800 and eclipsed grain within several decades.[19] Rice also assumed increasing popularity. Other new crops included clover, assorted legumes, and, in southern Europe, maize. Other gains came from mixed farming, winter feeding of animals with clover and root crops and improved farm implements, including various early forms of mechanisation.

Once again, as with the first agricultural revolution, population increases occurred. In Britain the population trebled in size between 1700 and 1850 without significant import of food. The consequent migration of surplus population to the cities provided the workforce for the industrial revolution, and also created a new pressure on food supplies – including the need to feed rapidly increasing numbers of city-dwellers. This provided a new impetus for technological developments in food processing, transport and preservation. It also caused waves of migration from England and other parts of Europe to the new frontiers of North America, South America and Australasia, and, relatedly, an increase in food imports to Europe from those and other colonial sources. The cultivation of beef – an environmentally bruising activity – surged at home and abroad, as a preferred high-status food of the European upper class and bourgeoisie.

4.3.3 Modern times – and modern diseases

Along with increased production and improved distribution (especially canning and long-distance refrigeration) of food late last century came enormous technological changes in food processing. Further, the selective breeding of plants and animals was intensified. Hence, the nutrient profile of the modern diet in industrialised societies is far removed from those of the hunter-gatherer and peasant-agriculture phases of human existence. This new abundance and security of food caused major public health gains by eradicating starvation and nutritional deficiencies. It also foreshadowed a new type of diet with a much-changed nutrient composition. The traditional working-class diet of bread and cereal, with few dairy products, negligible meat and only occasional fresh vegetables, was transformed. The per capita consumption of fat and sugar increased ten-fold to twenty-fold in England over the past 250 years, while the consumption of complex carbohydrates and dietary fibre has declined substantially.[13,20] Between the eighteenth and nineteenth centuries the per-person consumption of meat in Europe increased by a remarkable one hundred-fold.[8] By 1900, fleets of refrigerated ships were delivering a cornucopia of beef to England from deforested Argentina and Australia. Against the timescale of human history and, particularly, biological evolution, these were very dramatic and rapid changes in the population's food supply.

As other countries have developed, the diet of wealthy subgroups has changed in the direction described above.[13,21] The immediate health benefit of this assured supply of varied food has been the elimination of starvation, the near-elimination of micronutrient deficiency diseases and a reduction in childhood susceptibility to infectious diseases. However, while the affluent diet has increased life expectancy (primarily by reducing childhood deaths), it has also caused increases in a range of chronic (degenerative) diseases: heart disease, stroke, diabetes, various cancers, dental caries, gallstones, chronic bowel disorders and various bone and joint disorders.[13] Voluminous epidemiological research between and within populations has established the importance of diet in the most common causes of premature death in developed countries, namely cardiovascular diseases and cancer.[13]

These diet-induced changes in the pattern of chronic diseases have occurred widely in various developing countries, where the chronic diseases typically increase first in the urbanising middle-class minority. The tendency to abandon traditional (predominantly agrarian-based) diets is exacerbated by inappropriate community perceptions of what is a 'good' diet and a counter-traditional expectation that every day can now be a feast day. This dietary transition, which is extreme in some small countries such as Nauru (where 70% of persons aged over fifty have diabetes), is evident in parts of Africa, South America, Polynesia and Asia.[13,20] A huge country-wide study of diet and health across a spectrum of population groups in China, conducted collaboratively by Chinese, English and American epidemiologists, has provided strong evidence that the traditional near-vegetarian agrarian Chinese diet, with 10–15% of (an otherwise sufficient total of) calories coming from dietary fat, is generally good for human health – being associated with very low rates of death from heart disease, stroke, diabetes and many cancers.[22]

Within rich countries, the groups most affected by 'affluent' dietary imbalances now tend to be those at the social margins, who have acceded to such diets relatively late. Further, in the absence of knowledge, money and healthy food choices, they often eat a more extreme version of the affluent diet, high in refined carbohydrates (sugar and white flour), processed foods, fatty meat and alcohol – hence their typically higher levels of obesity, blood pressure and blood cholesterol. In Australia, Aboriginal people living on the urban fringe or in reserves, and therefore removed from their traditional lifestyle and diet, have unusually high rates of adult-onset (obesity-related) diabetes mellitus, high blood pressure and heart disease. Strong corroboration of this causal relationship comes from studies done by Kerin O'Dea, showing that young Australian Aboriginal males with early-stage diabetes undergo clear improvements in their prediabetic metabolism after 'going bush' and reverting to traditional hunter-gatherer diets for a month or so.[23] Overall, there is abundant evidence that the modern problem of overnutrition from an excessive, energy-dense, affluent diet represents a health-endangering evodeviation – although recent downturns in heart disease in many rich countries may, in part, reflect the benefit of a newly reattained dietary diversity.

In summary, over most of human history dietary differences between societies have reflected differences in local environment, climate and associated cultural traditions. However, rapid developments in the technology of agriculture, animal husbandry and food processing in Western populations in recent centuries have radically changed the type and amount of food available. The main benefit has been the general elimination of undernutrition and the assurance of a regular food supply. The accompanying disadvantage has been the onset of widespread overnutrition from the abundant energy-dense foods and the consequent rise of chronic degenerative diseases as the main causes of premature death in developed countries (see Fig. 4.2). Those diseases are now also increasing, in both absolute and proportional terms, in urban middle-class populations of the developing world.

4.3.4 Loaves, fishes and fatted calves: the ecological consequences

There are two sides to the coin linking food, health and environment. While risks to health result from deviations from the dietary nutrient profile for which human biology evolved, so the production of food in an overcrowded and unequal world poses risks to the environment.

The world's poorest one billion people have barely enough food for subsistence; many are malnourished, including the estimated 30 million currently facing severe malnutrition in sub-Saharan Africa. They eat mostly cereal grains, with occasional seasonal fruit and vegetables. The three billion other people in developing countries have more secure supplies of the same, plus regular but small amounts of meat, fish and dairy products. Fat typically contributes less than 20 % of their total calories. The 1.3 billion people in industrialised countries eat diets with a much higher content of meat, fish and dairy products, with regular and abundant supplies of cereal grains, fruit and vegetables, and with a range of ever-present processed and packaged foods. The rich meat-eaters obtain 35–40 % of their calories from fat. In 1989, the average per-person consumption of red meat was 96 kilograms in East Germany, 76 kilograms in the USA, 21 kilograms in China, 12 kilograms in Egypt, and 1 kilogram in (predominantly Hindu) India.[24]

The poorest one billion often have to farm in marginal and deteriorating soils. They have few resources with which to manage and protect their

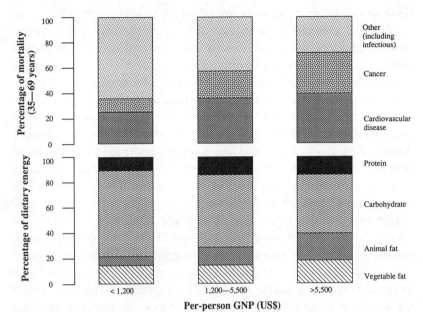

Figure 4.2. Relationship between nutrient composition of diet and profile of adult mortality by major causes, for countries grouped by per-person GNP. (Data source: WHO, 1990 – ref. 13)

farmland. Fuelwood for cooking is in rapidly shrinking supply. Their day-to-day challenge is to sustain life in the short term, not to sustain their resource base in the long term – and so their farmland and woodland becomes degraded. At the other end of the scale, the meat-eating quarter of humanity poses enormous demands on the world's environment. Forty per cent of the world's grain production is required to feed and fatten the livestock. As along all food chains, there is a loss of energy-efficiency: 1 kilogram of American beef requires 5 kilograms of grain – not to mention the required extrinsic energy equivalent of 9 litres of gasoline, the contribution of belching bovines to methane emissions, and the environmental impacts of intensive mechanised agriculture.[25] We will return to these issues in chapter 8.

4.4 Energy use: environmental sources and impact

4.4.1 Introduction

Nearly all of our energy comes from the Sun. Solar energy is available as sunlight, as kinetic energy through wind and falling water and as heat.

Sunlight can be used indirectly, by eating plants, fermenting them or by waiting for millions of years and then burning fossil fuels. Today, photovoltaic technology (as used in solar-powered cars or calculators) enables us to use sunlight directly. There are several non-solar energy sources. Tidal power depends on the movement of the moon around Earth. Uranium, a very heavy element, was formed by nuclear fusion reactions within ancient large stars, which, when they exploded, scattered uranium and other elemental cosmic debris – some of which coalesced into planets. Atoms of uranium are unstable and gradually undergo radioactive decay to lighter atoms, which entails the emission of high-energy radiation (either as waves or particles). Radioactive decay in the bowels of the Earth sustains the heat contained in geothermal energy.

Life on Earth can be thought of as a way of *intensifying* the flow of energy. In nature, the molecular building blocks, or 'nutrients', cycle endlessly between soil, water, air and living species. It is the energy input that drives the cycle and enables the assembling of molecules into integrated organic structures, capable of self-replication – i.e. 'life'. Before the advent of humans, all life-forms had lived in balance with the solar energy arriving at Earth's surface. Plants trap incoming sunlight; animals get their energy by eating plants or one another; and so energy passes up the food chain (see also Fig. 2.1). Then, several hundred thousand years ago, humans began to use wood-fire (releasing short-term *stored* solar energy) to cook meat, keep warm and modify the local woodland. Then, having spread into temperate climes where they first decimated the local large-animal supplies of dietary energy, agrarian humans began to nurture their own energy-trapping plants and edible animals, to harness beasts of burden, to use water power and to create hotter fires for the smelting of metals. Sea-faring communities began using wind energy. These uses of deliberately harnessed energy enabled human colonisation and trans-formation of less familiar, more distant environments.

Within the last thousand years or so, a quantum-like leap was taken when combustible coal was discovered (apparently first in China). Suddenly, the reserves of energy available were not just the petty cash accumulating from yesterday's or last year's capture of solar energy, but were the fossilised fortunes stored up from millions of years ago. Human ecology began to change fast, and the environment with it. The industrial revolution has transformed the power of Western populations to extract and use materials, to create wealth and physical comfort and to migrate and colonise. Subsequently, the partial spillover of energy-based tech-nology into Third World populations, applied particularly to improved sanitation, control of infectious diseases and the storage and distribution of food, underwrote the unprecedented explosion in their population numbers that has occurred over the past fifty years. We will take up that story in the next chapter.

More recently, we have discovered how to tap into another part of the universe's stored energy, the energy emitted during radioactive decay (nuclear fission) of uranium. Optimistic nuclear scientists imagine that we are on the brink of achieving nuclear fusion of simple, light-weight, atomic nuclei – just as happens in the core of stars – to produce heavier elements and the release of energy. We may yet learn to create domesticated mini-suns with which to generate more energy!

4.4.2 Societal patterns of energy use

Industrialisation and economic development have required large and concentrated sources of energy. Early industry used on-site generation of energy, usually from fossil fuels. Around the world, wherever industrialisation has occurred, fossil fuels have been the dominant source of energy. These fossil fuels (initially coal, then more recently oil and gas) have been cheap because the true costs to the environment, ecosystems and human health have, as the economists say, been 'externalised'; the costs have been passed on to other species, distant human populations (as with acid rain) or future generations. Of all the energy consumed commercially in today's world, 88% comes from fossil fuels, 7% from hydropower and 5% from nuclear power and other minor sources.[26]

The discovery of electricity, last century, brought a new flexibility in energy supply. Today, electricity generation accounts for about one-third of all the primary energy that we extract from the environment. Three-fifths of electricity is produced from fossil fuels, one-fifth each from hydroelectric power and nuclear fission, and less than 1% from renewable energy sources. Households in rich countries use a mix of electricity and fossil fuel combustion (gas, heating oil and some coal), while those in poor countries rely on dwindling supplies of biomass fuels (wood, crop residues and animal dung – which is thus diverted from soil fertilisation).

The per-person consumption of energy in the USA is approximately 100 times higher than in sub-Saharan Africa. More generally, there is a ten-fold difference in per-person energy consumption between the rich and poor countries. The WCED estimated that, for the whole world to live at the current level of energy usage of North America, about five times more energy would be required.[27] This, it argues, is technologically and ecologically unachievable. Even if we extrapolate current trends in energy use to 2030, including continuation of current disparities between rich and poor, there will be an estimated four-fold increase in total energy use, including a seven-fold increase in electricity use. While it is estimated that increases in conservation and end-use energy efficiency could reduce the *per capita* demand for electricity in developed countries by around 50 % the net demand for electricity in the rapidly industrialising parts of the world, with their more rapidly growing populations, will grow substantially over the next few decades.

The fact that energy, including electricity, is the basic currency of technical and industrial development poses a dilemma, since energy-intensive human activity is a fundamental source of the overload that we are now placing on the planet's functional capacities. Coal-fired power plants and petroleum-fueled vehicles contribute mightily to the total emission of carbon dioxide, a major greenhouse gas. Indeed, fossil fuel combustion returns to the atmosphere – unnaturally – the carbon dioxide that was removed from the atmosphere of a warmer world several hundred million years ago. Much of the ongoing debate about society's energy futures reflects the struggle between the two titans of the power-generating industry – coal and nuclear power. However, if Chernobyl (1987) was a major strike against nuclear power, the greenhouse effect is a potentially crippling blow to the burning of fossil fuels, particularly coal.

As the global demand for electricity and other forms of energy increases, risks to Earth's natural systems will increase. It is the sheer scale of our energy use that may trigger planetary responses not seen at lower levels of aggregate consumption. Those responses, directly or indirectly, will have various adverse effects on human population health. Longer-term problems will arise because of the heat-trapping effects of excess atmospheric carbon dioxide; shorter-term problems are already arising because of the regional spread of acid rain.

4.5 Acid rain

4.5.1 Introduction

The term 'acid rain' was coined in the 1850s in industrial Manchester, Engel's 'masterpiece of the Industrial Revolution', when sulphuric acid was first found at high levels in rainwater. In the latter half of this century, acid rain has again been frequently in the news. Acidic precipitation – rain, snow or hail – results from the atmospheric accumulation of certain air pollutants, particularly the oxides of sulphur and nitrogen. Coal-fired, and to a lesser extent oil-fired, power plants have long been the major source of these pollutants, and their high chimneys assist their dispersal into the atmosphere.

Unlike other air pollution, acid rain is mostly a trans-boundary problem, with distant and indirect effects upon the environment and health. During the 1950s, 60s and 70s, lakes and streams in England and in Scandinavia were found to have become acidified. Many had become devoid of all fish life. Controversy about the cause of surface water acidification was eventually resolved by research which showed that the concentration of certain acid-sensitive diatoms (the remains of minute living organisms) in historical layers of lake sediment changed quite suddenly after 1800, and coincided with the first appearance of tiny particles of soot in the sediment. Clearly, acidification was an unnatural process. Further confirmation came from finding that a halving in acid content in rainfall over much of Scotland since the early 1970s was followed by a 30% reduction in the acidity of some lakes.

Rain is naturally a little acidic, reflecting the presence of low concentrations of carbonic acid which forms from carbon dioxide and water vapour. The normal pH is 5.6, relative to a neutral value of 7.0. Since the pH scale is logarithmic, a change of 1.0 units equals a ten-fold change in acidity. (Therefore a pH of, say, 2.6 is 1000 times – i.e. 10^3– more acidic than the normal rainfall pH of 5.6). In the early episodes of acid rain in southern California, Scotland and the Appalachian Mountains in West Virginia, the pH was in the range 2.2–3.0 – i.e. up to several thousand times more acidic than normal. In the 1980s, in Western Europe and North America, rainfall of pH 4.5 or lower was widespread. The main source of this additional acidity has been the gaseous pollutant sulphur dioxide, which combines with water vapour to form sulphurous and sulphuric acids. Human-made sulphur dioxide emissions, globally, rose from seven million tons annually in 1860 to 155 million tons annually in 1985. In recent years, however, sulphur emissions in OECD countries have declined, in response to a tightening of environmental standards, and the mantle has passed to other poorer industrialising countries, particularly China.

In Europe, the prevailing winds blow most of the sulphur dioxide, to be deposited on adjoining eastern neighbours; some ends on the Asian steppes of the USSR. Acid rain has also been a major concern in North America in recent decades. The Canadians have long been aggrieved that acid air pollutants from the USA have greatly damaged the maple syrup industry and killed fish in many of their lakes. In the coal-mining provinces

of southwest China, the levels of acid pollution now approach those
recorded in the more polluted parts of the USA, while acid rain from
industrial northeast China (Manchuria) drifts eastwards to Japan. More
localised zones of acidification have been reported elsewhere – e.g. in
Brazil, Venezuela, South Africa and Australia. Acid rain illustrates well
how environmental problems can, and increasingly do, transcend national
boundaries.

4.5.2 Effects on fish and forests

Forest dieback in central Europe has long been attributed to acid rain –
although not without controversy.[28] An estimated one-third of Europe's
forest has been damaged by acid precipitation, including the loss of about
one-sixth of the potential timber harvest. In the worst-affected areas, such
as around Usti Nad Ladem in the old Bohemian region of Czechoslovakia,
whole forests have been decimated. Nearby conifer forests in mountain
resort areas near the Polish–Czechoslovakian border are visibly dying.
Elsewhere, however, physical appearances can be misleading. Instead of
large unbroken swathes of dead trees, there may be small clumps of dead
trees. More often the telltale signs are of dead or dying tree-tops or of
unhealthy foliage, especially of the non-deciduous pine trees.[29]

Acid rain may leach certain nutrients (e.g. magnesium) out of the forest
soil thus causing nutritional deficiencies. Acidification of soil and water
increases the bioavailability of metals (e.g. aluminium – a plant toxin –
and the heavy metals). By mobilising the sulphates and nitrates that have
accumulated in soil since industrialisation began, acid rain may directly
damage root growth. It may also act directly on tree foliage. As with most
complex biological systems, the causal mechanism is unlikely to be simple.
If nutritional deficiencies and root damage cause weakness in the tree, then
subsequent stresses may suffice to damage or kill the tree. The vulnerability,
and therefore the rapidity of damage, is influenced by the natural acid-
buffering qualities of the soil – much of which, in Europe and North
America, was depleted of its surface alkaline content during ice-age
glaciation.

Large quantities of acid precipitation can fall on lake systems for
decades with no discernible impact. However, once the lake's buffering
capacity (i.e. its capacity to absorb and neutralise) has been exhausted
rapid acidification and ecological deterioration can occur. This is another
example of the non-linearities in ecosystem dynamics, discussed in chapter
2. The ensuing changes include the mobilisation of toxic metals – such as
aluminium, copper, lead and mercury – from surrounding soils into the

water. Plants, bacteria, invertebrates, amphibians and fish are all affected. The tiny plant species that capture solar energy by photosynthesis and create the base of the aquatic food web are replaced by a less photo-synthetically active and more acid-tolerant spectrum of vegetation. The ecosystem becomes simplified and less energetic – and fish supplies for human consumption decline.

4.5.3 Effects upon human health

Acid rain can affect human health indirectly, by contamination of drinking water due to the leaching of toxic metals or by impairment of food production. However, there is little evidence of direct adverse effects on respiratory health caused by inhalation of acid aerosol or particles. (This contrasts with the acidic pollutant *gases* in air – especially sulphur dioxide and the oxides of nitrogen – which contribute to the respiratory illness burden caused by air pollution.) Occupational exposure to mists of sulphuric acid may increase the risk of laryngeal and lung cancers,[30] but those exposures are markedly higher than that due to acid rain.

Acidification of drinking water is likely to have various adverse effects upon human health, although these are difficult to quantify.[31] Mercury accumulation in edible fish from acidified lakes – such as has been detected in acidified lakes in Sweden – may present a health problem, particularly as a neurotoxin. Greatly increased concentrations of some contaminants, particularly acid-leached metals such as aluminium, lead and cadmium, have been found in acidified drinking water in Scotland and the USA. Aluminium, a known neurotoxin, may be involved in Alzheimer's disease – although the evidence for such a link remains rather low-grade. Environmental lead exposure causes impaired intellectual development in young children[32] and is a well-documented contaminant in acidic drinking water (such as in parts of Scotland) which passes through pipes either made of lead or joined together with lead solder (see also section 10.2.5). Cadmium, likewise, can be dissolved out of piping solder by acidic water, and may accumulate in, and damage, kidney tissue. Any increase in the acidity of surface waters would therefore exacerbate these processes.

The adverse effects of acid precipitation upon food crops are three-fold: (i) direct chemical damage to crops, (ii) acidification of the soil, which reduces fertility and (iii) the leaching out of toxic metals such as cadmium which may then enter the crop foods at above-average concentrations. There is little systematic information available about these problems. One other effect may be beneficial; the deposition of sulphates on sulphur-poor soils provides an unintended fertiliser effect!

4.6 Environmental pollutants and ecosystems

Microbes and chemical toxicants have long been focal points of public health activity. The biological effects of these environmental agents are relatively straightforward in that they act directly upon the exposed person. Because microbes and chemical toxins are part of the natural environment, human (and other mammalian) biology has evolved defence mechanisms against them, principally the immune system and the various detoxifying and excretory metabolic pathways. Typically, these agents do not pose risks to whole populations or to ecosystems.

Consider the interplay between human biology and the vast array of chemicals naturally present in plant foods. Nearly all such chemicals are complex organic molecules made by the plant for its own biological purposes: proteins, carbohydrates (complex and fibrous, or simple and sweet), oils, natural pesticides, scents, ultraviolet-blocking chemicals and so on. Some of these plant chemicals are useful to animals as energy sources (carbohydrates and oils); others supply amino acids or essential fatty acids (proteins and oils); yet others that we call 'vitamins' have been found, via the trial-and-error of natural selection, to facilitate important metabolic pathways in animals; and elements such as iron, iodine, magnesium and selenium have been evolutionarily recruited into critical roles in important molecules in animal metabolism (e.g. iron in hae-moglobin and iodine in thyroid hormone). But many other plant chemicals are not useful, and some are potentially toxic. Therefore, animals have evolved ways of detoxifying, sidelining or excreting them. Where biological evolution has not furnished adequate metabolic defences, painful ex-perience (transmitted as culture) has enabled humans to learn to avoid eating known poisonous foods – and to feed them only to political or personal rivals.

Thus, the human body has a long familiarity with dealing with microbes and toxic chemicals. Since life on Earth never stands still, there are always new microbes to respond to, such as the seemingly endless repertoire of 'Asiatic' strains of influenza virus that sally forth to temperate climates each year. Likewise, new chemicals occur in evolving plants. While our mammalian biology has a sound basic set of defences against these noxious environmental agents, these defences may be overwhelmed by excessive exposures in combination with heightened susceptibility. Urban crowding and associated socially-disordered behaviours have opened up ecological opportunities for many microbes. Industrialisation, via occupational exposures and environmental pollution, has hugely increased the range of chemicals to which human populations are now exposed.

This increase in global chemical pollution has been largely due to the growth in industrial activity. The world manufactures almost 100 times more goods each year now than it did in 1900. Today, the most rapid growth in industry, and in industrial pollution, is in the developing countries. Indeed, a further five-fold to ten-fold increase in industrial output and energy use will be needed just to raise the level of consumption in developing countries to that of today's developed countries by 2050, when world population growth is expected to begin flattening out. In the words of the WCED: 'Pollution problems that were once local are now regional or even global in scale. Contamination of soils, ground-water, and people by agrochemicals is widening and chemical pollution has spread to every corner of the planet. The incidence of major accidents involving toxic chemicals has grown.'[27] Coastal ecosystems are deteriorating widely because of increasing contamination by poorly controlled industrial, agricultural and domestic waste disposal. If effective counter-measures are not taken, long-term damage to many of our ecosystems may result.

Since Rachel Carson published *Silent Spring* in 1962 (see chapter 2), the US production of synthetic organic chemicals has increased five-fold. Although some impressive gains have been made in developed countries in reducing toxic emissions and wastes, the total global output has increased. Few species or lands now escape the effects of this pollution. Pesticide residues are found in birds and mammals that live their whole lives in the remote Antarctic. Heavy metals such as lead occur with increasing concentration in the polar ice; the lead concentration has increased over twenty-fold since 1800. The concentration of lead, mercury and copper in the tissues of Inuit eskimos in Greenland today is 4–8 times higher than samples from some of their well-preserved, naturally refrigerated ancestors of five centuries ago. Because some of these chemical exposures are thought to impair reproduction in birds; because the dumping of certain wastes (including sewage) reduces oxygenation of rivers and lakes and thus endangers fish life, as does eutrophication by 'nutrient' pollutants; and because the mobilisation of heavy metals such as lead into the biosphere is something that only occurs marginally in nature, so these pollution problems may also disrupt ecosystems.

Thirty years after *Silent Spring* we are learning, in increasingly lurid detail, of the consequences of chemical pollution of the Soviet Union and its client Eastern European states. It is emerging as a disaster of biblical proportions, entailing decades of systematic neglect and poisoning of the environment and populace – expendable victims on the altar of ideological obsession and economic desperation. The catalogue of environmental and ecological devastation within the former USSR has grown: the poisoned Lake Baikal in Siberia; the shrivelled Aral Sea in Soviet central Asia; saturation of land and groundwater with pesticides in the cottonfields of Kazakhstan; the uncontrolled chemical industry emissions of Yerevan, Armenia; the air-polluting metal smelting and refining plants of northern Russia; and Chernobyl in the Ukraine. Much of the former USSR's surface water is badly polluted and not fit for drinking. The statistics on the blighted health of surrounding communities and on the declining productivity of soil and waterways, while still incomplete and of uncertain accuracy, outline the later chapters of this sorry story.[33] Whereas death rates in infancy and adulthood decreased in the 1950s and 60s in Eastern Europe and the USSR, subsequently the decline in infant mortality has stalled (at levels well above Western Europe) and adult male death rates from cardiovascular and respiratory diseases and from cancer have increased by 10–15 % (at a time when the rates in Western Europe declined by one-fifth).[34,35] In the most polluted areas of Czechoslovakia and Poland, life expectancy is several years less than in cleaner areas. Although the relative contributions to adult mortality of pollution, social conditions and personal behaviours (such as smoking) are unclear, for some adverse health effects in young children the association with environmental pollution is clearer – for example, official reports in the former USSR documented that, in regions where pesticide use is high, sickness rates in children are five times higher than those in relatively clean areas.[36]

As human populations and the scale of economic activity and consumption grow, so the problems due to chemical toxins can spill over from local backyard to affect whole ecosystems.

4.7 Summary

Ever since Hippocrates, medical science has known that human health is affected by the ambient environment – the airs and waters. More recently, we have learnt how the urban 'built environment' and social relationships influence the life chances and the health status of population subgroups. However, we are only now starting to view these things within an ecological framework, relating long-term population health to the carrying capacity of Earth's natural systems. Increasingly, we understand that human cultures have introduced many types of 'evodeviation' that have distorted ecological relationships and reduced the capacity of ecosystems to support health and life. Because of our powers of adaptation we have often been able to modify or defer such problems – and sometimes seemingly sidestep them. But it is in the nature of ecological systems that debts are finally called in.

Historically, various human civilisations have declined as their ecological support system has been eroded. The coming and going of infectious disease epidemics has reflected changes in humans ecology over the centuries. The health of populations has been profoundly influenced by changes in the security, the amount and the types of food available. Whereas the historical trend has been in the direction of more food and better survival, especially in childhood, more recently a range of chronic diseases due to dietary excesses and imbalances has appeared within the longer-living adult populations in richer countries.

The biosphere in which we live is essentially a closed and finite system. The UN predicts that the human population, currently 5.5 billion, will plateau between 10 and 14 billion later next century. In the meantime, as population growth continues, the probability of resource wars will almost certainly increase. The Gulf War of 1991 provided a pointer to how the world's dependence on fossil fuels can inflame international tensions. Wars over dwindling supplies of drinking water and arable land may occur increasingly frequently next century. The control of human population growth is a prerequisite for 'capping' the world's macro-environmental problems. As we shall see in chapter 5, population control is unlikely in the absence of a much fairer sharing of the world's assets between and within countries.

References

1. McNeill WH. *Plagues and Peoples*. New York: Anchor Doubleday, 1976.
2. Dols MW. *The Black Death in the Middle East*. Princeton: Princeton University Press, 1977.
3. McNeill suggests that new methods of keeping warm, coinciding with big increases in newly-available supplies of wool, might have lessened the (respiratory) transmission of the plague by nocturnal person-to-person huddling. See McNeill, ref 1 above.
4. Gribbin J, Gribbin M. *Children of The Ice*. Oxford: Basil Blackwell, 1990.
5. Bryson RA, Murray TJ. *Climates of Hunger: Mankind and the World's Changing Weather*. Madison: University of Wisconsin Press, 1977.
6. Johnson DL, Gould H. The effect of climate fluctuations on human populations: A case study of Mesopotamian Society. In: Biswas AK (ed).

Climate and Development. Dublin: Tycooly International Publishing, 1984, pp 117–38.

7. Jacobsen T, Adams RM. Salt and silt in Ancient Mesopotamian agriculture. *Science* 1958; **128**: 1251–8.
8. Ponting C. *A Green History of the World*. London: Penguin, 1992.
9. Boyden SV. *Western Civilization in Biological Perspective. Patterns in Biohistory*. Oxford: Oxford University Press, 1987.
10. Neumann J, Sigrist RM. Harvest dates in Ancient Mesopotamia as possible indicators of climatic variations. *Climatic Change* 1978; **1**: 239–52.
11. Culbert TP, Rice DS (eds). *PreColumbian Population History in the Maya Lowlands*. Albuquerque: University of New Mexico Press, 1990.
12. Diamond J. *The Rise and Fall of the Third Chimpanzee*. London: Radius, 1991.
13. WHO. *Diet, Nutrition, and the Prevention of Chronic Diseases*. Technical Report Series, No. 797. Geneva, WHO, 1990.
14. Powles JW. Global patterns and disadvantaged populations. In: McNeil JJ, King RWF, Jennings GL, Powles JW (eds). *A Textbook of Preventive Medicine*. Melbourne: Edward Arnold, 1990, pp 238–52.
15. Whiten A, Widdowson EM (eds). *Foraging Strategies and Natural Diet of Monkey, Apes, and Humans*. Oxford: Oxford University Press, 1992.
16. Eaton SB, Konner M. Paleolithic nutrition: a consideration of its nature and current implications. *New England Journal of Medicine* 1985; **312**: 283–9.
17. McKeown T. Food, infection, and population. In: Rotberg RI, Rabb TK (eds). *Hunger and History*. Cambridge: Cambridge University Press, 1985, pp 29–49.
18. While Ötze, the Tyrolean alpine 'iceman' (recently found preserved intact) was gathering-hunting around 3500 BC, great civilisations had already formed in southwest Asia, Egypt, China and the Indus Valley.
19. Knapp VJ. Major dietary changes in nineteenth century Europe. *Perspectives in Biology and Medicine* 1988; **31**: 188–93.
20. Burkitt D, Trowell HC (eds). *Refined Carbohydrate Foods and Disease: Some Implications of Dietary Fibre*. London: Academic Press, 1975.
21. The type of change is not an automatic progression. The evolution of the 'social diet', differentiated between urban rich, wage-labourers and peasants, is shaped by internal class relations and wider economic forces. The globalisation of food production and trade, dominated by international agribusiness, sets much of the food agenda – influencing preferences, shaping local supplies and often eroding local ecological balance and wisdom. See also chapter 8.
22. Chen J, Campbell TC, Li J, Peto R. *Diet, Lifestyle and Mortality in China: A Study of the Characteristics of 65 Chinese Counties*. Oxford: Oxford University Press, 1990.
23. O'Dea K. Marked improvement in carbohydrate and lipid metabolism in diabetic Australian Aborigines after temporary reversion to traditional lifestyle. *Diabetes* 1984; **33**: 590–603.
24. US Department of Agriculture. *World Livestock Situation*. Washington DC: USDA Foreign Agricultural Service, 1990.
25. Durning A. Asking how much is enough. In: *State of the World* 1991. *Worldwatch Institute Report*. New York: Norton, 1991, pp 153–69.
26. World Bank. *World Development Report* 1992. *Development and the Environment*. Oxford: Oxford University Press, 1992.

27. WCED. *Our Common Future*. Oxford: Oxford University Press, 1987.
28. Schwartz SE. Acid decomposition: Unravelling a regional phenomenon. *Science* 1989; **243**: 753–63.
29. The fact that only a minority of trees are visibly affected is not surprising, since the biology of forests and trees is complex. Humans afford an analogy: most smokers *appear* healthy enough, but in those who are biologically most susceptible or who are also affected by other noxious exposures, manifest disease occurs.
30. Soskolne C *et al*. Epidemiologic and toxicologic evidence for chronic health effects and the underlying biologic mechanisms involved in sub-lethal exposures to acidic pollutants. *Archives of Environmental Health* 1989; **44**: 180–91.
31. Nordberg GF. Human health effects of metals in drinking water: Relationship to cultural acidification. *Environmental Toxicology and Chemistry* 1990; **9**: 887–94.
32. McMichael AJ *et al*. Port Pirie Cohort Study: Childhood blood lead history and neuropsychological development at age four years. *New England Journal of Medicine* 1988; **319**: 468–75. And: Baghurst PA *et al*. Life-long exposure to environmental lead and children's intelligence at age seven: The Port Pirie Cohort Study. *New England Journal of Medicine* 1992, **327**: 1279–84.
33. Feshbach M, Friendly A. *Ecocide in the USSR: Health and Nature Under Siege*. Basic Books, 1992. The authors examine the Soviet Union's adversarial view of nature, in relation to the historical, political, social and economic complexities of Soviet life. Resources were treated as inexhaustible, human health as expendable and wilderness as adversary. Despite the lack of good data, there is much evidence of deleterious effects of environmental pollution upon health – along with anecdotal accounts of health disasters (e.g. that in areas of heavy pesticide usage, there were no healthy youths of military-service age).
34. Mexentseva E, Rimachevskaya N. The Soviet country profile: Health of the USSR population in the 70s and 80s – an approach to a comprehensive analysis. *Social Science and Medicine* 1990; **31**: 867–77.
35. Chopin K. Post-totalitarian medicine. *British Medical Journal* 1992; **304**: 1557–60.
36. French HF. Restoring the East European and Soviet environments. In: *State of the World* 1991. *Worldwatch Institute Report*. New York: Norton, 1991, pp 93–112.

5

Population increase, poverty and health

5.1 Introduction

The most spectacular aspect of this final decade of the twentieth century is the record-breaking annual increase in human population size. This figure, approaching 100 million per year, cannot go on for long without disaster, and one imagines that such growth will never recur. We are living through a demographic aberration. During the nineteenth century the annual increase was around 10 million; over the preceding five centuries it was around one million per year. Today, this massive surge in global population is being powered by declines in childhood death rates in most of the poor and populous countries as they struggle to get on to the escalator of 'development'. Those countries will contribute 90% of world population increase over the coming half-century. So long as their relatively high birth rates do not match their newly-reduced death rates, then this imbalance will persist until, somehow, a new equilibrium is established. The longer the world procrastinates over population control, the less likely it is that such equilibrium will be attained without widespread starvation, environmental devastation, social disruption and war.

There is an intimate connection between population growth rates and systemic poverty. In the world's poor and populous countries almost one-quarter of humanity live in 'absolute poverty', with their most basic needs for food, clothing and shelter neither met nor likely to be. The criterion of 'absolute poverty' used by the World Bank is an annual income below around US$450. Robert McNamara, when president of the World Bank, described it as 'a condition of life so limited by malnutrition, illiteracy, disease, squalid surroundings, high infant mortality and low life expectancy as to be beneath any reasonable definition of human decency'.[1]

There have been marginal reductions in world poverty in percentage

Table 5.1. *Poverty in the developing world, 1985–90.*

Region	Percentage of population below the poverty line		Number of poor (millions)	
	1985	1990	1985	1990
All developing countries	30.5	29.7	1,051	1,133
South Asia	51.8	49.0	532	562
East Asia	13.2	11.3	182	169
Sub-Saharan Africa	47.6	47.8	184	216
Middle East and N. Africa	30.6	33.1	60	73
Eastern Europe[a]	7.1	7.1	5	5
L. America and Caribbean	22.4	25.5	87	108

Note: Poverty line used here = $370 annual income per capita in 1985 purchasing power. In 1990, the poverty line was approximately $420 annual income per capita.
[a] Does not include the former USSR.
Source: World Bank, 1992[2]

terms, but the absolute number of people living in poverty is still increasing (see Table 5.1).[2] During the 1980s, per-person incomes actually declined in more than forty poor countries; all measures of poverty worsened in sub-Saharan Africa, the Middle East, North Africa and Latin America.[2] This has made the distribution of income around the world increasingly skewed. The average personal income is around fifty times higher in the rich (OECD) countries than in the three-fifths of world population living in poor countries – and the gap is still growing as the rising tide of world poverty continues.[3] Poverty *within* countries is also unevenly spread. Some of the world's greatest gradients of inequality are in Mexico and Brazil, with twenty- to thirty-fold differences between the richest and poorest fifths of the population.[2] Likewise, in many affluent countries there are large gaps between rich and poor. In the USA during the 1980s, the richest 1 % increased their share of the nation's wealth from around 31 % to 37 %. However, poverty in rich countries generally has a different scale and texture, with most of the poor having basic shelter, clothing and food, and – because of the social mobility that extends towards society's lowest reaches – having, or seeming to have, some prospect for material betterment.

Do high birth rates in today's poor countries perpetuate poverty, or does being poor cause people to have more children? Opinions differ over this complex issue. Biologists tend to view high birth rates as the prime problem; an uncontrolled basic drive that overloads resources, keeps people illiterate, poor, hungry and illness-prone and damages ecosystems. Social scientists argue that poverty, social inequality and political powerlessness lead to high birth rates because of the parents' perceived present or future need for supportive children or because of sheer helplessness. The social biologist, comparing hunter-gatherer cultures with settled cultures, may point out that once society's structure changes, once personal survival depends on economic rather than ecological factors and once population density and dynamics change sufficiently, then traditional restraints on population growth recede. Those restraints include killing or abandoning the new-born and the senile, prolonged breast-feeding, ritualised abstention from intercourse, abortion, periodic famine, inter-tribal warfare and physical accidents.[4]

The direction of causal relationships aside, overpopulation and poverty have adverse and intertwined environmental impacts. Some impacts result largely from population overload – for example, the consequences of prodigious coal-burning in China, rice-field methane emissions in India and China, and widespread deforestation throughout the Third World. Others result primarily from poverty (both national and local) – for example, poor sanitation, use of marginal farmlands, soil erosion from cash-cropping, excessive logging and burning of forests and uncontrolled industrial pollution. The Ehrlichs have summarised the determinants of environmental degradation algebraically: the aggregate environmental impact (I) of humans is the product of the number of people (P) times their level of affluence (A) times the type of technology and waste disposal (T).[5] That is: $I = P \times A \times T$. The World Bank has used this same formula, elaborated to take account of the input-output balance and the energy-efficiency components of 'T'.[2] Although the basic formula readily describes the key influences upon environmental degradation in rich countries, it is less attuned to the impact associated with Third World poverty. In poor countries the meagre 'A' causes environmentally destructive forms of 'T'. Desperately poor farmers in Ethiopia clear and plough the stoney hillside land, and most of the soil is soon washed away. China, striving to raise the 'A' of more than a billion people, intends to double its combustion of coal this decade.

In all of this arithmetic there looms a recurring multiplier that strains the biosphere's carrying capacity – the burgeoning human population ('P'). Demographers tell us that even if the world's birth rate were immediately reduced to replacement level, the demographic momentum of the on-

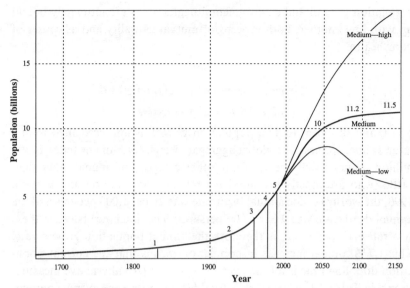

Figure 5.1. Recent increase in world population, and UN projections to 2150. (Data source: UN, 1992 – ref. 6)

coming preadult population would still cause a 60% increase in world's population, from 5.5 billion to nine billion, by around the year 2030 (see Fig. 5.1). The UN's current best estimate, the 'mediant' variant, is that a plateauing will occur between 11–12 billion late next century.[6] It is important to note that these population projections make assumptions about likely changes in fertility rates in relation to specified scenarios of economic growth, education, child mortality and family planning. The 'mediant' estimate assumes that the replacement level of fertility (2.1 children per woman) will be attained by around 2050. There is therefore much uncertainty in such estimates, and those of the UN and World Bank for the year 2100 range from a 'low' of around six billion to a 'high' of 20-plus billion.

The ecological problem posed by population growth over the coming century is not just a Malthusian more-people-than-food concern. This massive population growth will multiply the destruction of farmland and forest, the contamination of the global commons (air and water), the disruption of climate and the extinction of species. In recent decades, much of the increased emission of methane, a potent greenhouse gas, has come from population-dependent increases in irrigated rice-fields. Likewise, the increased carbon dioxide emissions of industrialising Third World countries, such as China and India, largely reflect population size. Poverty, population growth and environmental degradation are intimately bound up with one another, both in webs of mutual causality and as causes of poor health.

5.2 Human ecology and population growth

5.2.1 Historical perspective

The graph of recent human population growth is unprecedented in any large animal species. The global aggregate weight of humans today is 300 million tonnes – well ahead of any other category of animal other than cattle (who are mostly produced for human consumption and use). In 1990, the world's crude annual birth rate was 26 per 1,000 persons and the annual death rate was 9 per 1,000 persons.[2] The difference between these two rates, 17 per 1,000, is therefore the annual population growth rate. This 1.7% annual increase in population means that the current population doubling time is a little over forty years.[7] (An alternative measure, shown in Table 3.1, is the world's total fertility rate – the average number of children per woman during her reproductive years. This was 3.4 in 1990.)

When agriculture evolved, beginning about 10,000 years ago, there were an estimated five million people. Subsequent population growth milestones have been: 250 million at the birth of Christ, 500 million in the fourteenth century, one billion in 1830, two billion in 1930, five billion in 1988, and a projected 9 billion at 2030 (see also Fig. 5.1). It took our species about 200,000 years to reach the first billion; it will have taken just 200 years to add another eight billion. By any reasonable definition, that qualifies as 'swarming'. Indeed, to a biologist it resembles part of the familiar boom-and-bust pattern followed by plagues of mice, swarms of locusts and weevils in the flour.

Unconstrained populations always grow explosively. When a species of animal or plant is placed in an unfamiliar but supportive environment, it may proliferate unchecked. Examples are legion: rats in refuse-rich urban environments, and rabbits in rural Australia. Rabbits were introduced into Australia in 1854 by Thomas Austin, a pastoralist, who wanted to recreate the gentrified English hunt in his newly adopted homeland. Within a few short years he and his neighbours were paying hunters to try and eradicate the local infestation of rabbits; but to no avail. Austin's twenty-four imported rabbits multiplied into millions of rabbits within decades. This century they have munched their way through much of the natural habitat of Australia's small and unassertive marsupials. Likewise, a species whose

natural environment becomes hyper-enriched may proliferate spectacularly. Sporadic algal blooms along the nutrient-rich polluted coasts of
Italy, eastern USA and the Baltic Sea illustrate this. So too did the choking
bloom of blue-green algae within the mighty Murray-Darling river system
in rural Australia, in late 1991, as agricultural fertiliser run-off and
untreated sewage from river towns provided an unnatural, evodeviationist
banquet for the algae.

But these population booms cannot last indefinitely. Nature's two basic
constraints are, first, limits to the food supply and, second, being some
other species' food supply. Epidemic disease and, among animal species,
physiological responses to overcrowding[4] are also important. Further,
within a closed system, as for bacteria in a laboratory culture medium, the
build-up of toxic wastes becomes a constraint. Closed-system population
dynamics exemplify catastrophe theory – i.e. when a critical point is
reached the initial growth trend is breached dramatically and the
previously-proliferating population 'crashes'. Pessimistic commentators
argue that this may yet be the destiny of the human population.[8]

Over many millenia, the human species has responded to new environmental opportunities by migration and proliferation. For a million
years or so, our hominid ancestors roamed their ecological patches in small
nomadic bands. Anthropological observations made on contemporary
hunter-gatherers – the 'bushmen' (!Kung) of the Kalahari in southern
Africa, the Hadza of eastern Africa, the Australian Aborigines, and the
pygmies of the Congo Basin – suggest that the early bands of humans
typically comprised around 20–40 persons.[9] Much of their time was spent
in finding food, mostly plant food (gathered by women) and some animal
food (hunted by men). The food supply was usually very varied and
seasonal. Infant mortality rates were high. Life-expectancy at birth in the
Kung is 30–35 years, with about a 20% mortality in the first year of life.
From childhood onwards, disabling injury or debilitating sickness constantly threatened personal survival; neither was compatible with nomadic
existence. Fertility rates were around 30 births per 1,000 persons per year.
Women typically bore around five children, with fertility being under a mix
of cultural and biological control, including suppression of ovulation by
breast-feeding. Since death rates closely matched birth rates, a 'steady
state' existed, and population growth was, at most, slow. Typically,
population density was less than 10 persons per 100 square kilometres,[10]
compared to 400 persons per square kilometre today in Bangladesh,
Holland and South Korea.

With the onset of agriculture, population densities initially increased by about twenty-five-fold in response to increased food supplies. Wild cereal grasses (wheat and barley) were cultivated; wild sheep and goats were herded. The security of agrarian life enhanced some aspects of health. The disabled and debilitated had a chance to recover, and the physically weak were better able to find a socially useful function. On the other hand, infectious diseases were spread more easily. New infectious agents (e.g. smallpox and measles) were sustained within the larger populations. The food supply, being less varied, was more vulnerable to food shortage through crop failure. There were major demographic impacts too. Fertility rates increased by around 50% and women typically bore up to ten children. Because death rates (particularly due to infection) kept pace, a new steady state was established, albeit at a much greater population size and density. Subsequent population growth therefore remained relatively slow.[9]

The ensuing evolution of forms of human society culminated in larger political and ethnic groupings. Human settlement spread into uninhabited regions. As trade, warfare and conquest occurred, religions and civilisations rose and fell. From around the time of Christ to the seventeenth century, the world population increased slowly, and populations were vulnerable to occasional catastrophes, especially famines and pestilence. The next great and sustained population growth was triggered by the early industrial revolution, beginning in eighteenth-century Europe. Over the past century, the transmission of *some* of the fruits of that revolution – particularly the technologies of sanitation, vaccination, malaria abatement and (latterly) oral rehydration – to developing countries has laid the foundations for the contemporary population surges in those countries.

5.2.2 The demographic transition – in transition?

Austin's exponentially breeding rabbits were unrestrained by considerations of life-choices and peer pressure. However, the human species is different in one important respect; human culture modulates basic biological drives. As social progress occurred last century, in the wake of the early industrial revolution, so population growth rates slowed in the richer countries. This 'demographic transition' occurred in Europe and North America between the late eighteenth century and mid-twentieth century.[11] The process, as classically described, comprised three stages. The initial, preindustrial stage was characterised by high birth rates and high death rates, in approximate equilibrium. The second stage, occurring in response to economic and social gains, brought a reduction in death

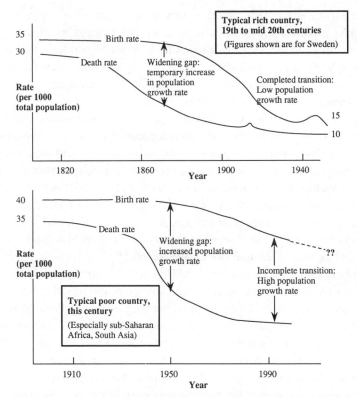

Figure 5.2. The demographic transition. The sequential fall in death rates and then birth rates in (today's) rich countries, which substantially occurred last century, has not been mirrored by poor countries this century. That 'incomplete' transition is fuelling the Third World's population increase.

rates – especially of infants and children – while high birth rates persisted. In the third stage, occurring several decades later, a catch-up (catch-down?) reduction in birth rates occurred as social aspirations changed, female literacy and status increased, desired family size declined and contraceptive skills increased (Fig. 5.2).

This demographic transition, however, does not occur automatically – or even readily – in response merely to falls in death rates. (Indeed, demographers now point out that, historically, this staged sequence often did not apply to today's rich countries. Falls in mortality and fertility can occur, if the right social and cultural conditions exist.) In retrospect, a particularly favourable, perhaps fortuitous, sequence of events occurred in today's developed countries when they went through their demographic transition. They had a strong base of improved nutrition, education and civic infrastructure, reflecting their new wealth from industrialisation and colonial exploitation. Birth rates were further reduced by migration of young adults to North America, South America, Australasia and South Africa. Social and technical control of fertility therefore occurred readily and soon after the downturn in death rates.[11] By contrast, much of Africa is currently stalled in the middle of the transition; the second stage has occurred – but not the third. Death rates have fallen in recent decades in the wake of some early post-independence economic growth, coupled with 'imported' vector (especially mosquito) control, some improved sanitation and programmes of child immunisation and oral rehydration for diarrhoea. Smallpox has been wiped out; yellow fever is under partial control. But these mortality declines have not yet been followed by a compensatory reduction in birth rates. Accordingly, Africa's rate of annual population growth has increased over recent decades and is currently around 3% (i.e. a doubling time of around twenty-five years).[2,6]

Populations thus suspended in mid-transition, and vigorously expanding, are likely to exceed the carrying capacity of their local ecosystems – especially populations living in the more fragile tropics. Prolonged non-completion of the transition could herald a disaster, eventually entailing sharp increases in mortality as the food-producing ecosystem collapses. Maurice King, a pioneer of primary health care in Africa, has bluntly referred to this scenario as the 'demographic trap', and suggested in 1990 that such a collapse may have already begun in Ethiopia.[12] He argues that this potentially disastrous uncoupling of the second and third stages of the demographic transition raises important questions about the socioeconomic prerequisites for effective population control. Parental motivations for a large net family size are primarily economic and social. In poor rural populations children, particularly sons, are the equivalent of a tractor and an old-age pension – for people that have neither. Children, later, can migrate to the city and eke out a wage to share with parents. Within an impoverished and improvident society, children are thus an economic asset. Therefore, to curb fertility, King and others argue that economic gains must bring widespread reduction of poverty, increased literacy and marriage age in women, and social development that makes a large family a burden rather than a survival kit. That is a very tall order for impoverished countries, further enfeebled by corruption and social disintegration. Other evidence suggests, however that some fertility reductions are achievable by better access to contraception in the presence of continued poverty (e.g. liberal-Muslim Bangladesh) and by the process of urbanisation (e.g. Thailand).

In the worst-case scenario, with persistent high fertility (see Fig. 5.1), a dreadful dilemma could arise. Does the time-honoured public health imperative of working to reduce childhood death rates merely sow the seeds of eventual ecological collapse from population overload? Under the banner of public health imperatives, WHO and UNICEF estimate that over 40% of the 14 million child deaths each year could be prevented through vaccination and low-cost interventions such as oral rehydration therapy.[13] But child salvage, without complementary social gains and contraceptive services that enable fertility decline, may be no real long-term kindness. King's unsettling question, 'Are ecology and compassion incommensurate?', spawned angry correspondence in medical–scientific journals – but it reminds us that nature is entirely dispassionate in re-establishing the equilibrium of disrupted ecosystems. If rich countries are to help solve the public health and demographic problems of poor countries then they will have to increase hugely the scope and generosity of their assistance – and poor countries will need to introduce heroic measures to stem population growth.

Meanwhile, another symptom of this demographic tension is looming. As population and ecological pressures increase, precipitating the collapse of subsistence agriculture in some countries, so the numbers of inter-country refugees have swollen. The world total of refugees grew by nearly three-quarters during the 1980s, rising from around 8 million to 15 million in 1990.[14] Of that total, 4.6 million refugees are in Africa, 7.3 million in Asia, 1.5 million in Europe, 1.3 million in North America, and 0.3 million in Latin America. Displaced Palestinians number another 2.3 million. If intra-country refugees are included, then the global total doubles to over 30 million. These numbers are likely to increase hugely over the coming decades.[15]

Most refugees are already poor, the countries to which they flee are no richer, and their resettlement circumstances are often bleak, makeshift and unhygienic. They are a vulnerable group, made more vulnerable by the emergency process and stress of displacement, and exposed to chronic deprivation and poor health. Malnutrition, child diarrhoeal disease, respiratory infections and the effects of inadequate shelter are rife among crowded refugees. Viewed from an ecological perspective, these environmental refugee populations often cause further overload on local, strained environments, and they tend to reintroduce diseases such as measles, polio, tuberculosis and leprosy into their host populations. Their existence reflects, in large part, the depletion of supportive environments in a world in which the last frontiers are rapidly closing off.

Table 5.2. *Estimated and projected population of the major areas of the world, medium-fertility extension (UN), 1950–2150.*

Area	Millions						
	1950	1990	2000	2025	2050	2100	2150
World	2,518	5,292	6,261	8,504	10,019	11,186	11,543
Industrialised	752	1,089	1,143	1,237	1,233	1,202	1,191
Europe	393	498	510	515	486	440	426
North America	166	276	295	332	326	314	308
Oceania	13	26	30	38	41	41	41
USSR	180	289	308	352	380	407	416
Third World	1,766	4,203	5,118	7,267	8,786	9,984	10,352
Africa	222	642	867	1,597	2,265	2,931	3,090
Latin America	166	448	538	757	922	1,075	1,117
China	555	1,139	1,299	1,513	1,521	1,405	1,389
India	358	853	1,042	1,442	1,699	1,870	1,949
Other Asia	465	1,121	1,372	1,958	2,379	2,703	2,807

Source: UN, 1992[6]

5.2.3 Profile of today's population problem

Thomas Malthus, the polymath parson, argued that population, unchecked, increases in a geometrical ratio, whereas food production increases only in an arithmetical ratio (see also chapter 1). In the middle years of this century, optimistic economists and agriculturalists dismissed the Malthusian spectre of a world in which population numbers would outstrip world food supply. Fantastic supportable populations of 50–100 billion, or more, were foreseen.[16] Predictions of ever-increasing food supplies were sustained by new agricultural technologies, including those of the early Green Revolution. Even so, renewed concern arose in the late 1960s over the world's accelerating population growth, and formal international attempts were made to disseminate new methods of birth control (including oral contraception, intra-uterine devices and vasectomy).[17] Although the current *absolute* annual increase in population is unprecedented, over the past 25 years global annual growth has dropped from 2.1% to 1.7%. Growth rates in some European countries are now at or even below zero. There is evidence that the 'inflection' point in population growth has now been reached in a number of developing countries, as they enter the third stage of the demographic transition. Growth rates in those countries should therefore begin a gradual decline over 20–30 years, prolonged by the demographic 'fly-wheel' effect of today's bottom-heavy population structures in which one-third of people are less than fifteen years old. Overall, however, the aggregate numbers still look formidable – and may not be sustainable.

The medium UN projections are that Africa's population of 750 million will increase four-fold before stabilising later next century, while for some individual African countries, including Ethiopia, Egypt and Nigeria, five-to eight-fold increases are forecast. For both India (currently 900 million) and Latin America (470 million) the projected increase is approximately two-fold (see Table 5.2). China, one of the power-houses of world population growth, has not yet stemmed its flow of new humans. The 1990 Chinese national census revealed a 2% excess over the planned total of approximately 1.12 billion, despite stern antinatalist policies in existence since the 1970s that limit each couple to one child, on pain of social sanctions. Nevertheless, China's achievement has been monumental, and it is expected that its population will peak at 1.5 billion around 2030 and then gradually subside. Many other developing countries are, for various reasons, hardly even trying to limit population growth. In much of Africa, West Asia, South Asia, and Latin America there has been little reduction in fertility. In Iran, where fertility rates were declining in the 1960s and 1970s, the advent of the fundamentalist Islamic Revolution has redefined women's role and status, lowered the legal marriage age of girls to nine years, and created a pronatalist social environment resulting in a 12% increase in fertility rates.[18] Even if the ecological consequences of unrestrained population growth were better understood, and even if there were sufficient economic resources to reduce fertility, there still remain the negative influences of pronatalist nationalism (now including many Islamic countries) and the reactionary opposition to birth control by the Roman Catholic Church.

Just how many people can the world feed, clothe and shelter, on an ongoing basis? Unlike other species, humans can deliberately and artificially extend the sustainable productivity base of their environment. This can only be done at the expense of other species, by commandeering natural habitat for human purposes. As we have seen, there have been three great surges in human population.[16] Systematic tool-making around several hundred thousand years ago potentiated the first surge; agriculture and animal husbandry beginning 10,000 years ago potentiated the second surge; and the mechanisation of agriculture, the industrial revolution and the recent transfer of childhood death-averting technologies from rich to poor countries have potentiated the ongoing third surge. The absolute increases in each of the first two surges were trivial compared to the might of the current surge. The central question, to be addressed in the following chapters, is whether this third surge is propelling our demands upon the world's natural systems beyond sustainable limits.

The world's food supply is perhaps the most compelling focus of concern. Given today's global food production, if everyone ate a simple, basic vegetarian diet, there would be enough food for around six billion people – but only enough for 2.5 billion to eat in the high, meat-laden, style of North America and Europe.[19] If we deduct the food currently produced with unsustainable farming methods (which includes an estimated one-sixth of US agricultural production) then the world-feeding figures look even bleaker. As we will see in chapter 8, after forty years during which the growth in global food production has exceeded population growth, the pendulum may now be starting to swing in the other direction. This faltering of ecological support systems is prompting a fourth phase in attitudes towards 'the population question'. Mid-century optimism, followed by neo-Malthusian fears in the 1970s, led to a revisionist renewal of faith in human numbers and technological adaptation. Humans are themselves an intellectual asset, argued the enthusiasts; they are the source and stimulus of the economists' stock-in-trade solutions: technical innovation, materials substitution and resource discovery.[20] However, this 'the-more-humans-the-better' argument is now falling under the shadow of global ecological overload, a new dimension to the population problem.

Questions of 'carrying capacity' aside, there is no intrinsic merit in the human species having a very large population size. The notion that maximising the total throughput of human individuals would somehow maximise the sum total of happiness and fulfilment is a theological absurdity. Hence, the real issue for *Homo sapiens* is grounded in ecological fundamentals. Will it be possible some time later next century to achieve a cooperatively managed, gradual and equitable reduction in global population – or will there be a demographic catastrophe? While biological theory may suggest the latter, we can hope that those things that are special about the human species will enable us to achieve the former.

5.2.4 Fertility, poverty – and education

During this decade, the four-billion-strong Third World population will grow by almost one billion, while that of rich countries will grow by little over 50 million.[6] Countries with annual growth rates of at least 3 % during the eighties, and therefore with population doubling occurring in around twenty years, include Kenya (3.8 %), Zambia (3.7 %), Ethiopia, Somalia,

Tanzania, Congo, Ghana, Nigeria, Cote d'Ivoire, most of the Arab countries (including several exceeding 4.0 %), Pakistan and several Latin American countries.[2] By contrast, the overall population growth rate in developed countries is 0.5 %, and includes some countries with negative or zero growth rates – such as Denmark, Germany, Belgium and Italy. Rapid declines in fertility rates have occurred in Japan, China and the 'newly industrialising' countries of East Asia: Hong Kong, Taiwan, South and North Koreas. Fertility rates have also turned down in Indonesia, Philippines, Sri Lanka and Thailand.

High fertility accompanies high child mortality, for reasons I have already mentioned. Thirteen million of the world's children aged under five years die annually, and 98 % of these deaths occur in developing countries. The dominant causes of child death are diarrhoeal diseases and malnutrition (approximately four million each), acute respiratory infections (three million), other vaccine-preventable diseases and malaria (around one million each). All are closely related to poverty, overcrowding, poor hygiene and lack of education. Demographers have long observed that child mortality falls as the education and status of women increases – and that this is particularly the case in the contemporary Third World.[21,22] Figures from a cross-section of countries show that where no females are enrolled in secondary education the average woman has seven children, compared with an average of three children in societies where 40 % of females have secondary education.[2]

The world's literacy rates have risen considerably since the 1960s, particularly in Third World countries in the wake of post-colonial independence. However, the *absolute* number of illiterate adults – particularly women – has increased. The rates of illiteracy are highest in Africa and in West Asia, being approximately 65 % in females and 40 % in males.[2] In the poorest countries, such as The Gambia in West Africa, the figures exceed 80 % and 60 %, respectively; the figures in South Asia (predominantly India) are little better. While school enrollment rates are expected to rise in the 1990s, the number of children missing out on primary schooling (particularly girls in conservative traditional communities) will remain at over 100 million in the less developed countries in 2000. Indeed, the recent worldwide recession has caused school enrollment rates to drop in one in five developing countries. Education also suffered in the 1980s in some countries because of the externally-imposed 'structural' economic reforms (see also chapter 12).

In many parts of the world, traditional attitudes and the burden of domestic work result in many more daughters than sons being withdrawn from school. Further, precious education money is often spent pursuing higher, prestige-seeking, levels of education for small (male) minorities, where there would be much greater social benefit from increasing the proportion of young girls who extend their schooling. In Bangladesh, India, Nepal and Papua New Guinea, the best-educated 10% (predominantly male) have received over half of all governmental expenditure on education.[13,23] The dismal end-result is that in many poor countries the perpetuation of maternal illiteracy militates against child survival and the reduction of fertility rates. In its 1992 World Development Report, the World Bank acknowledged observation that money spent on improving the literacy of women in poor countries is likely to be the most effective investment of all in social and economic development and environmental protection.[2]

5.3 Controlling population and alleviating poverty

5.3.1 Population control

Humans, through their cultural developments, have moved far away from the ecologically-based natural constraints on population growth – and must now seek corrective action through that same cultural–political medium. The UN has estimated that spending on family planning, currently around US$2 billion annually, must double by the year 2000 if population growth early next century is to be kept at manageable levels. Greatly increased family planning, particularly in the non-industrialised poor countries, will be needed just to achieve the UN's projected 'medium variant' in population growth.[6,17]

However, there have been powerful political forces ranged against population control. Since the mid-1980s, conservative US Administrations have officially withdrawn much of their substantial funding support from international family planning organisations presumed to support abortion.[17,24] Birth control programmes in developing countries thus stalled because strident political pressures on government from affluent right-to-life activists caused the International Planned Parenthood Federation and UN Family Planning Association to lose crucial funds. The Vatican, resonating to the papal encyclical, *Humanae vitae* (1968), remains adamantly opposed to family planning as a contravention of God's will – and argues, obliquely, that since the highest fertility rates occur in non-Catholic countries (e.g. Kenya) therefore the forbidding of contraception is not the problem. The resurgent political influence of the Roman Catholic Church in post-communist Poland has led to moves to outlaw the sale of contraceptives and, in 1992, the Polish General Medical Council proscribed abortion as a professionally 'unethical' procedure. Islam, too, frowns on at least some types of contraception. Another insidious – and evodeviationist – influence has been the unprincipled commercial promotion of bottle-feeding of Third World babies with First World formula, not only thus exposing them to health risks (infection and loss of the assorted biological benefits of mother's milk) but also removing the natural (albeit short-term) contraceptive effect of breast-feeding.[17] A further political obstacle in international birth control is that family planning programmes proposed by rich countries for poor countries are often seen by the latter – or are portrayed – as discriminatory; even genocidal. Some Third World leaders imply that rich countries fear the prospect of hordes of hungry, dispossessed migrants and are more concerned to preserve whales and elephants than impoverished starving humans. Racial tensions within some developing countries also present obstacles. The Malaysian government, for example, is overtly encouraging higher birth rates in Malays, for long economically subordinate to Chinese and Indian Malaysians.

Ideological and political barriers aside, family size is now declining worldwide, particularly in southern and eastern Asia. In Thailand, for example, a vigorous, community-based, family planning programme has caused the total fertility rate to drop from 7.5 in 1960 to 2.5 in 1990. Family size remains large in Africa, but is anticipated to fall from six children to three by 2025. Women throughout the world, including those in many Muslim countries, want fewer children than before.[17] Within various Third World countries, the desired family size fell by a quarter between the 1970s and 1980s.[2,6] Over half of the world's one billion reproductive-aged married couples now use some form of contraception, compared to a figure of around one-third in the 1970s and 10 % in the 1960s. In China over 70 % of married women use contraception, while for the rest of the developing world the figure is 45 %. In East Asia and Latin America, the proportion has gone up from around 30 % in 1970 to 65 % in 1990. Many of these couples rely on sterilisation, usually of the woman. The IUD and the contraceptive pill are becoming more widely used; although the use of condoms may increase with the spreading threat of HIV infection. The economic arguments in favour of birth control are becoming more apparent. For example, in Mexico the total expenditure on family planning education, the provision of tax incentives and child education subsidies for couples who restrict their child-bearing has been found to be much less than the cost of supporting the averted population increase.[13]

Despite these recent advances, the demand for contraception greatly exceeds supply. Consequently, there are an estimated 250 million couples worldwide (i.e. about one in every four reproductive-age couples) who currently want but cannot obtain contraceptives. The resultant unwanted pregnancies contribute to the tragic toll of around 150,000 women who die every year of complications from unsafe abortions.[17] Women in every age and culture have resorted to abortion to control unwanted pregnancy. In today's world, where overpopulation poses unprecedented social and ecological threats, it is disastrous that community poverty and restrictive laws deny many women access to safe abortion. In approximate terms, of the 220 million pregnancies that would now be occurring annually *if* the 1960s fertility rate applied, we prevent 30 million with contraception and terminate another (estimated) 50 million with abortion. To attain the UN's medium population projection will require reducing annual births below today's 140 million – and that means extending contraceptive use to at least half the people that currently lack it.[17] We, the clever species, have loosened the yoke of premature mortality. We now need, urgently, the insight, intelligence and enlightened generosity to ease the burden of fertility in the Third World.

5.3.2 *Poverty and inequality*

Much of the Third World's poverty and, in turn, its uncontrolled population growth and environmental degradation, arises from past and present inequitable relationships between rich and poor countries. The regressive influence of the structure of international trade and aid is considered further in chapter 12. These structural problems are exacerbated by the widespread, sometimes massive, corruption and inefficiency of Third World non-democratic governments. Even allowing for all international aid, the net flow of money is now – once again – from poor to rich countries.[2] During the 1980s downturn in worldwide economic growth, external debt increased dramatically in many of the poorest countries as prices for their main commodities fell. This heightened disparity of living standards ensures that environmental degradation will worsen in the world's poor countries. Population growth compounds that degradation, as environmental protection and natural resource management are ignored by financially desperate governments.

As we noted early in this chapter, there are some major differences in the impact of rich and poor countries upon environmental degradation. The rich countries, long conditioned by the spoils of empire, no longer depend on their own renewable resources (land, forests and biomass fuels) for economic viability. They have the wealth, the technological know-how and the access to global resources to obtain energy-dense fossil fuels (and/or nuclear power) as required. Through their increasingly rich, versatile and politically powerful transnational companies, operating in an era of economic deregulation, they have access to the world's stocks of non-renewables (metal ores and oil for petrochemicals and plastics) and they obtain renewables (food, fibre and forest products) from the poor countries, at low cost. Their energy-intensive lifestyle produces a high proportion of the world's industrial toxic wastes and air pollution, including most of the greenhouse gases.

The poor countries, with much less industry and with precarious economies, depend more on subsistence agriculture, supplemented with cash crops and the sale of natural resources, both renewable and non-renewable, to rich countries. A high proportion of their energy comes from essentially non-polluting renewable sources (biomass and hydropower) – although China's accelerating coal combustion is clouding the picture. Exacerbated by continuing population growth and the encroachments of mechanised export-oriented production, traditional agricultural prudence is abandoned and marginal land is pressed into service. The overuse of arable land, the exposure of vulnerable land to farming, and the mismanagement of irrigation, cause exhaustion and salinity of soils. This and deforestation leads to soil erosion. As farmland becomes non-productive, as food runs out, and as economies stall, so the pressures build up for people to emigrate as environmental refugees, or to resort to civil strife or war. In 1987, the Prime Minister of Zimbabwe said:

Those who are poor and hungry will often destroy their immediate environment in order to survive ... They [the developing countries] are faced with falling commodity prices, a rise in protectionism, a crushing debt burden and dwindling or even reverse financial flows. If their commodities bring little money, they must produce more of them to bring in the same amount or at times even less. To do this, they cut down trees, bring under cultivation marginal land, overgraze their pastures and in the process make desert out of previously productive land. But in these actions the poor have no choice. They cannot exercise the option to die today so as to live tomorrow.[25]

In poor countries, awareness of environmental problems is usually low. Environmental and occupational exposure standards are difficult to enforce, because of a lack of professional training, monitoring resources, and – often – political will. Further, governments of those countries may not want to jeopardise the interests of overseas investors. Besides, in circumstances of economic hardship, desperately needed foreign earnings from exported commodities or products – no matter how produced – take precedence over other societal and ecological niceties.

Many commentators consider that the problems of overpopulation in the Third World and its associated environmental destruction are the result of entrenched inequality.[26] Murdoch, an American political economist, argues that the 'structural poverty' of rural populations determines the pattern of hunger and population growth: 'In most developing countries, this segment of the population is the largest, the poorest, and has the highest birth rates, so it is the dynamo that drives population growth.'[27] 'Structural' refers here to poverty caused by the political and economic framework of arrangements and institutions that sustain inequitable patterns of land ownership, of access to credit, capital and modern technology and the form of the relationship between rich and poor nations. It implies that poverty and overpopulation are *not* simply the result of poor soils, poor climate, inadequate natural resources, or other physical or biological conditions, but that they are determined primarily by social and political structures. Technical and economic assistance, from rich to poor, will not work within an unjust and insecure social framework.

5.3.3 Prospects for economic development

The UN, in 1990, assessed two of the structural problems facing the global economy thus: 'If current patterns were to continue unchanged, the world distribution of income between countries would worsen. ... The risk of serious deterioration of the physical environment will increase unless patterns of production and consumption are radically altered.'[23] Meanwhile, we contemplate business as usual. For industrialised countries, including those in Eastern Europe and the former USSR, economic growth rates in the range 2–4% per year have been predicted during this decade.[28] Continued rapid growth is expected in some Southeast and East Asian countries. However, growth will be 'negative' in sub-Saharan Africa and West Asia, and most Latin American economies continue in a precarious situation.[25] So the existing, yawning, gap widens. Most of the newly-created 'wealth' from this century's unprecedented twenty-fold expansion of the global economy, and the fifty-fold expansion in industrial production, has accrued to the minority of rich countries, and continues to do so (see Fig. 5.3).

Africa now presents a special problem.[13,29] Of the forty countries scoring lowest on the UN's index of political-economic development, thirty-two

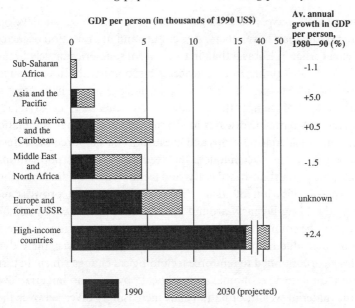

Figure 5.3. Current, recent and projected per-person Gross Domestic Product (GDP), by category of country. (Source: World Bank, 1992 – ref 2)

are in Africa. Over the past three decades, Africa's share of the world's aggregate GNP has fallen by over one-third. There are few jobs outside agriculture in Africa, and attempts at industrialisation have not been very successful. The international terms of trade have been increasingly unfavourable, and domestic economic performance has been encumbered by top-heavy state intervention and often a bloated public service. Thus, the current crisis in sub-Saharan Africa combines high population growth, severe poverty, political instability, drought and environmental degradation of agricultural and pastoral land. The chronic political troubles of southern Africa and of Somalia, Ethiopia and the Sudan have seriously damaged the social infrastructure, wasted precious resources, devastated farmlands, and countless refugees have been created. Since sub-Saharan Africa's international debt now approaches its entire collective GNP, most countries have had to enter into 'structural adjustment' economic reform programmes with the World Bank (discussed further in chapter 12). In Latin America, despite similar structural adjustment imposed by international agencies since the 1980s debt crisis, there has been little increase in economic development, as debt repayment continues to dominate the balance sheets.

Taking the longer historical view, human populations living in favoured environments have, on average, undergone technological development more rapidly than have others. Agriculture and urbanisation occurred two millenia earlier in Eurasia than in tropical Mesoamerica largely because of the lack of cultivatable crops and accessible, manageable arable land in the latter region. Further, both China and, later, Europe have enjoyed the combination of temperate climates, rich soils and ecosystems with relatively few parasitic diseases to sap the vigour of human populations or their domesticated food-crops and livestock. This favoured the diffusion of productive farming techniques, the feeding of urban populations, the development of urban-based crafts and technologies, trading, mobility and the cross-fertilisation of ideas. China's technological prowess receded during the Ming Dynasty around five centuries ago when, by edict of the Emperor, the country turned inwards, stopped its sea-faring and commerce and awaited due homage from all.[30] In Europe, however, these dynamic forces have continued to reinforce technological change which, in turn, has conferred military, economic and political power – and imperial leverage. So the material wealth of nations has increasingly diverged over the past millenium – and particularly in recent centuries, as those cultures which had remained relatively static became subordinate to the colonising, technologically dynamic, cultures. Thus were sown most of the seeds of economic disparity between today's rich and poor nations.

While the complete explanations are, of course, more complex, the outcome is clear enough. The modern world economic system, created within the framework of Western capitalism, systematically tends to shift wealth to those most fluent with the 'rules' and in control of the resources – i.e. from the poor to the rich. The wealth gap is now so wide that there is little prospect of poor nations achieving social–economic 'lift-off' without massive assistance from the rich world. Without that lift-off, poverty will persist, fertility will remain high – and ecological strain will grow.

5.4 Summary

Pressures of population growth and the hand-to-mouth needs of Third World poverty are fundamental threats to the integrity of ecosystems – particularly via the aggregate loss of fertile land and climate-stabilising, biodiversity-nurturing forest. Population growth is grounded in the biology of birth, fertility and death. But the surging and receding population tides, over the millenia, have been intimately bound up with human culture and the rides of the Four Horsemen of the Apocalypse – pestilence, famine, warfare and conquest. Population growth is closely related to poverty: each influences the other. There is a pessimistic view that some of today's impoverished Third World masses are becoming enmeshed in a 'demographic trap' – a trap from which there will not be quick rescue by the sort of systemic decline in population fertility that completed the demographic transitions in earlier, more favoured populations.

Poverty is, in some respects, a relative notion. Historically, the notion of poverty can be traced to settled communal living, when goods and chattels begin to accumulate, when 'property' becomes identifiable, when social stratification becomes explicit, and when the amenities of daily living focus on tangible assets. In such a setting, a personal lack of goods and chattels makes for the real experience of relative deprivation. This, then, was one of the most profound consequences for human society of the emergence of agriculture. History aside, in today's world, the widespread and grinding poverty that afflicts over a billion people in the poor countries is starkly obvious. That poverty and its associated powerlessness are the down-side of the wealth, control and exploitation that continue to shape dealings between rich and poor countries – and that widen the gap between them. The statistics on health and survival for the world's poor are grim, in both rural and urban-slum settings. Infant and child mortality, while declining, remains high; infectious diseases are rife. The recent increases in malaria, cholera, tuberculosis and AIDS – particularly in impoverished Africa – reflect the combined influences of poverty, crowding and environmental decline.

World population growth is expected to tail off later next century. By drawing graphs, one can plot a soft asymptotic plateauing. Meanwhile, the reality is a continued population growth in those countries least able to afford it and an accelerating environmental decline by overload, depletion and pollution. On these current trends, as populations continue to grow next century, many parts of the world will become less healthy places in which to try and live.

The next five chapters examine the major areas of ecological disruption. Each is examined in an historical, social and technical context, paying particular attention to the consequences for human population health. Of course, there are many interfaces and overlaps between the topics of these seven chapters. The problems are neither independent in their origin nor in their processes and impacts.

References

1. McNamara RS. *The McNamara Years at the World Bank: Major Policy Addresses of Robert S. McNamara* 1968–1981. Baltimore: Johns Hopkins University Press, 1981.
2. World Bank. *World Development Report* 1992. *Development and the Environment*. Oxford: Oxford University Press, 1992.
3. UN Development Program (UNDP). *Human Development Report*, 1991. Oxford: Oxford University Press, 1991.
4. Wynne-Edward VC. Self-regulating systems in populations of animals. *Science* 1965; **147**: 1543–8.
5. Ehrlich PR, Ehrlich A. *The Population Explosion*. New York: Simon and Schuster, 1990.
6. UN Department of International Economic and Social Affairs. *Long-range World Population Projections. Two Centuries of Population Growth* 1950–2150. ST/ESA/SER.A/125. New York: United Nations, 1992.
7. To calculate doubling time, the rule-of-thumb is to divide 70 by the percentage increase per unit time. Thus, 70 divided by 1.7 gives 41.
8. Last JM. Global Environment, Health, and Health Services. In: Last JM (ed). *Maxcy–Rosenau–Last. Public Health and Preventive Medicine*. Norwalk, Connecticut: Appleton-Lange, 1992, pp 677–84.
9. Powles JW. Global patterns and disadvantaged populations. In: McNeil JJ, King RWF, Jennings GL, Powles JW (eds). *A Textbook of Preventive Medicine*. Melbourne: Edward Arnold, 1990, pp 238–52.
10. Heathcote RL. *The Arid Lands: Their Use and Abuse*. New York: Longman, 1983.
11. McKeown T. *The Modern Rise of Population*. New York: Academic Press, 1976.
12. King M. Health is a sustainable state. *Lancet* 1990; **336**: 664–7.
13. World Bank. *World Development Report* 1991. *The Challenge of Development*. Oxford: Oxford University Press, 1991.
14. Population Division, Department of International Economic and Social Affairs. *Population Newsletter Number* 51. New York: UN, 1991.
15. Myers N. Famine and refugees. *Climate Alert* 1992; **5(1)**: 6–7.
16. Deevey ES. The human population. *Scientific American* 1960; **203(3)**: 195–204.
17. Potts M, Rosenfield A. The fifth freedom revisited: I, background and existing programmes. *Lancet* 1990; **336**: 1227–31; and: Potts M, Rosenfield A. The fifth freedom revisited: II, the way forward. *Lancet* 1990; **336**: 1293–5.
18. Aghajanian A. Population change in Iran, 1966–86: A stalled demographic transition? *Population and Development Review* 1991; **17**: 703–15.
19. Smith T. The population bomb has exploded already. *British Medical Journal* 1990; **301**: 681–2.
20. Ridker RG. Population issues. In: Darmstadter J (ed). *Global Development and the Environment. Perspectives on Sustainability*. Washington, DC: Resources for the Future, 1992, pp 7–14. Ridker points out that this 'revisionist' view coincided, in America, with the reduced support for birth control programs – see section 5.3.1.
21. Caldwell J. Major new evidence on health transition and its interpretation. *Health Transition Review* 1991; **1**: 221–9.

22. Powles J. Changes in disease patterns and related social trends. *Social Science and Medicine* 1992; **35**: 377–87.
23. United Nations. *Global Outlook 2000. An Economic, Social and Environmental Perspective*. UN Publications ST/ESA/215/Rev 1, New York: United Nations, 1990.
24. Horgan J. Exporting misery. *Scientific American* 1991; **265(2)**: 8–9. (Note: In 1993, the Clinton Administration moved quickly to reverse this retrograde policy.)
25. United Nations, Department of International Economic and Social Affairs. *World Population Monitoring 1989. Population Studies No.* 113. New York: United Nations, 1990.
26. Johnston RJ, Taylor PJ (eds). *A World in Crisis? Geographical Perspectives*. Second edition. Oxford: Blackwell, 1989.
27. Murdoch WM. *The Poverty of Nations. The Political Economy of Hunger and Population*. Baltimore: Johns Hopkins University Press, 1980, pp 6–7.
28. Some commentators argue that late-stage industrial capitalism now faces 'structural unemployment', because production processes have increasingly substituted energy-intensive, sophisticated technology for human labour. As the wage-paid workforce declines, so too does the domestic consumer market. Thus, the self-sustaining, wealth-creating system of wage-labourers-as-consumers falters – which, in turn, heightens the drive for export markets.
29. Morrow L. Africa: The scramble for existence. *Time* 1992; **7(36)**: 46–55.
30. Merson J. *Roads to Xanadu – East and West in the Making of the Modern World*. Sydney: Child and Associates, 1989.

6

Greenhouse warming and climate change

6.1 Climate change

6.1.1 Earth's atmosphere

Describing the sky seems easy enough: blue during the day, black at night; clouds, sun, moon or stars. However, beyond this – the stuff of literature, legend and ancient religions – the atmosphere is a complex part of the biosphere. Via the natural greenhouse effect it regulates Earth's temperature; it absorbs much of the incoming ultraviolet radiation; it mediates the weather; it provides carbon dioxide for plant photosynthesis and oxygen for aerobic metabolism; and it participates in the life-supporting circulation of water, carbon, nitrogen, phosphorus and many other elements.

Earth's average surface temperature is a moderate 15°C. Yet, the diaphanous atmosphere of gravity-weak Mars and the luxuriant cloud-laden atmosphere of Venus are accompanied by planetary temperatures well below the freezing point (-53 °C) and well above the boiling point of water (460 °C), respectively. Unlike Earth, neither of our neighbouring, lifeless, sibling planets have more than a trace of oxygen or a few percentage points of nitrogen; nor any methane or ozone. The atmospheres of both are at least 95% carbon dioxide, whereas the carbon dioxide concentration in Earth's atmosphere is now less than 1%.

Earth's atmosphere extends outwards for about 100 kilometres. The troposphere, the lowest stratum, is about 10 kilometres thick and contains about 90% of the total mass of the atmosphere. Within that layer our daily weather occurs – winds, clouds, rain and snow. The troposphere comprises mostly nitrogen (78%) and oxygen (21%); the other 1% is made up of argon, water vapour and traces of carbon dioxide, methane, ammonia, hydrogen and other minor gases. Without the replenishing actions of life

132

on Earth, this distinctive and chemically improbable composition would not be sustainable, since oxygen, methane and other gases interact with one another in the presence of sunlight. The next, less dense, stratum is the stratosphere, which extends from approximately 10–50 kilometres altitude and contains the ozone 'layer'.

The atmosphere's role in regulating Earth's temperature and weather is complex. The creation of hot and cold air masses, following heat exchanges with land and sea, causes day-to-day variations in temperature, wind and rain. Seasonal changes in the climate reflect the annual, orbital cycle of change in the orientation of Earth's hemispheres to the Sun. On an increasingly grand timescale are four other geological and astronomical influences on Earth's temperature:

1. The approximately eleven-year cycle of solar flares ('sunspots') entailing bursts of extra solar radiation.
2. The hierarchy of astronomical cycles (the Milankovic cycles) due to changes in the geometry of Earth's orbit around the Sun and in the gyroscopic wobble (precession) around its axis of rotation.[1]
3. Vastly longer trends due to the drift of continents around the globe (leading to prolonged sets of ice ages every several hundred million years).
4. The long-term increase in the Sun's temperature.

The atmosphere freely admits short-wave solar radiation (including visible light) to Earth's surface, but then traps some of that energy when it is reradiated outwards by Earth as longer-wave infrared radiation (heat). This heat retention, technically called 'radiative forcing', causes the natural greenhouse effect (see Fig. 6.1). The naturally-occurring greenhouse gases – predominantly water vapour, carbon dioxide and methane – each absorb particular wavelengths of infra-red radiation. (Smaller two-atom gaseous molecules, like carbon monoxide and nitric oxide, do not absorb infra-red radiation.) Consequently, much of this radiant energy is retained within the lower atmosphere, thus raising its temperature. The resultant average global surface temperature, 15 °C, is about 34 °C warmer than if all the reradiated heat escaped back into space. Without this atmospheric heat blanket the world's water would freeze and life would not be possible. The corollary, of course, is that the greater the concentration of these gases the 'thicker' the blanket, and the bigger the warming effect within the lower atmosphere.

Over the past 160,000 years, fluctuations in the atmospheric concentrations of carbon dioxide and methane have been closely correlated with

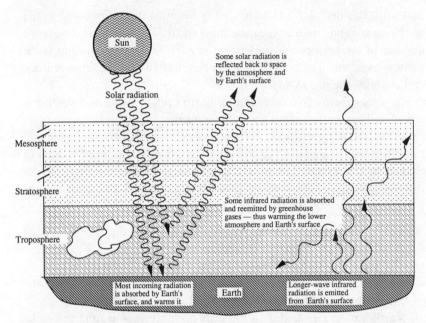

Figure 6.1. A simplified diagram of the greenhouse effect. The greater the concentration of heat-trapping gases in the lower atmosphere the greater the heat retention close to Earth's surface.

changes in the average global temperature. This period, shown in Fig. 1.1 and spanning more than one full glacial–interglacial cycle, has entailed temperature variations of 5–7 °C. The measurements of carbon dioxide concentration and of the temperature-dependent ratio of oxygen isotopes (oxygen atoms with different atomic weights) have been made on chronological layers of ice from the one-kilometre deep Vostok 5 Ice Core, drilled by Russian scientists in the Antarctic. The correlation suggests that variations in atmospheric carbon dioxide contributed to variations in global temperature. However, the relationship could have also been in the opposite direction: variations in Earth's surface temperature could have influenced the emission of carbon dioxide and methane. As with much of the science of climate change, there are uncertainties.

6.1.2 The biosphere's carbon cycle

Earth's 'metabolism' is powered by the incoming solar radiation, with assistance from subterranean thermodynamic forces. The two great cycles of the biosphere are the hydrological (water) cycle and the biogeochemical cycle. The hydrological cycle, driven by solar-induced temperature gradients and the evapotranspiration of forests, influences the distribution of water and its evaporation and precipitation. The biogeochemical cycle circulates essential elements – particularly carbon, nitrogen, phosphorus and sulphur – and is driven primarily by the solar energy used by living organisms to acquire and, later, to release these elements.

Carbon and water, along with solar energy, are the basic building blocks of plants, upon which all animals depend for food. All life on Earth is based on carbon, a very versatile element. Chains or rings of carbon atoms provide the central framework of organic molecules in plants, animals and their derivative fossil fuels. Carbon also exists as carbon dioxide gas in the atmosphere, dissolved in the oceans, solid limestone carbonate sediment and fossil fuel. Decaying organic matter, fires and exhalations by plants and animals all naturally contribute carbon dioxide to the air. Meanwhile, it is continually removed from the atmosphere by Earth's two immediate 'sinks': plant photosynthesis and absorption into the oceans (see also Fig. 2.1). These natural processes maintain an equilibrium with one another from year to year. Over the lifetime of the Earth, however, there has been a massive reduction in the amount of carbon dioxide in the atmosphere and an equivalent increase in the amount buried below Earth's surface.

There are around 40,000 giga-tonnes (Gt = one billion tonnes) of mobile carbon circulating through the biosphere (Fig 6.2). This comprises three sources of actively circulating carbon – each of 600–1000 Gt – in the atmosphere, plants, and the surface layer of the oceans. Animal life accounts for a tiny 1–2 Gt. Next, there is a slow-moving 1,500 Gt of carbon in the topsoil as terrestrial detritus, and an even slower 38,000 Gt in the deep oceans. In addition to all this naturally-circulating carbon, there is a vast store of carbon locked away after its removal from the Earth's adolescent atmosphere, either in sedimentary limestone deposits or, of great importance to this chapter, in fossil fuels (an estimated 10,000 Gt of carbon). Hence, by burning 'fossil' fuels we release the carbon and the solar energy trapped long ago by primitive plant life. The combustion converts the organic, energy-containing, hydrocarbon molecules to oxides of carbon, hydrogen and, in lesser amount, nitrogen – i.e. carbon dioxide, water vapour and nitrous oxide.

Within the finely balanced natural exchanges of the biosphere's carbon cycle, approximately 100 Gt of carbon moves from land-sources (predominantly plants) to the atmosphere each year, and vice-versa. Likewise, approximately 90 Gt moves in each direction between oceans and atmosphere, with photosynthesis by phytoplankton accounting for most

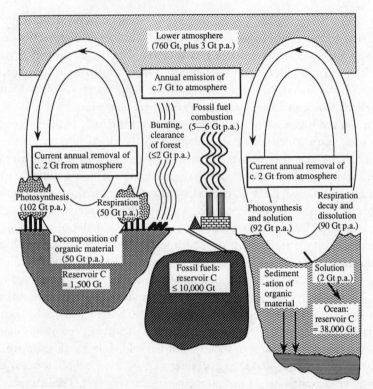

Figure 6.2. The carbon cycle, showing natural and anthropogenic movement of carbon into and out of the atmosphere. The net effect of emitting 7 gigatonnes (Gt) of anthropogenic carbon to the atmosphere each year, of which approximately 4 Gt is removed by natural 'sinks' (ocean and forest), is to add 3 Gt annually. (Adapted from diagrams in Leggett J, 1990 (ref. 12), and UNEP, 1991 (ref. 13, chapter 7))

of the ocean's uptake.[2,3] (In passing, we should note, perhaps with surprise, that about as much photosynthesis occurs at sea as on land.) Now, however, human activities are causing a net annual addition of 3 Gt of carbon to the atmosphere. This derives from the carbon that we release by burning fossil fuel (6 Gt) and burning other biomass, particularly forests (1–2 Gt) – over half of which is (currently) absorbed by the planet's ocean and forest 'sinks'.

Carbon dioxide is very important to the topic of climate change because of its stability. Unlike other long-lived heat-trapping gases, carbon dioxide does not break down. Hence, elevated atmospheric carbon dioxide levels tend to persist long after anthropogenic emissions are curtailed. If all such emissions ceased in the year 2000, global temperatures would probably continue to rise until around 2025 (as the backlog of trapped energy

Table 6.1. *Profile of the main anthropogenic greenhouse gases.*

Characteristic	Carbon dioxide	Methane	Nitrous oxide	CFC-11[a]	CFC-12[a]
Atmospheric concentration (by volume):[b]	ppm	ppm	ppb	ppt	ppt
Preindustrial (1750–1800)	280	0.8	288	0	0
Present day (early 1990s)	356	1.74	311	290	500
Current rate of change per year	0.5%	0.9%	0.25%	4%	4%
Atmospheric lifetime (years)	50–200	10	150	65	130
Global warming potential (GWP):					
Direct	1	11	270	3,400	7,100
Indirect	none	positive	?	negative	negative

[a]CFC = chlorofluorocarbon
[b] parts per million/billion/trillion
Source: IPCC (1990, 1992)

becomes realised warmth), after which, because of the long slow removal of excess carbon dioxide from the atmosphere, most of that temperature excess would still be present in the year 2100.[3] By comparison, if excess emissions of methane ceased in 2000, that methane would be removed within two decades and its temperature effect would fully dissipate by around 2050.[2]

6.1.3 The human-made greenhouse effect

The composition of Earth's atmosphere has changed much over the past four billion years, particularly in response to the evolution of life. Whereas those changes occurred on a grand timescale, human activities have very recently introduced further, rapid changes. Polar ice cores show that emissions of various greenhouse gases into Earth's atmosphere have increased substantially since the early industrial revolution 200 years ago, and markedly so since 1950. Methane concentrations have doubled since around 1800. Carbon dioxide concentrations have gone up by almost one-third, with well over half of that increase occurring since the 1950s.[4]

The main anthropogenic greenhouse gases are carbon dioxide, methane

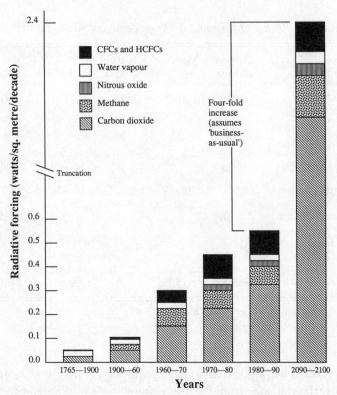

Figure 6.3. Changes across time in the contributions to enhanced radiative forcing (heat retention) by the major types of anthropogenic greenhouse gas, 1765–1990. Estimates for 2090–2100 are based on projection of current trends. (Data source: IPCC, 1990 – ref. 3)

(sometimes called 'marsh gas'), nitrous oxide and the entirely synthetic chlorofluorocarbons (CFCs). These trace gases vary enormously in their longevity and heat-trapping capacity (see Table 6.1). Within a 100-year time horizon, methane is approximately 11 times more heat-absorptive than carbon dioxide (per unit weight), while nitrous oxide is 270 times and the CFCs several thousand times more heat-absorptive.[5] The estimated contributions to increased 'radiative forcing' during the 1980s were: carbon dioxide 55%, CFCs 24%, methane 15% and nitrous oxide 6% (see Fig 6.3).[3] Within 50 years, however, methane could become the main greenhouse gas because, as the climate warms, the voluminous peat-like permafrost in the sub-arctic tundras may release much more methane.

We should look further at the story about carbon dioxide told by the archaeological air in those polar ice cores. The recent, brief, energy-intensive phase of human history has entailed a hundred-fold increase in the global use of energy since around 1800. This rate of increase has surged since around 1950, causing an approximately hundred-fold increase in the *rate* of annual global production of carbon dioxide.[6] Previously, the atmospheric concentration of carbon dioxide had increased only slowly, from around 200 parts per million (ppm) after the last ice age to 280 ppm by around 1800. Subsequently, the concentration of carbon dioxide has surged to almost 360 ppm. The most developed countries produce about half the carbon dioxide, while Eastern Europe and the populous Third World countries (particularly Brazil, China and India) produce one quarter each. Emissions of carbon dioxide have increased more than ten-fold in Asia over the past four decades. Three-quarters of anthropogenic carbon dioxide emissions comes from the combustion of fossil fuels, particularly coal, but with an increasing amount coming from the exhaust of cars and commercial vehicles. Most of the remainder comes from the burning of rainforest and, to a lesser extent, the production of cement. Globally, increases in agricultural and industrial emissions of other greenhouse gases are also accelerating. Anthropogenic methane derives from irrigated agriculture, flatulent cows, mines, gas pipelines and rubbish tips; its atmospheric concentration has increased by 50 % since 1950. Nitrous oxide comes from fossil fuel combustion and from use of nitrogenous fertilisers. In the late 1980s, the top five countries for total greenhouse gas emissions were the USA (18 %), USSR (12 %), Brazil (11 %), China (7 %) and India (4 %).[7]

Clearly, greenhouse gases are accumulating – but do we yet *know* whether global temperature is increasing because of these emissions? Predictions of global warming are not new. Late last century, a Swedish scientist, Svante Arrhenius, predicted that the build-up of atmospheric carbon dioxide from the burning of fossil fuels would increase the retention of Earth's heat, causing global warming. Further scientific debate occurred in the 1920s, but then largely subsided until the 1980s. One difficulty is that the relationship between *heat* and *temperature* is complex. The Earth's deep oceans are natural 'sinks' that divert heat away from the atmosphere, thus deferring any surface temperature increase.

Nevertheless, we know that the global temperature has increased since around 1900 (see Fig. 4.1).[3] From recorded measurements made at thousands of stations around the world, adjusted for altered circumstance and technique, global temperatures seem to have increased by 0.5 °C over the past 100 years.[8] The International Panel on Climate Change (IPCC), established by the UN Environment Programme and the World Meteorological Organization in 1988 as an expert body of over 300 scientists, estimates that the global mean surface temperature has increased by 0.3–0.6 °C over the last 100 years. This increase, however, is well within the range of natural climate variability predicted by climate-change computer models when applied to historical data. This could be due to other influences counteracting global temperature increase, and there has been speculation that the rapid increase in urban and industrial air pollution since World War II and an unusual succession of volcanic eruptions may each have contributed a cooling effect.[9] The scattering of solar radiation by the haze of fine-particle industrial air pollutants could have accounted for several tenths of a degree of cooling in recent decades. Further, a temporary lengthening of the sunspot cycle in mid-century may have marginally decreased solar heat input. One thing is clear in all of this: the signal-to-noise ratio is not yet very high!

On a very much longer timescale, Earth is presumably entering the latter half of the current interglacial period and its temperature will begin edging downwards sometime in the next ten thousand years, before plunging (by 5–10 °C) into yet another ice age. The several previous stable interglacials have lasted, on average, around 20,000 years (see Fig. 1.1). On an even broader timescale, Earth has experienced three long warm 'greenhouse' eras over the past billion years. Each of these lasted about 250 million years, with a longer first phase and shorter second phase separated by a 'brief' cooling. If the current, third, 'greenhouse' era conforms, then we may be approaching the end of the intervening cooling (which has culminated in a series of ice ages over the past several million years), and Earth may get warmer again over the coming 50 million years.

Meanwhile, our task is to appraise a near-instantaneous, apparently anthropogenic, blip in temperature. The world is on a short-term warming trend that may well be of our own doing. During 1980–91, there occurred eight of the twelve hottest years since record-keeping began last century; indeed, 1990 and 1991 were the two warmest years on record. If the world does warm by several degrees over the coming century, then the *rate* of temperature change would be far faster – and very much sooner – than any interglacial cooling that might occur in future.

Estimating climate change

There is now general agreement among climatologists that the main influence on the world's climate in the near future will be the warming effect of anthropogenic greenhouse gases.[3] The IPCC predicts that these gases will enhance the greenhouse effect and cause additional warming,

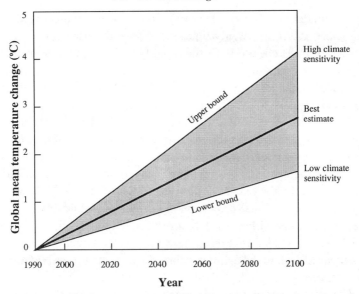

Figure 6.4. Estimated changes in average global temperature during the coming century, by IPCC, assuming the most likely economic scenario. Estimates for three different climate sensitivities to atmospheric heat retention are shown. The probable cooling effects of sulphate pollutant aerosol and stratospheric ozone depletion have not been included (both effects are likely to diminish in coming decades). (Source: IPCC, 1992 – ref. 5)

with the average global temperature rising by around 1 °C by 2025 and by 2.5–3 °C by the end of next century (see Fig 6.4).[3] Temperature increases will be greater at high latitudes, and, because land surfaces warm more quickly than oceans, those increases may be much greater in the northern hemisphere. Although some scientists argue that the computer models used to make such predictions are defective, support has firmed for the predictions of temperature rises made with increasingly complex computer models.[10] Even so, because of the difficulty in distinguishing a sustained trend from background fluctuations in temperature, the IPCC cautions that the unequivocal detection of an enhanced greenhouse effect is not likely for a decade or more.[3] This delay in getting a clear signal-to-noise ratio resembles the difficulty that economists have in detecting the onset of a recession. Certainty comes with hindsight.

The predicted rate of increase in global temperature would be several times faster than the cooling of around 1 °C at the end of the warm mediaeval era (Fig 4.1) as the higher latitudes slipped towards the Little Ice Age. It would occur much faster than did the approximately 5 °C rise over several thousand years after the last ice age (equivalent to a rate of approximately 1 °C every 700 years). Today we anticipate a predicted 1 °C rise every 35 years – twenty times faster than the rise experienced by our pre-agrarian forebears. Both the rapidity and the absolute magnitude of temperature increase are important. Rises of 2–3 °C may not seem much, but small changes can have dramatic ecological consequences.[11] An increase of just 2 °C would result in temperatures not seen on Earth for 125,000 years. A 3 °C rise, as predicted by the IPCC, would make the world hotter than it has been since the early emergence of the *Homo* genus two million years ago. (See the mystery dot at bottom-right of Fig. 1.1.). However, there are many uncertainties underlying these predictions. While a rise of 3 °C by the year 2100 was the IPCC scientists' initial best estimate for a business-as-usual projection of current economic and demographic trends, the range within which that rise could plausibly lie is 1.5–4.5 °C. In 1992 they concluded that recent research results 'either confirm or do not justify alteration of the major conclusions of the first [1990] IPCC Scientific Assessment'.

The debate about feedback mechanisms is complex – and somewhat speculative. Positive feedback reinforces and therefore destabilises, whereas negative feedback dampens and stabilises. If further greenhouse warming eventuates then positive feedback could occur, for example, because of the increased release of methane by warmed permafrost in high northern latitudes. Positive feedback could also occur from reductions in the planet's heat-reflecting snow and ice cover. Negative feedback could arise from the greater reflectivity (albedo) of an increase in cloud cover, particularly the formation of low-level stratus clouds over the tropical regions when the ocean surface temperature rises. Clearly, complexities abound and the net effect of feedback mechanisms, known and unknown, remains uncertain. Some scientists predict that positive feedbacks will outweigh negative feedbacks, leading to 'runaway' warming.[12] We could, as the IPCC has pointed out, be in for some surprises.

Uncertainty also arises from limitations in computer models. These models are not 'statistical', like economic models, but seek numerical solutions to basic physical equations. Modelling climate change is a daunting task. There are many non-linear relationships, feedback loops and interactive effects between factors that amplify or suppress one another's effects. The widely-used 'general circulation models' combine information about energy radiation with equations of motion, thermodynamics and heat transfer in oceans and atmosphere – but take little account of other complexities of oceans and clouds. While 'coupled' ocean–atmosphere models are now being introduced, there are other uncertainties about how living organisms affect the climate, and how the extra trapped heat would translate into temperature rise, water evaporation, cloud formation and so on. Further, modelling regional effects is difficult, and high-resolution estimates for grids of less than several hundred kilometres may not be possible for a decade.

Two final points should be made in this introductory section. First, although we are still struggling to make confident estimates, some irreversible forward momentum in global heating is already underway. The world is committed to *some* further warming, because of the excess long-lived greenhouse gases we have already launched. Further, the oceans act as a vast thermal sink, and the extra absorbed heat will take a while to reach the biosphere. The IPCC estimates that, as the greenhouse effect proceeds, the world's *attained* temperature at any one time is about two-thirds of its *committed* temperature. Second, despite our arbitrary references to the year 2100, the problem will not magically terminate at that arbitrary time-horizon. We are dealing with processes that could continue – and might worsen – over a longer period. Meanwhile, it seems prudent to assume that we are heading for a change in global climate greater than any other experienced by humans since the advent of settled agriculture 10,000 years ago. That would have a number of effects, mostly adverse, on human health.

6.2 Climate change and human health

6.2.1 Introduction

Climatic changes due to greenhouse warming would have both direct and indirect effects upon human health (Table 6.2).[13,14] Direct effects would be due to thermal extremes, respiratory effects and more 'natural' disasters. Indirect effects would arise from altered distribution of parasites and infectious disease vectors, reduced agricultural productivity, demographic shifts and social disruptions.

In general, it is very difficult to identify and quantify these health effects in advance. Many of the anticipated environmental changes fall outside the range of recorded human experience and their health impacts are uncertain. For example, in its 1990 'Green Plan', the Canadian Government identified nine types of impact of global warming, shown in (Fig. 6.5).[15] Further, there would be great variations in impact upon the health and wellbeing of different segments of the world's population. Impoverished populations, such as those of Bangladesh and sub-Saharan Africa, with

Table 6.2. *Main effects of global climate change on population health.*

Direct:	Deaths, illness and injury due to increased exposure to heatwaves
	Effects upon respiratory system
	Climate-related disasters (cyclones, floods, fires, etc.)
Indirect:	Altered spread and transmission of vector-borne diseases (malaria, etc.)
	Altered transmission of contagious diseases (cholera, influenza, etc.)
	Disturbance and impairment of crop production
	— Effects on soil, temperature, water, pests
	Various consequences of sea-level rise
	— Inundation, sewerage disruption, soil salinity, etc.
	Demographic disruption, environmental refugees

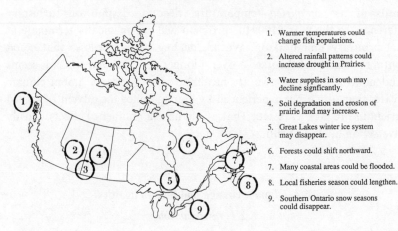

1. Warmer temperatures could change fish populations.

2. Altered rainfall patterns could increase drought in Prairies.

3. Water supplies in south may decline signficantly.

4. Soil degradation and erosion of prairie land may increase.

5. Great Lakes winter ice system may disappear.

6. Forests could shift northward.

7. Many coastal areas could be flooded.

8. Local fisheries season could lengthen.

9. Southern Ontario snow seasons could disappear.

Figure 6.5. Predicted impacts of global warming on Canada. (Source: ref. 15)

limited resources, will be much more vulnerable to adverse climatic events than will the rich nations. Within countries, socioeconomically disadvantaged subgroups will generally be the most vulnerable.

6.2.2 *Direct effects*

Temperature, especially thermal extremes

Humans cope with wide extremes in environmental conditions much better than other species, by combining physiological and behavioural adaptation. Culturally, we have developed various adaptive behaviours that enable us to cope with a range of climate for which we are not biologically well equipped by evolution. The temperate climate of Europe, into which our early ancestors eventually migrated from their presumed evolutionary origin in eastern Africa led to the adaptive donning of copious clothing. Other subsequent adaptations include housing, heating devices, patterns of food and drink, and styles of recreation.

Minor and gradual changes in temperature and humidity evoke both physiological and behavioural responses. Healthy persons have an efficient bodily heat regulatory mechanism that copes with a moderate rise in ambient environmental temperature. Thus, within certain limits, thermal comfort can be maintained. Acclimatisation, particularly via increased perspiration, develops after several days, and heat stress will be minimal.[16] With continuing exposure to heat stress, a process of physiological adaptation occurs. But frail or ill individuals with lesser physiological resilience will not adapt as well. In general, therefore, temperature increase is a greater health hazard in people with disorders of the cardiovascular, respiratory, kidney or immune systems, in infants with immature regulatory systems and in the elderly frail.[13,16] The main thermal hazards of global warming, however, will come not from a gradual drift upwards in temperature, but from an increased frequency and severity of thermal extremes. Acute heat stress in an unacclimatised urban population takes an immediate toll; the first day or two of a heatwave is a well-known 'killer'. Subsequently, protracted extremes of heat can overwhelm the body's physiological adaptive mechanisms.

A small increase in the *average* temperature means that there will be many more days each year when temperatures are extreme. Heat waves, in which the daily ambient temperature exceeds the normal bodily temperature (38 °C or 98.6 °F) for a number of days in a row, will occur more often. For example, it is estimated that Washington, DC, which currently averages one day each year with a temperature over 38 °C, will average twelve such days by around 2050.[17] Relatedly, there would be an increase from around ten to thirty days with temperatures over 32 °C. Similarly, extreme warm *years* would occur more often. In England, the frequency of extraordinarily hot summers like that of 1976 could increase 100-fold by 2030, to occur once in every 10 years.[18]

Higher summer temperatures in temperate and tropical countries will increase the rates of serious illness and death from heat-related causes. For example, during the 1980s in Missouri, USA, upwards fluctuations of 2–3 °C in the average summer temperature caused a severalfold increase in the annual number of heat-related deaths, while the major heatwave of 1980 caused a much greater increase (See Fig. 6.6).[19] From experience such as

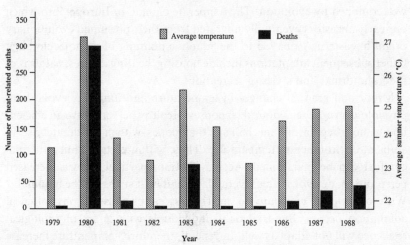

Figure 6.6. Annual fluctuations in average summer temperature and heat-related deaths in Missouri, USA, 1979–88. (Data source: ref. 19)

this, it is estimated that the approximately 3 °C temperature increase anticipated from next century's doubling of atmospheric carbon dioxide (or equivalent) would, without other adaptive change, cause a substantial increase in heat-related deaths; some American figures suggest a six-fold increase.[20] Old people are at particularly high risk; in heatwaves in developed countries around three-quarters of deaths and hospital admissions are in persons aged over 60. An 85-year-old is about six times more likely to die from the effects of a heatwave than is a 60-year-old, while babies and young children are 5–10 times more likely to die than are older children. The profile of increased hospitalisations for cardiovascular problems during heatwaves resemble that for mortality.[21]

The most direct thermal effect is heat stress, which can lead to heat exhaustion and heatstroke. Heat stress results from a breakdown in the balance between the competing demands of body temperature regulation and control of blood pressure. Since the control of body heat is mainly achieved by the cardiovascular system, heat stress, which necessitates increased blood flow through the skin to enable loss of body heat, readily exacerbates preexisting heart and blood pressure disorders.[16] Heat exhaustion, characterised by dizziness, weakness and fatigue, is not due to failed bodily temperature control, but is caused by an imbalance in bodily fluids and salt. It typically occurs after several days of high temperatures and perspiration. Heatstroke, however, is a serious condition in which the inner body temperature exceeds 41 °C (106 °F), and in which confusion, stupor and finally unconsciousness occurs. The outcome is often fatal. Heatstroke is typically preceded by falling blood pressure, faintness, dehydration, salt depletion and cramps.

The pattern of summertime deaths in New York City shows a threshold of 33 °C above which the overall death rates increase.[20] Only about half of heatwave-related deaths are caused by heatstroke or heat exhaustion; many are caused by heart attacks or cerebrovascular 'stroke' (quite different from heatstroke). From American studies, it has been estimated that, *within* the temperature range of −5 °C to +25 °C, death rates from heart disease and stroke decrease as the temperature increases, whereas at temperatures above +25 °C or below −5 °C death rates increase as temperatures become more extreme. At very high temperatures, death from cerebrovascular stroke is common in older people.[22] The fact that heat stress makes the blood more likely to clot could explain many of these cardiovascular deaths.[16]

Before the use of domestic air-conditioning in developed countries, major heatwaves would typically increase the death rate in persons aged over 50 by several-fold. In 1936, one of the hottest summers on record in the USA, 4,700 people were estimated to have been killed directly by heatstroke. This represented an increase of several per cent above the numbers of deaths expected during that summer period. In Melbourne, Australia, in 1959 there was a four-fold increase in mortality (145 excess deaths) above normal that was directly attributable to a prolonged heatwave; most of these deaths occurred during the first four days.[23]

There is some interesting information from three heatwaves that occurred in Los Angeles, in September, in 1939, 1955 and 1963.[24] All were of similar dimension, with 4–7 days above 100 °F, but the excess death rates were clearly greater in the first two episodes than in the 1963 episode. In the two earlier heatwaves, persons aged 50–69 years experienced a tripling in death rate, while persons aged over 80 had more than a five-fold increase. In the 1963 heatwave, the corresponding increases were much less: persons aged 50–69 years experienced about a 40% increase in death rate, while persons aged over 80 experienced a doubling. It has been suggested that the advent of domestic and commercial air-conditioning in the late 1950s made the difference. That is, via technical (and carbon-dioxide-generating!) adaptation, humans were able to reduce the risk to their health and survival. Excess heatwave-related deaths also occurred in the USA in 1966; the overall death rate in New York City rose by one-third in the week immediately following a severe heatwave, affecting elderly people particularly. A similar episode occurred in London, England, in

1976, when the hottest two weeks of the century caused the death rate among older people to increase by 50 % above the seasonal expected rate.[25] Again, these increases were less than in earlier episodes, presumably reflecting the health benefits of technical adaptation. Recent figures showing significant heat-related deaths in China confirm the potential global nature of this problem, especially in the absence of heat-protective resources.

The annual seven-day pilgrimage to Mecca, called the Hajj, provides an analogue for a warmer world in which air-conditioning is not available for most people. Approximately two million Muslims, mainly older people, make this pilgrimage. The temperature in Mecca during the Hajj is in the range 30–50 °C. The pilgrims are crowded in a confined space, body temperatures rise and heat-related illnesses are common.[26] In 1982, for example, over ten thousand cases of major heat-related illness occurred, including around two thousand cases of heatstroke, almost half of whom died.

Heatwaves also disturb psychological balance: tempers flare, crimes of passion increase in frequency and riots are more common. In crowded inner urban environments, thermal stresses may be amplified by the 'heat island' effect of living in treeless conglomerations of stone, concrete and asphalt, by the lack of insulation and air conditioning in poor housing, and by the lesser access to medical and hospital care. It is well documented that heatwave-related deaths are concentrated in these inner urban heat-traps.[16] The above-mentioned surge in heat-related deaths during the major heatwave in the central–southern USA in 1980 underscored the vulnerability of poor, high-risk communities.[27]

Global climate change would also tend to increase the occurrence of other heat-related health problems. Raised humidity would cause increases in fungal skin diseases and yeast diseases (such as candidiasis). During the Vietnam War, the occurrence of skin diseases among the US troops rose with increases in humidity (but not temperature).[28] Excessive temperatures can cause prickly heat and heat rash, both skin disorders that impair the skin's ability to 'breathe' and increase the thermal stress on the overheated individual. High temperatures can reduce fertility in men and may, in acute episodes, adversely affect early fetal development.[16]

Special problems will arise in certain occupations that already involve exposure to high levels of heat. In warm countries this would particularly include heat stress and heatstroke in those who work wholly or primarily outdoors. More generally, increases in temperature and humidity in the workplace will be likely to cause workers to discard their personal protective equipment – much of which is already unpopular because it is heavy and uncomfortable. Workplace studies have also shown that unsafe behaviour is minimal within the 'comfort zone' of 17–23 °C; an increase in thermal discomfort would cause more errors of judgement and hazardous impulsive behaviour, resulting in more occupational injuries.

Despite the undoubted health hazards of extreme heat and of rapid temperature increase, it is interesting that, in the world's temperate and cooler regions, death rates from heart disease, stroke and respiratory diseases are around 10% higher in the colder winter months than in the summer.[16] This tendency for death to occur in winter is greatest in the elderly, and includes the increasingly recognised problem of death from hypothermia (body-cooling). However, for persons aged below 45 years, the opposite is true: more deaths occur in summer than in winter. While cold temperature itself causes many of these winter deaths, particularly in infants and the elderly, the low humidity of winter favours transmission of certain infectious agents, particularly the influenza virus.[29] Since several other diseases, including juvenile-onset diabetes, are associated with winter infections, a greenhouse-induced reduction in severity of winter should reduce the incidence of these diseases. Likewise, the occurrence of the sudden infant death syndrome ('cot death') – although its causation still remains something of a mystery – would probably decline as one of its strongest correlates, very cold weather, is ameliorated. Thus, in some countries – such as the UK – increased summer mortality due to global warming might be partly offset by reductions in winter mortality.

Overall, there are many anticipated direct effects of greenhouse warming upon the health of populations, particularly those due to thermal extremes. These direct effects would increase only gradually, as temperature patterns change. Indeed, WHO has estimated that significant health effects of this kind will not be evident in less than a decade.[13]

Climate and the respiratory tract

Climate also affects the respiratory tract. Seasonal disorders may reflect air pollution in winter months and increased exposure to pollen, dusts and photochemical smog in summer months, whereas specific weather conditions such as cold fronts and rain may have direct effects.[30] People with chronic bronchitis and emphysema often experience exacerbations during winter, and the incidence of acute bronchitis peaks in winter. By contrast, asthma and hay fever tend to peak in the summer months, and the release of pollen is thought to be responsible. Hot dry summer weather increases the pollen count, whereas summer rain stimulates the release of certain fungi (*Sporobolomyces* and *Didymella*).[30,31]In eastern Australia, localised outbreaks of asthma seem to be associated with late springtime thunderstorms – perhaps because they prompt the sudden release of fungal spores or pollen into the air.

Many asthmatic individuals are not pollen-sensitive, however, and other factors – particularly those that increase bronchial responsiveness – may be important. For example, sudden falls in temperature, the build-up of respirable allergens in becalmed air, or summertime increases in ambient atmospheric ozone concentrations may potentiate asthmatic attacks. Sandstorms in Kansas (USA) and in the Sudan have been accompanied by increased illness and death from bronchitis and asthma.[30] Although the effects of climate change upon the respiratory system are less clearcut than those of temperature extremes upon the cardiovascular system, increases in temperature would probably cause a reduction in wintertime bronchitis and pneumonia and an increase in summertime asthma and hayfever.

Weather instability and natural disasters

Global warming may destabilise ocean and air currents, including the stratospheric jet streams. The resulting changes in temperature and in weather stability would alter the frequency and severity of bushfires, droughts, tropical cyclones, floods, landslides and coastal erosion. This, in turn, would increase the incidence of death, injury, stress-related disorders and the many adverse health effects associated with social disruption, enforced migration and settlement in urban slums. The impact of natural disasters, as ever, would be greatest on those who are most exposed and those with the fewest material resources and the least social and organisational support. (We call these events 'natural' disasters. However, if they are due to anthropogenic climate change, then in one sense they are 'unnatural'.)

The energy of tropical cyclones derives from warm moisture picked up from seas with surface temperatures above 27 °C. Greenhouse warming would cause the sea-surface temperature to rise, which may cause a greater frequency of cyclones. Low-lying and impoverished populations, such as in Bangladesh, are particularly vulnerable to increased cyclonic activity. Flash floods would occur more often in those regions that experience an increase in torrential rainfall. This would be exacerbated by upstream deforestation – such as is the case with Bangladesh, located downstream of the denuded Himalayan slopes. This, and storm surges, would be exacerbated by any associated sea-level rise. Dams and other water-control mechanisms will be more readily damaged, thereby increasing the likelihood of human disaster. Regions where there has been widespread forest clearance on hillsides adjoining flood plains will be particularly vulnerable because of the heightened run-off of water.

There is much that we still do not know about the workings of the oceans' heat-conveying currents, and their sensitivity to, and influence on, climate change. The vast east–west oceanic currents of warm water across the Pacific ('La Nina'), which periodically reverse and flow west–east as the 'El Nino' Southern Oscillation (ENSO), are a major determinant of weather and cyclonic activity along the coasts of the western Pacific and beyond. The severe 1991–92 drought in north-eastern Australia and in southern Africa was substantially caused by an El Nino event.

In the world today, the most serious effects of climatic extremes on human health are malnutrition and starvation due to severe drought.[32] The protracted drought of the 1980s in Sahelian Africa, from Somalia to the Atlantic west coast, has caused millions of deaths already, and has brought further tens of millions to the edge of starvation. The gradual southern encroachment of the Sahara may reflect one of nature's grand cycles, or it may be the result of incipient global and regional climatic change, exacerbated by forest clearance, land degradation and loss of local moisture. The food shortages have been chronically compounded by soil erosion, civil war, poor transport, and a meagre social-political infrastructure.

Bushfires, particularly in the wooded and warmer regions of Australia and the USA, are predisposed to by drought, heatwaves, low humidity and high winds. The widespread 'Ash Wednesday' bushfires in Australia in 1983 were partly due to the preceding drought having dried out the vegetation and soil. (Further, as the peri-urban populations of cities expand into surrounding, often more hilly and wooded areas, so the risk of death, injury and costly damage from bushfire increases.) Trauma on the roads may also increase, at least sporadically, because of increases in rain, wind, damaged roads and related sources of driver stress such as heatwaves. Some coastal roads would be vulnerable to various adverse effects of sea-level rise.

There are many psychosocial effects of natural disasters, affecting both victims and rescuers. The 'post-disaster syndrome' encompasses depression, withdrawal, mania, chronic anxiety and frank psychosis. In the year after the above-mentioned Ash Wednesday fires in Australia, an excess of mental disorders and stress-related disorders such as hypertension and ulcers occurred. A major dam collapse at Buffalo Creek in the USA in 1972 caused 125 deaths and left 4,000 people homeless.[33] Post-traumatic neurotic reactions occurred in most survivors, with persistent unresolved grief, shame at having survived and feelings of hopelessness and impotent rage. Many children in affected families showed developmental problems over ensuing years. In some individuals, post-disaster reactions, including the emotionally numbing guilty-survivor syndrome, persist indefinitely.

6.2.3 *Indirect effects*

The indirect effects of climate change upon population health are those that do not entail a direct-acting causal connection between a climatic factor (such as heat or humidity) and human biology. Since climate-induced changes in the distribution and behaviour of 'vectors' that transmit infectious disease organisms entail a relatively straightforward effect, I will deal with it first in this section.

Vector-borne diseases

Insects and microbes generally thrive in warmth and moisture, and many are responsive to small changes in these conditions; likewise parasitic worms and flukes. For them, the coming century could therefore be auspicious.[34] This is not good news for humans, particularly since the great majority live in coastal regions where rainfall increases are expected to be greatest.

In 'contagious' infectious diseases, the infective agent spreads directly from human to human by skin contact, air, food or water. However, the vector-borne (three-factor) infectious diseases need an intermediate 'vector' organism to pass the infectious parasite between humans. Malaria, endemic in over 100 countries, provides a key example of a vector-borne disease – and an ominous one, given that its impact around the world is now increasing. The 'four-factor' infectious diseases include a vertebrate animal host within the transmission loop. Plague is an example (rats, fleas, bacteria and humans); so is Lyme disease, discussed below. But, first, consider malaria.

Mosquito species of the *Anopheles* genus host four species of the malarial parasite, *Plasmodium*, during one of its several developmental life-stages. The pregnant female mosquito becomes infected (with 'gameto-cytes') by ingesting the blood of an infected human and, later, transmits the malarial parasite ('sporozoites') to other humans while feeding off their blood. Each such vector-borne infectious disease in humans requires an adequate population of the vector (typically an insect, snail or crustacean) and favourable environmental conditions (temperature, humidity, surface

Table 6.3 *Global status of major vector-borne diseases and possible changes in their distribution due to climatic change.*[a]

Disease	Populations at risk (millions)[b]	Prevalence of infection (millions)	Present distribution	Possible change of distribution as a result of climate change[c]
Malaria	2100	270	tropics/subtropics	+ + +
Lymphatic filariases	900	90.2	tropics/subtropics	+
Onchocerciasis	90	17.8	Africa/L. America	+
Schistosomiasis	600	200	tropics/subtropics	+ +
African trypanosomiasis	50	25,000 new cases/yr	tropical Africa	+
Leishmaniasis	350	12 million infected, plus 400,000 new cases/year	Asia/S. Europe/Africa/S. America	?
Dracunculiasis	63	1	tropics (Africa/Asia)	0
Arboviral diseases				
Dengue	No estimates		tropics/subtropics	+ +
Yellow fever	available for		Africa/L. America	+
Japanese encephalitis	these		East/Southeast Asia	+
Other arboviral diseases				+

[a]Source: WHO 1990.[13]
[b]Based on a world population estimated at 4.8 billion.
[c]o = unlikely; + = likely; + + = very likely; + + + = highly likely; ? = not known.

water) for both vector and parasite. Slight changes in climate can alter the viability and geographical distribution of vectors. For example, the malarial mosquito can only survive where the winter mean temperature is above approximately 15 °C.[35] Temperatures between 20 °C and 30 °C and humidity of at least 60% are optimal for the mosquito to survive long enough to acquire and transmit the infection. Further, a small increase in average temperature accelerates the passage of the parasite through its mosquito-borne 'extrinsic incubation' developmental stage, reducing it from 3 weeks to as little as 1 week.

Estimations by WHO of possible changes in the distribution of the major vectors caused by global warming are shown in Table 6.3. The geographic distributions of many protozoal (unicellular) parasites are closely correlated with temperature.[36] These include trypanosomiasis (sleeping sickness), filariasis (better known as the disfiguring condition of elephantiasis, in which the microfilamentous parasites block the lymphatic channels), onchocerciasis (river blindness), schistosomiasis, leishmaniasis, hookworm, guinea worm and various tapeworms. Likewise, the distributions of various vector-borne viral infections, such as dengue and yellow fever, are affected by temperature and surface water distribution. As with malaria, temperature affects both the vector's viability and the rapidity of the parasite's life cycle. For example, the extrinsic incubation period for the mosquito-hosted stage of the yellow fever virus can vary ten-fold, from three days to several weeks, depending on local temperature. Likewise, the mosquito-hosted reproduction of filarial larvae is sensitive to water temperature.

In tropical countries, these vector-borne diseases are a major cause of illness and death. For example, malaria and schistosomiasis pose health risks to 2,100 and 600 million people, respectively, and the total numbers of infected persons are approximately 270 million and 200 million. Over 1 million deaths from malaria occur each year. In sub-Saharan Africa, which accounts for around 95% of all malaria, half a million young children die from the infection each year. In eastern Africa, a relatively small increase in winter temperature would enable the malarial zone to extend 'upwards' to engulf the large urban highland populations that are currently off-limits to the mosquito because of the cooler temperature at higher altitudes – e.g. Nairobi (Kenya) and Harare (Zimbabwe). Indeed, such populations around the world, currently just outside the margins of endemic malaria, would provide early evidence of climate-related shifts in the distribution of this disease. Climate aside, malaria is now resurgent in many regions of the world. In the never-ending cut-and-thrust of biological evolution, the malarial parasite has become increasingly resistant to antimalarial drugs and it remains an elusive, and antigenically labile, target for the many different types of vaccines that have been tried out.[37] Likewise, the malarial mosquito has acquired, in fast-forward evolutionary fashion, various forms of metabolic and behavioural adaptation to insecticides.[38]

Ancient Rome and mediaeval Europe were malarious. The disease was then attributed to 'mal' air (an ancestral concept for later notions of the pervasive 'miasmas' thought to spread cholera and tuberculosis). Until recent decades, parts of today's developed world were malarious, including the USA, Italy and northern Australia. Throughout the early nineteenth century, outbreaks of the most widespread form of the parasite, *Plasmodium vivax*, occurred in Scandinavia, New York, Ottawa and England. It persisted in parts of Europe – such as around the Kent marshes in eastern England – during the first half of this century. There is now concern that, if the global climate changed sufficiently, malaria may return to some of these currently unaffected non-tropical areas, as well as extending its spread and seasonality in tropical countries (particularly the dangerous *Plasmodium falciparum* species). The vulnerability of newly-affected populations would initially be high because of their lack of natural acquired immunity.

Schistosomiasis, or 'bilharzia', has doubled in prevalence since mid-century, largely because of the enormous expansion of irrigation systems in hot climates. In this disease, blood flukes (schistosomes) infect the blood vessels of the intestines and urinary bladder, causing inflammation and tissue damage. The parasite's eggs are then excreted in faeces or urine and often end up in surface water (especially the shallow, slow-moving water of canals, ditches and irrigation channels) where they infect water snails. Climatic change would influence this cycle. In Egypt, for example, the water snails tend to lose their schistosome infections during the winter months (January–March). However, if temperatures increase, snails may spread schistosomiasis throughout the year, thus increasing the already heavy parasite burden in rural Egypt.[13,36]

A numerically much less important consequence of any increase in either malaria or schistosomiasis would be an increase in certain cancers. In Africa, malaria is associated with Burkitt's lymphoma, a cancer of the body's lymphatic tissues that usually occurs in early life. The malarial infection stimulates proliferation of the lymphatic system's B-cells, which produce antibodies to the infection. Cancer researchers think that this cell proliferation increases the probability that in one of them there will occur the particular genetic mutation (a chromosomal translocation) that initiates this cancer. Schistosomiasis increases the occurrence of cancer of the urinary bladder, presumably because of the increased turnover of epithelial cells lining the chronically inflamed bladder. Bladder cancer is already the most frequent cancer in Egypt, where schistosomiasis is widespread among workers in the well-watered fields.

Viruses are another important category of vector-borne infectious disease agent. The arboviruses (*ar*thropod-*bo*rne) have an insect vector. More than 100 arboviruses affect humans and half are transmitted by mosquitoes. Arboviral infections span a wide clinical spectrum, from those that cause mild feverish illness or subclinical infections to those causing severe and often fatal encephalitis (brain inflammation) or haemorrhagic fever. Under favourable environmental conditions, an arboviral disease can become epidemic (population-wide) from a local endemic base – in much the same way that cholera has recently broken out from its endemic enclave to sweep through much of the tropical world. The distribution and abundance of vectors are influenced by various physical factors (temperature, rainfall, humidity, groundwater and wind) and biological factors (vegetation, host species, predators, parasites and human interventions).[13]

Dengue and yellow fever, both viral diseases transmitted by the *Aedes aegypti* mosquito, will tend to increase in locations that become sufficiently wet and warm – although this species of mosquito depends on confined, human-made, sites of pooled water (such as cans, coconut shells and car tyres) for much of its breeding. Dengue causes a febrile illness which may be severe and painful in children. In Australia, where dengue occurs endemically in the north-east, the habitat of the mosquito is restricted to areas with a mean mid-winter temperature of more than 10 °C. Research in Mexico, based on population samples from 70 localities, has shown that an increase of 3–4 °C in the average temperature doubles the rate of transmission of the dengue virus.[39] Other experimental research shows that the viral incubation period within the mosquito is halved by a temperature increase of this size. Meanwhile, annual epidemics of dengue have returned to the Americas over the past decade (as they did about twenty years ago in Asia) and, in Mexico, dengue has spread to previously unaffected higher altitudes, above 1,200 metres. Pin-pointing the influence of climate upon such changes remains a tantalising task.

Increased temperature and rainfall in Australia would allow various vector-borne diseases to extend to higher latitudes (or higher altitudes).[35] For example, mosquitoes and other insect vectors would spread southwards, enabling increases in arboviral infections such as Murray Valley encephalitis (which often causes brain damage and can be fatal), Ross River virus (Australia's dominant arbovirus, causing multiple, often long-lasting joint inflammation) and dengue. In 1991–2 in Perth, Western Australia, a six-fold increase in Ross River virus infections appears to have been related to the dramatic increase in late spring and summer rainfall. Mosquitoes may increase their range if heavier summer rainfall occurs over a wider area, with more pooled water. Malaria, no longer endemic in Australia, could thus be reestablished by climatic change. These ecological intricacies are well illustrated by the commentary about Murray Valley encephalitis in a recent report by Australia's National Health and Medical Research Council:

The main vector is the mosquito *Culex annulirostris* which is present throughout Australia . . . In southeastern Australia, *Culex annulirostris* becomes active from October, reaches its peak in mid-summer [February], and becomes undetectable by May. In central and northern Australia, and southern Queensland, *Culex annulirostris* is active throughout the year. One hypothesis is that the growth of *Culex annulirostris* populations starts when the mean daily spring temperatures exceed 17.5 °C and declines rapidly when mean daily temperatures fall below this . . . A higher prevalence of this virus is strongly linked with the creation of large non-immune host populations (particularly bird populations, and perhaps mouse plagues) such as occurs after a drought breaks. Increased waterbird breeding will mean more potential carriers, extended area of endemicity, and increased potential for outbreaks in tropical Australia, the Northern Territory, the Murray-Darling Basin [which waters the southeast quartile of the continent], and South Australia . . . The new distribution and incidence will depend not only on vector-vertebrate host redistribution but also on the degree of man-vector contact. It is probable that the disease will migrate south.[35]

Vector-borne diseases are now relatively rare in developed countries. However, health authorities in the USA have identified various vector-borne diseases that might arrive or increase in the USA because of warmer temperatures.[28] St. Louis viral encephalitis, dengue, leishmaniasis and rabies (borne by the vampire bat, which is very temperature dependent) might extend into the southern USA. Two species of the dengue-transmitting *Aedes* mosquito genus are already well established in the southern USA, and are expected to extend north if temperatures continue to increase. Bat-borne rabies has severely depleted cattle herds in parts of South America, and now appears to be moving northwards. Malaria, yellow fever and Rift Valley fever might also enter the US population, perhaps assisted by immigrants from tropical countries. Computer modelling, using a model developed for tropical countries, indicates that malaria could increase in various eastern cities such as Nashville, Atlanta and Richmond.[28]

The tick-borne (four-factor) infectious diseases Rocky Mountain spotted fever and Lyme disease, both now well-established in the USA, would probably extend their current territory. Rocky Mountain spotted fever, which is due to a rickettsia (bacteria-like organism), is widespread in the eastern and south-eastern states. Lyme disease, which is due to a spirochaete (bacterium), has a more restricted and fragmented distribution in several parts of the country – although this appears to be increasing. Introduced to the USA within the past several decades, Lyme disease is now that country's leading vector-borne disease. Both diseases cause similar initial symptoms: high fever, chills, headache, backache and profound fatigue. Rocky Mountain spotted fever can produce bleeding ulcers, whereas Lyme disease can cause residual damage to brain, heart and joints. The ticks that spread these diseases are affected by climatic factors directly and via their intermediate hosts. The distributions of the intermediate hosts (e.g. the white-tailed deer, mice and birds) are climate-affected. Further, the temperature must be sufficiently warm for completion of the tick's life cycle, sufficiently humid to avoid the drying out of eggs and cold enough in winter to initiate the resting phase of the life-cycle. Higher temperatures enhance proliferation of the infectious agent itself within the infected host animal, but excessive heat would reduce the survival of ticks. Further, continued clearance of forests would deplete the numbers of deer, the intermediate host of Lyme disease. Once again, we note the complexity of predicting the impact of climate change upon ecological systems. Computer modelling of the potential impact of climate change on Rocky Mountain spotted fever indicates that tick populations would shift northwards in the USA, while being virtually eliminated from southernmost locations. However, small changes in the model's weather inputs caused a four-fold variation in predictions. Changes in the density of the intermediate host animals were also shown to affect greatly the predictions.

There could well be climatic effects upon the global and regional distribution of non-vector-borne infections, such as the common cold virus, influenza and pneumonia. However, these are not yet predictable. More tangibly, the hot European summer of 1990 – the third above-average summer in a row – caused a reported proliferation of insects in England, from cat fleas, wasps and cockroaches to, less certainly, tropical mosquitoes (including *Aedes aegypti*, which carries dengue and yellow fever). However, the important general point here is that increases in warmth and humidity are likely to promote the range, numbers and activity of a wide variety of human disease parasites and vectors. As with other aspects of climate change that disrupt ecological systems, there is much uncertainty about the likely timing and altered spread of these parasitic diseases. Unexpected consequences, such as the activation of residual pockets of plague in temperate regions, could occur.[34,40]

Overall, it seems clear that climate change, particularly warmer temperatures, would result in increases in many vector-borne infectious diseases in human populations. Control methods exist for many of these diseases, and it is likely that developed countries would be able to minimise the impact on their populations. However, it is likely that the quicker 'turn-over' of the parasite's life cycle at warmer temperatures will increase its chances for evolving greater resistance to pesticides and other control methods. There are other associated possibilities. Toxic effects upon humans could occur if increased amounts of chemical pesticides were used for controlling vectors, particularly if warmer climates cause greater volatilisation of these chemicals. Increased exposure to ultraviolet radiation because of ozone layer depletion might cause some immune suppression in humans – thus increasing their susceptibility to infection. As ever, when ecological systems are disrupted complex consequences may follow.

Water-borne infectious diseases

Global warming will affect rainfall and, therefore, the distribution of surface water and the pattern of flooding. Since this would cause less hygienic water quality in many settings, diarrhoea, cholera and dysentery epidemics could spread, particularly in developing countries. Recent studies on the cholera organism, *Vibrio cholerae*, have revealed how its survival within the environment, including within saline waters, is greatly enhanced by its sheltering beneath the mucous outer-coat of various algae and other phytoplankton, which are themselves very responsive to climatic conditions and nutrients in wastewater (see also chapter 10).[41] Diarrhoeal diseases can be caused by a large variety of bacteria (e.g. *Salmonella* and *Shigella*), viruses (e.g. *Rotavirus*) and protozoa (e.g. *Giardia lamblia*). Many of these organisms can survive in water for many months, and increased rainfall would enhance their transport between groups of people, while warmer climates would enhance their survival in the environment. An increase in diarrhoeal disease is most likely to occur within communities with insufficient clean water and poor sanitation. Poor countries are therefore at most risk, and there have been outbreaks of diarrhoeal disease after flooding in many such countries in the past. There would also be risks of outbreaks in developed countries among temporary settlements of communities displaced by flooding. As discussed in the next section, the disruption of sewage disposal by rising sea-levels would compound this problem.

Greenhouse warming could also create a water-quality problem by causing extra heating of above-ground piped water supplies. In parts of Australia, for example, there is already a seasonal problem of amoebic meningoencephalitis (inflammation of the brain and its lining membrane) caused by the *Naegleria fowleri* amoeba which proliferates in long-distance overland water-pipes in high summer temperatures.[35] Such problems would be exacerbated by further increases in temperature. Meanwhile, offshore, if the oceans become warmer, toxins produced by temperature-dependent dinoflagellates – including those that cause 'red tides' – may occur more often in seafood (e.g. paralytic shellfish poisoning and ciguatera poisoning in reef fish).

Sea-level rise

As the world warms, the thermal expansion of ocean water, augmented by melting ice, is expected to raise the sea-level. The IPCC estimated, in 1990, a rise of approximately 20 centimetres (eight inches) by 2030 and of 65 centimetres (two feet) by 2100.[3] As with estimates of global warming itself, there is considerable statistical uncertainty about these estimates. Some scientists estimate that the increased evaporation caused by global warming will greatly increase snow-fall at the poles, particularly in the Arctic region, and that this would withdraw some water from the oceans.[42]

The level of the seas is always changing, usually on a grand, slow, timescale. Over millions of years, the Mediterranean Sea has emptied and filled like a huge bathtub, as the sea-level has fluctuated by a 100 metres or more. During ice ages, particularly those occurring periodically over the past million years, the shift of water from ocean to extended polar ice-cap has created land bridges, enabling the migration of species – including that of *Homo sapiens* from Asia to Australia and from Siberia to Alaska. When Earth was pulling out of the last ice age, the seas rose rapidly by around one metre every century for several thousands of years. The total rise was of the order of 50–100 metres.[43] As occurred during each of those vast changes in sea-level stretching over many thousands of years, the world now faces a rise of half a metre or more over the next century. That, too, would be a rapid change, and faster than anything experienced by human populations since settled agrarian living began.

During the intervening millenia, human settlements have clustered on the world's coastal lands. A sea-level rise of half a metre during the coming century would threaten many such coastal communities. A rise of one metre (the IPCC's 'high' estimate) would inundate an estimated five million square kilometres of the world's lowlands, destroying one-third of all cropland and creating fifty million environmental refugees. This would include inundating one-quarter of Bangladesh, one-seventh of its cropland, and displacing 15–35% of its population.[17] Likewise, a one-metre rise would inundate 12–15% of Egypt's cropland; indeed, much of the heavily populated Nile Delta would be flooded if the seas rose just 50 centimetres.[44] The resultant displaced Egyptian farmers would have few fallback options, since coastal lands account for much of the national total of arable land.

There are many other low-lying coastal regions (for example, the Netherlands, eastern England, parts of Indonesia, the Florida Everglades, much of southern Louisiana and parts of the northeast coast of Latin America) that would be vulnerable to severe flooding or inundation. Some small island-states such as The Maldives and Vanuatu, at present only a few metres above sea-level at their highest point, could eventually be at risk of total immersion. A one-metre rise would obliterate several hundred Pacific atolls (after having earlier obliterated much of their underlying capsule of fresh water).

The projected rise in sea-level would be 5–10 times faster than the rise since around 1800, and would be likely to overwhelm many of our social and technical resources. A one-metre rise would increase the estimated probability of the dykes in the Netherlands being breached from once in every 10,000 years to once in every 100 years. Some of the world's great cities are already vulnerable to flooding, including Shanghai, Bangkok, New Orleans, Alexandria, St. Petersburg and Venice. All of these stand beside major rivers, on estuaries, or on low-lying land barely above sea-level. Further, some cities, such as Bangkok and Venice, are actually subsiding, either because of depletion of underlying groundwater caused by industrial and domestic consumption or because of natural land movements.

While inundation is the obvious adverse consequence of rising seas, there are others. In parts of Florida, the infiltration of the subterranean water table by sea water, in porous rock formations, is encroaching upon supplies of drinkable water. Rising seas would disrupt stormwater drainage and the disposal of sewerage and waste, causing increased exposures to toxic chemicals and infectious agents. Local damage to coastal roads, bridges and coastal protection structures would result from increased flooding, erosion, storm surges, wave impact and seawater intrusion (including rust). The biggest such danger to large coastal populations would come from storms, combining weather instability with the precariously high sea-level. Poverty and the absence of social infrastructure – as in Bangladesh – would compound the public health consequences of storms, disruption of sanitation and displacement of coastal dwellers in many such populations.

Rising seas would cause saltwater to encroach upon freshwater estuarine and tidal areas – often used as municipal and industrial freshwater supplies. Increased salinity of water tables would damage farmlands. Wetlands that nourish the world's fisheries, and the natural tide-aided hatcheries for fish fry, would be damaged. Around Subang, in Indonesia, for example, the predicted encroachments of the sea would reduce greatly the annual harvest of fish and shrimp, destroying the livelihood of tens of thousands. The world's wetlands are vital to the ecology and economy of many coastal areas. Not only is their biological productivity per unit of energy input very high, but they have been the natural source of various food crops, including rice and palm plants, and are a continuing source of genetic stock for future crop development.

Even greater rises in sea-level would occur, albeit over several centuries, if polar ice sheets melted. Particular concern has been focused on the western Antarctic ice sheet which, if melted, would cause the ocean to rise by around six metres. However, that huge slab of 'unstable' ice apparently survived the world's previous interglacial period, when temperatures were a little hotter than those predicted next century. Satellite photographs indicate that parts of the Antarctic ice shelf have shrunk substantially since the mid-1960s. However, since there has probably been progressive shrinking of polar ice during the current interglacial period, the significance of these observations is unclear.[45] Finally, because of the menace of unexpected positive feedbacks triggered by global warming, we should note the worst-case scenario: if the ice covering Greenland and Antarctica were to melt, the world's oceans would rise by not one, or six, but tens of metres!

Agricultural productivity

All human civilisations have known that climate is fundamentally important in growing food. Religions, ancient and modern, have fixated on gods of the weather and the seasons. Stravinsky's pulsatingly pagan *Rites of Spring* encapsulates the age-old dependency relationships between agrarian humans and their weather gods – to Whom appeasement, sacrifice and gratitude must be directed. These seasonal vicissitudes aside, most of the world's food production methods have developed over a relatively brief 10,000-year period, during which there has been little change in the global climate. Today, however, we must contemplate the effects upon agriculture of potentially dramatic climatic changes.

Climate change could affect agriculture both by long-term changes in ecosystems and by a possible increased frequency and severity of extreme events: heatwaves, droughts, flooding, cyclones and outbreaks of plant diseases and infestations. The 1988 experience in North America, when a continent-wide heatwave and drought reduced the grain crop by 30 % (and significantly increased the world price of grain), illustrates well the potency of climatic variation. However, the great uncertainties, including those due to limitations in computer models, make it difficult to estimate the impact of climatically-altered agriculture upon human population health other than in qualitative terms.

Global warming would alter regional temperature and rainfall, two of the major constraints on agriculture. In the view of Martin Parry (who chaired the IPCC's agricultural-effects working group): 'Relatively small changes in the mean values of rainfall and temperature can have a marked effect on the frequency of extreme levels of available warmth and moisture. For example, the number of very hot days which can cause damaging heat stress to temperate crops and livestock could increase significantly in some regions as a result of a 1–2 °C increase in mean annual temperatures.'[46] Similarly, he argues, reductions in soil moisture due to increased evapotranspiration could substantially increase the number of days below a minimum threshold of water availability for given crops. In Parry's view, while climate change might increase productivity in, for example, the northern regions of North America, Europe and Asia, such gains at high latitude may not compensate for quite possibly large losses in the middle and low latitudes.[46]

Scientists predict that global warming will cause an overall global increase in rainfall, reflecting increased evaporation from the oceans. However, here the plot thickens again. Increased rainfall will occur mostly as increased rainfall *intensity* (rain per rain day).[47] For example, warmer temperatures in the current temperate and cold zones will cause more of the total precipitation to fall as rain, rather than as snow, resulting in quicker runoff and less natural stockpiling for the summer months. Although equatorial monsoon rains may extend to higher latitudes, drenching the drought-prone regions in Africa's Sahel and northwest India, much of that increased rain could come as more intensified events rather than as a sustained increase, thus exacerbating flooding and soil erosion. The overall global increase in rainfall will not be evenly distributed, and some regions may have a reduced rainfall. Further, there will be seasonal shifts in the annual pattern of rainfall. For example, in central North America, the IPCC predicts that, along with temperature increases of around 3 °C by 2020, rainfall will increase by up to 20% in winter but decrease by 5–10% in summer – which could reduce soil moisture by 10–15% during much of the growing season.[44]

Insufficient soil moisture (which reflects the balance of rainfall, runoff and evaporation) is the main climatic limitation upon agriculture in many countries, particularly developing countries.[46] The extent of rainfed agriculture is already very limited in some regions, particularly southwest Asia, and any further constraint due to climate change could create serious food shortages. These vulnerable places are mainly in the cooler tropics (the Andean region, the Mahgreb in North Africa and the mountain regions of southwest Asia), the Sahel and the Horn of Africa, the Indian subcontinent and central and southeast Asia. A reduction in rainfall, or at least soil moisture, is likely in some of those areas.

Climatic change would have other effects upon agriculture. It would allow new types and combinations of food parasites to emerge. Warmer and more humid conditions would enhance the growth of bacteria and moulds on stored foods, increasing spoilage and posing some specific health hazards – as with the *Aspergillus flavus* mould that grows on stored ground-nuts in tropical climates and produces a potent toxin (aflatoxin) that causes liver cancer. In Europe, warmer winters would allow various crop pests that normally die out annually to survive year-round and flourish. In the light of research done in the warm early 1990s, Britain's Institute of Arable Crops Research predicts that aphids, which spread many important plant virus diseases, will enjoy an extended season of activity in a warmer world.

More debatably, the ongoing carbon dioxide-enrichment of the atmosphere may benefit agriculture via a 'fertilisation' effect. There are two major categories of plants: the C3 plants (e.g. wheat, rice and potatoes) and the C4 plants (e.g. millet, sorghum and maize). The C3 plants evolved before the C4 plants and use different metabolic pathways to convert carbon dioxide to plant material. Today's atmosphere, low in carbon dioxide relative to past aeons when C3 plants first evolved, limits the photosynthesis rates of those plants, but not of the 'contemporary' C4 plants. Hence, the C3 plants (which include many weed species) respond more vigorously to an increase in atmospheric carbon dioxide. One study showed a doubling in the yield of C3 sunflowers in response to a doubling of carbon dioxide to 700 ppm, whereas C4 maize did not respond.[48] However, other scientists argue that this thesis is ecologically simplistic, since it overlooks the likely proliferation of pests in a warmer climate, the need for adequate rainfall, the fact that temperature increases will reduce soil moisture, and the sensitivity of plants to increases in temperature (e.g. the accelerated maturation of wheat under warmer temperatures reduces the time for the grains to fill out). Scientists in the Atmospheric Research Division of Australia's major governmental research organisation (CSIRO) have simulated the *combined* effect of increased carbon dioxide (to 560 ppm, as expected by 2150), 3 °C temperature increase, altered rainfall and soil moisture, and conclude that there would be a halving of yield in some early-maturing wheat strains – but increases of around 20 % in late-maturing strains.[49] On balance, it seems unlikely that increases in carbon dioxide would cause large long-term gains in agricultural productivity.

The IPCC scientists, in 1990, were uncertain whether climate change would produce a net increase or decrease in the world's agricultural production. Certainly, there would be marked regional differences. For the northern hemisphere, most computer models predict that global warming will result in mid-latitude, mid-continent, summer dryness. Thus, in developed countries, a 2 °C rise in temperature could cause wheat yields in Europe and North America to fall by 3–17 %.[44,46] The IPCC concluded that warming is likely to reduce cereal productivity in North America and southern Europe, but increase productivity in northern Europe. While the desert in southwestern USA would shift north and encroach upon the American mid-west grain belt, that belt would itself 'migrate' to northern USA and Canada (where, however, the soils are naturally poorer). Similarly in Europe, while Greece and Italy may suffer declines in harvests, Britain and Scandinavia might benefit.

Of the rest of the world, the IPCC concluded that food security could be seriously threatened in the poorer countries in the semi-arid and humid tropics.[44] Poor countries, already struggling with overpopulation and marginal climatic conditions, are the most vulnerable. Reduced rainfall would lead to a decline in soil moisture and an increase in water loss from plants by evapotranspiration. These, in turn, would reduce the effectiveness of fertiliser use and irrigation, respectively. The overall result could be a substantial reduction in agricultural yields. Starvation and malnutrition may follow, causing increases in infectious diseases and widespread demographic disruption. Already in Africa, there are over 100 million people who are 'food insecure', many of them in the arid Sahel region, home to approximately 35 million people. FAO has assessed that this nutritionally precarious region could be severely affected by global warming and reduced rainfall.

Reduced rainfall could also be problematic outside the arid zones. One of the world's most important cereal crops is rice – the staple food of China, India, Indonesia and many other smaller tropical countries. In some of those countries where the management of water resources, including irrigation, is poorly developed, such as in India and Myanmar, rice yields may be very vulnerable to changes in temperature and rainfall. A UN-funded study of the impact of climate change upon Southeast Asia has estimated that yields in the main rice-growing region of Malaysia would decline by 12–22 % because of shorter maturation periods caused by higher temperatures early next century, while maize yields throughout many parts of Southeast Asia would be halved.[50]

The cost of food on the world markets would increase if food production declined in the world's mid-latitude 'bread-basket' regions: the US Great Plains, the Ukraine, Mediterranean and north European lowlands, Australian wheat belt and Argentinian pampas. The one-fifth of the world population who already suffer from malnutrition would then face an even greater threat to survival from agricultural failure and rising food costs. The IPCC concluded that: 'food production at the global level can be maintained at essentially the same level as would have occurred without climate change; however, the cost of achieving this is unclear'. Parry estimates that, for moderate climatic impacts on agriculture due to carbon dioxide doubling, the price of maize and soybeans would increase by about 10%.[46] Rice prices would fall because the world's main rice-producing regions would benefit from a small amount of warming combined with moderate increases in rainfall.

If, by international action, we limit global warming to tolerable proportions over the coming century, then adjustments in land use and management will provide some alleviation. In naturally highly variable climates, farmers may be more adaptable than those in more equable climates. But in poorer countries, and particularly in areas of marginal agriculture, the capacity for adaptive change would be much less. Current disparities would therefore tend to widen.

Other agriculture-related impacts

There are several other agriculture-related impacts of climate change that will also bear on human health and survival, including effects on animal productivity, fuelwood supplies and availability of freshwater. Further afield, warming of the seas and changes in ocean currents may alter the distribution and productivity of fish species, a major source of protein for many poor countries.[51]

With respect to adverse effects on livestock, the US Environmental Protection Agency has identified several infectious diseases – such as the horn fly in beef and dairy cattle, and insect-borne anaplasmosis infection in sheep and cattle – which might well increase in response to climate changes.[52] African horse sickness provides an instructive example. In the early 1990s, a series of unusually warm winters enabled the *Culicoides* midge which carries this disease to survive in Spain, killing many horses there.

A rise in temperature could also have significant effects on the growth and health of farm animals. Young animals are less tolerant of a wide range of temperature than are adult animals. For example, hens can tolerate a 10 °C range, while chicks can tolerate only a 1 °C range. The differences are even more striking for pigs and sheep, where the adults can tolerate 15–20 °C ranges while their young can only tolerate a 1 °C range.[53]

The world's forests, a source of fuelwood, food and timber, will face some particular problems. Because their growth and maturation spans several decades, changes in climate will lead to declines in the productivity of many forests.[44] Population pressures and unresolved poverty will exacerbate unsustainable patterns of use of dwindling forest supplies. Little can seem more urgent in life than cooking today's food – unless it is obtaining today's food.

Surface water runoff from natural watersheds, particularly in arid and semi-arid regions, is very sensitive to small changes in climate. Indeed, the IPCC estimates that a 1–2°C temperature increase coupled with a 10% reduction in rainfall could reduce annual runoff by 40–70%.[44] The regions where water collection resources necessary to sustain populations are at greatest risk are in northern and Sahelian Africa, western Arabia, southeast Asia, India, Mexico, Central America, parts of Brazil and the Mediterranean margin of Europe.

6.3 Politics of climate change

There is a skein of moral, political and scientific problems associated with human-induced climate change. There is the general moral problem mentioned in the Preface – that of bequeathing to future generations a negative legacy, an ecologically damaged world. The second moral problem is that, while the rich countries have caused most of the increase in greenhouse gas emissions to date, the whole world will experience the consequences of climate change.

The major political problem is that, not surprisingly, the poor countries insist on the right to follow the same general path of industrial development that was previously followed by rich countries. Yet, notions of historical equity aside, the control of greenhouse gas emissions is already becoming a major international concern, as evidenced by the priority given to the Climate Convention at the 1992 Earth Summit. Because greenhouse gases have long half-lives and because the oceanic heat sink will take some time to fill (and then to empty), protracted delays in achieving emissions reduction may commit the world to substantial climate change next century. However, basic economic development entails an increase in energy use. With current technology, and end-use practice, that would mean more greenhouse emissions. Even if substantive cuts in emissions are technically possible in rich countries, most developing countries will not be able to afford such technology without massive financial aid (which should perhaps be tied to commitments to achieving population control).[54] Without such aid, there will be large increases in greenhouse gas emissions from those poor countries (especially China) as they embark on rapid industrialisation.

The major scientific problem relates to the uncertainties inherent in the available data, theory and computer models of climate change. Unlike stratospheric ozone depletion, land degradation and loss of biodiversity – all of which are manifestly occurring and are therefore measurable – scientists must still address climate change primarily via forecasts. Not surprisingly, the politics and the science have become intimately bound up with one another. This was evident at the 1990 Second World Climate Conference, where delegations from several of the industrialised countries cast doubt on the quality of the IPCC's 'science', and therefore on the need for cutbacks in greenhouse gas emissions. The IPCC has estimated that stabilising carbon dioxide emissions at current levels would reduce the rate of increase in temperature from 0.3 °C to 0.1 °C per decade, and that the industrialised nations *could* actually cut emissions by at least 20 % by the year 2000, assuming an immediate start. Indeed, in Eastern Europe and the former USSR, for reasons of painful structural economic reform, carbon emissions have declined by 10–20 % since 1987. However, the USA (which contributes the most emissions) has persistently argued that the science is inconclusive and, anyway, that cutting carbon emissions would be unaffordable.

The countries and regions that currently emit the most greenhouse gases – North America, Europe, USSR, Japan – will tend to experience the greatest warming since they lie furthest from the equator, while poorer countries, mostly below 35° latitude north and south, will experience less warming. However, for many of the richer, non-tropical, countries, this may entail a tolerable – even, in some locations, an agriculturally beneficial – increase in temperature, along with some increase in rainfall. For many of the poorer, hotter, countries, the adverse effects may include exacerbation of agricultural difficulties, with further heating and drying of farmland. Rathjen, an American economist, argues that, where rich countries may be able to adapt, the prospects for poor countries are bleak:

For many in the developing world, adjustment to climatic change may be more difficult; indeed, for some, impossible. Poverty implies less mobility and less flexibility to otherwise adapt to change. For the Dutch, building higher dikes as a hedge against rising sea-levels and greater frequency of storms is a realistic possibility. Not so for the Bangladeshis. Their choices are likely to be between attempting migration to an already overpopulated, and hence a probably resistant, India, or death by drowning or starvation – not all at once or suddenly, for there is no basis for projecting a sudden, dramatic rise in sea-level – but as a result of intermittent flooding of increasing severity and perhaps frequency.[55]

The nexus between economic growth (GNP) and energy use is more flexible than was previously thought. In the most developed countries, economic activity is evolving towards an increased emphasis on providing services (education, communications, entertainment, recreation, health and welfare) rather than further material consumption. There are also mounting social and political pressures to improve the efficiency of energy consumption. We should take some encouragement from knowing that in poorer countries, and in the hugely inefficient Eastern European countries over the past four decades, the amount of energy used per unit of economic productivity has been up to ten times greater than in the most energy-efficient rich countries (e.g. Switzerland, Sweden and Japan). The World Bank estimates that renewable energy technologies will be widely afford-able by 2020, when carbon dioxide emissions could stabilise at 150% of 1990 levels – rather than the 350% predicted from business-as-usual trends.[56]

Finally, there may be some appropriate 'technical fixes' that could at least buy us some time. For example, carbon dioxide emissions from power plants could be diverted to 'bioreactor' ponds in order to feed photo-synthesising algae, which could then be converted to stable products such as plastics. Even if feasible, however, such antidotal responses would only reduce, but not remove, this escalating global problem. The problem is deeply embedded in the energy-intensive model of economic development – and it will require radical social responses beyond technical adaptation and gains in energy efficiency.

6.4 Summary

Climate change caused by our enhancement of Earth's greenhouse effect will have a wide range of health effects. The direct effects, via temperature change, thermal extremes and increased natural disasters, are easier to predict than are the various indirect and delayed effects. However, it is likely – and it is important to note this in relation to the general issue of ecological disruption and human health – that the indirect effects will, in aggregate, outweigh the direct effects. Alterations in patterns of vector-borne infectious diseases, reductions in agricultural productivity and the social disruption caused by sea-level rise and associated disasters could all become major public health problems. Regional declines in agriculture will accelerate the flight to the cities by impoverished rural dwellers.

Climate change presents a more fundamental challenge than does preserving the stratospheric ozone layer. While the latter may be essentially within the realm of 'technical fixes', the avoidance of global warming is much more politically and economically complex. The solution lies in controlling world population growth, weaning societies off cheap fossil-fuel energy and redistributing international wealth to obviate the need for inefficient industrialisation and the destruction of rain forests. Human population growth and economic activity are combining to alter the gaseous composition of Earth's unique, life-supporting atmosphere. This, in turn, is widely predicted to cause changes in global climate larger than any that have occurred since settled agriculture began 10,000 years ago – and more rapid than any that have occurred since humans first walked on Earth.

References

1. The strength of the cycle's influence on temperature is, incidentally, in the same rank order as the periodicity: eccentricity of orbit (c. 100,000 years), obliquity/plane of orbit (c. 40,000 years), precession (c. 20,000 years).
2. National Institute of Public Health and Environmental Protection (Langeweg IF, ed.). *Concern for Tomorrow. A National Environmental Survey* 1985–2010. Bilthoven, Netherlands: National Institute of Public Health and Environmental Protection, 1989.
3. Houghton JT, Jenkins GJ, Ephraums JJ. *Climate Change. The IPCC Assessment.* Cambridge: Cambridge University Press, 1990.
4. The rate of increase in carbon emissions has declined a little since the late 1980s. This might reflect the massive industrial slow-down in the former USSR. See: Flavin C. Carbon Emissions Steady. In: Brown LR, Flavin C, Kane H (eds). *Vital Signs: The Trends That Are Shaping Our Future.* New York: Norton, 1992, pp 60–1.
5. Intergovernmental Panel on Climate Change, Submission from Working Group I. 1992 *IPCC Supplement. Scientific Assessment of Climate Change.* Geneva: WMO/UNEP, 1992.
6. Boyden S. *Western Civilization in Biological Perspective. Patterns in Biohistory.* Oxford: Oxford University Press, 1987.
7. World Resources Institute. *World Resources,* 1990–91. Oxford: Oxford University Press, 1991.
8. Jones PD *et al.* Assessment of urbanisation effects in time series of surface air temperature over land. *Nature* 1990; **347**: 169–72.
9. Some temporary cooling is expected in the early 1990s because of atmospheric pollutants (ash and sulphur dioxide) from the massive eruption of Mount Pinatubo in the Philippines in June 1991. Global temperature in 1992 was 0.2 °C below that of the preceding two years.
10. Wigley TML, Raper SCB. Implications for climate and sea-level of revised IPCC emissions scenarios. *Nature* 1992; **357**: 283–300.

11. Broecker WS. Unpleasant surprises in the greenhouse? *Nature* 1987; **328**: 123–6.
12. Leggett J. The nature of the Greenhouse Threat. In: Leggett J (ed). *Global Warming. The Greenpeace Report*. Oxford: Oxford University Press, 1990, pp 14–43.
13. WHO. *Potential Health Effects of Climate Change: Report of a WHO Task Group (WHO/PEP/90/10)*. Geneva: WHO, 1990.
14. Haines A, Fuchs C. Potential impacts on health of atmospheric change. *Journal of Public Health Medicine* 1991; **13**: 69–80. And: Ewan C, Bryant EA, Calvert GD. Potential health effects of greenhouse effect and ozone layer depletion in Australia. *Medical Journal of Australia* 1991; **154**: 554–8.
15. Government of Canada. *Canada's Green Plan*. Ottawa: Ministry of Supply and Services. 1990.
16. Kilbourne EM. Illness due to thermal extremes. In: Last JM, Wallace RB (eds). *Maxcy-Rosenau-Last. Public Health and Preventive Medicine* (13th edition). Norwalk, Connecticut: Appleton Lange, 1992, pp 491–501.
17. Leaf A. Potential health effects of global climatic and environmental changes. *New England Journal of Medicine* 1989; **321**: 1577–83.
18. Climate Change Impacts Review Group, Department of the Environment (UK). *The Potential Effects of Climate Change in the United Kingdom*. London: HMSO, 1991.
19. Centers for Disease Control. Heat-related deaths – Missouri, 1979–1988. *Morbidity and Mortality Weekly Report* 1989; **38**: 437–9.
20. Kalkstein LS *et al*, 1986 – cited (ref. 31) in: Haines A, Fuchs C. Potential impacts on health of atmospheric change. *Journal of Public Health Medicine* 1991; **13**: 69–80.
21. Sontaniemi E, Vuopala V, Huhti E, Takkunen J. Effect of temperature on hospital admissions for myocardial infaction in a subarctic area. *British Medical Journal* 1970; **4**: 150–1. And: Gill JS, Davies P, Gill SK, Beevers DG. Wind-chill and the seasonal variation of cerebrovascular disease. *Journal of Clinical Epidemiology* 1988; **41**: 225–30. See also reference 27 below.
22. Rogot E, Padgett SJ. Associations of coronary and stroke mortality with temperature and snowfall in selected areas of the United States 1962–1966. *American Journal of Epidemiology* 1976; **103**: 565–75.
23. Rankin DW. Mortality associated with heat wave conditions in the Melbourne Metropolitan Area, January and February 1959. *Australian Meteorological Magazine* 1959; **26**: 96–100.
24. Goldsmith JR. Three Los Angeles heat waves. In: Goldsmith JR (ed). *Environmental Epidemiology: Epidemiological Investigation of Community Environmental Health Problems*. Boca Raton, Florida: CRC Press, 1986, pp 73–81.
25. Landsberg HE. Climate and health. In: Biswas AK (ed). *Climate and Development*. Dublin: Tycooly International Publishing, 1984, pp 26–64.
26. Khogali M, Hales JRS. *Health and Temperature Regulation*. Sydney: Academic Press, 1983.
27. Jones TS *et al*. Morbidity and mortality associated with the July 1980 heat wave in St. Louis and Kansas City, Missouri. *Journal of the American Medical Association* 1982; **247**: 3327–31.
28. Longstreth JA. Human health. In: Smith JB, Tirpak D (eds). *The Potential Effects of Global Climate Change on the United States*. EPA-230–05–89–050. Washington DC: US Environmental Protection Agency, 1989, pp 219–36.

29. Curwen M. Excess winter mortality: A British phenomenon? *Health Trends* 1990; **22**: 169–75.
30. Ayres JG. Meteorology and respiratory disease. *Update* 1990; **40**: 596–605.
31. Editorial. Asthma and the weather. *Lancet* 1985; **i**: 1079–80.
32. Escudero JC. Health, nutrition and human development. In: Kates RW, Ausubel JH, Berberian M (eds). *Climatic Impact Assessment. Scope* 27. Chichester: John Wiley & Sons, 1985, pp 251–72.
33. Titchener JL, Frederic TK. Family and character change at Buffalo Creek. *American Journal of Psychiatry* 1976; **133**: 295–9.
34. Shope R. Global climate change and infectious diseases. *Environmental Health Perspectives*. 1992; **96**: 171–4.
35. National Health and Medical Research Council. *Health Implications of Long Term Climatic Change*. Canberra: Australian Government Printing Service, 1991.
36. Gillett JD. Direct and indirect influences of temperature on the transmission of parasites from insects to man. In: Taylor AER, Muller R (eds). *The Effects of Meteorological Factors Upon Parasites*. Oxford: Blackwell Scientific Publications, 1974, pp 79–95.
37. Marshall E. Malaria parasite gaining ground against science. *Science* 1991; **254**: 190.
38. Recent research in The Gambia, in West Africa, has shown that the timing of mosquito bites has changed in response to the increasing use of bed-nets. This reflects the irrepressible hand of natural selection wherein those mosquito strains that feed at times when bed-nets are not in use will survive – and breed – best.
39. Koopman JS *et al*. Determinants and predictors of dengue infection in Mexico. *American Journal of Epidemiology* 1991; **133**: 1168–78.
40. The outbreak of Black Death in Europe may have been influenced by climatic changes in the steppes of Asia associated with the post-mediaeval cooling in the fourteenth century. Epidemic plague infestation probably occurred in climatically-stressed wild rodents – the natural host to the plague organism. The organism could then have subsequently passed, via fleas, to the black 'European' rats travelling with humans by overland caravan between China and Europe. This theory, admittedly speculative, illustrates well the complex ways in which ecological perturbations can affect human health.
41. Epstein P. Cholera and the environment (Letter) *Lancet* 1992; **339**: 1167–8.
42. Schneider S. Will sea-levels rise or fall? *Nature* 1992; **356**: 11–12.
43. Gribbin J, Gribbin M. *Children of the Ice*. Oxford: Basil Blackwell, 1990.
44. IPCC (Intergovernmental Panel on Climate Change). *Policymakers Summary of the Potential Impacts of Climate Change. Report Prepared for the IPCC by Working Group II*. World Meteorological Organization and UN Environment Programme: Geneva and Nairobi, June 1990.
45. Doake C. Rapid disintegration of the Wordie Ice Shelf in response to atmospheric warming. *Nature* 1991; **350**: 328–30.
46. Parry M. *Climate Change and World Agriculture*. London: Earthscan Publications, 1990.
47. Gordon HB, Whetton PH, Pittock AB, Fowler AM, Haylock MR. Simulated changes in daily rainfall intensity due to the enhanced greenhouse effect: implications for extreme rainfall events. *Climate Dynamics* 1992; **8**: 83–102.

48. Hunt R, Hand DW, Hannah MA, Neal AM. Response to carbon dioxide enrichment in 27 herbaceous species. *Functional Ecology* 1991; **5**: 410–21.
49. Wang YP, Handoko JR, Rimmington GM. Sensitivity of wheat growth to increased air temperature for different scenarios of ambient carbon dioxide concentration and rainfall in Victoria, Australia – a simulation study. *Climate Research* 1992; **2**: 131–49.
50. Hunt P. Storm warning over Southeast Asia. *New Scientist* 1992; **1810**: 12–13.
51. Glantz MH (ed). *Climate Variability, Climate Change and Fisheries*. Oxford: Oxford University Press, 1992.
52. Rosenzweig C, Daniel MM. Agriculture. In: Smith JB, Tirpak D (eds). *The Potential Effects of Global Climate Change in the United States*. EPA-230–05–89–050. Washington DC: US Environmental Protection Agency, 1989.
53. Bianca W. The significance of meteorology in animal production. *International Journal of Biometeorology* 1976; **20**: 139–56.
54. Developed countries have discussed with China the substantial gains in the efficiency of energy production from coal burning that could be achieved by insulating and improving the nearly one million industrial generators around the country. If money is paid to China, then this upgrading of technology could be undertaken, as an alternative to burning yet more cheap coal.
55. Rathjen GW. Energy and climate change. In: Mathews JT (ed). *Preserving the Global Environment. The Challenge of Shared Leadership*. New York: Norton, 1991, pp 154–86.
56. World Bank. *World Development 1992. Development and the Environment*. Oxford: Oxford University Press, 1992.

7

The thinning ozone layer

Nationally determined action programmes ... should undertake, as a matter of urgency, research on the effects on human health of the increasing ultraviolet radiation reaching the earth's surface as a consequence of depletion of the stratospheric ozone layer.

UN Conference on Environment and Development, 1992[1]

7.1 The ozone layer

7.1.1 Introduction

Conventional passenger aircraft, cruising at 10,000 metres altitude, are at the ill-defined boundary between troposphere and stratosphere. The stratosphere contains a diffuse 'layer' of ozone which filters out much of the incoming solar ultraviolet radiation. Most of this ozone exists in the lower stratosphere at an altitude of 15–25 kilometres, at very low concentrations of up to 10 parts per million. This mantle of stratospheric ozone began forming 1–2 billion years ago from newly-accruing atmospheric oxygen – a waste product of photosynthesis in aquatic algae and plants.

Sunlight-powered photosynthesis in plants converts the basic 'building blocks', water and carbon dioxide, into complex energy-containing organic molecules. In ages past, the waste oxygen (O_2) released as a byproduct of photosynthesis was mopped up by the chemically 'reducing' Archean environment, in which it combined with oxidisable elements such as iron and sulphur in the surrounding sludge. Later, oxygen began to spill over into the atmosphere (see also Fig. 1.1), while carbon dioxide was progressively removed by the increasing profusion of photosynthesising algae and aquatic plants. The atmospheric accumulation of oxygen, a thermodynamically reactive element, had profound consequences for the evolution of life. It enabled the development of aerobic (oxygen-using)

174

metabolism in animals and plants. It also accelerated, by oxidative weathering of rocks, the release of mineral elements many of which, through evolution, became important parts of complex molecules and metabolic pathways in multicellular organisms.

A third, equally important, consequence for life on Earth was the conversion of some of the oxygen in the upper atmosphere to ozone (O_3). This ozone, which was formed by the high-energy photolytic action of the sun's ultraviolet radiation, subsequently absorbed much of that same incoming radiation – thus making the shallow aquatic environments and dry land safer places for live organisms. Within the spectrum of solar radiation, ultraviolet radiation (short wavelength) and infrared radiation (long wavelength) lie at either end of the visible 'light' band. Of solar radiation reaching Earth's surface, 55 % is infrared, 40 % is visible, and 5 % is ultraviolet radiation. The ultraviolet includes wavelengths from 200 to 400 nanometres and is subdivided into three bands: A, B and C. The ozone layer blocks out nearly all of the highest energy, shortest wavelength, ultraviolet (UV-C) band, approximately half of the next-highest energy band (UV-B), but only a small part of the UV-A. Because of this ultraviolet-attenuating effect of ozone, the 'migration' of plant life from its ancient watery habitat onto land was facilitated about 400 million years ago, and was followed by various worm-like creatures and rudimentary limbed fish (the early amphibians) – or, more correctly, these creatures were better able to survive in the increasingly shielded environment when caught short by receding waters. And so the evolution of terrestrial life-forms accelerated.[2]

7.1.2 Damage to the ozone layer

Scientific debate about human-made threats to the stratospheric ozone layer began in the 1970s. In particular, there were fears about nitrogen oxides in the exhaust of high-flying supersonic aircraft (such as Concorde). Although very few such aircraft were built, that debate highlighted the theoretical vulnerability of the ozone layer to chemical assault. Ozone is constantly being created and destroyed by natural photochemical processes that are in dynamic equilibrium, as follows:

$$O_2 \rightarrow O + O \text{ (energised by UV-C)}$$
$$O + O_2 \rightarrow O_3$$
$$O_3 \rightarrow O_2 + O \text{ (energised by UV-B)}$$

However, the introduction of foreign catalytic molecules can disrupt this equilibrium. Because the 'half-time' for ozone regeneration is 3–4 years, any such disruption can cause a sustained deficit in the ozone layer.[3] Unwittingly, since mid-century, industrialised countries have been releasing a synthetic family of long-term ozone-destroying chemicals into the atmosphere – the halocarbons, in particular the chlorofluorocarbons (CFCs).

In 1974 two American scientists, Molina and Rowland, theorised that free chlorine 'radicals' (Cl·) would be released from CFCs by the action of ultraviolet radiation at extremely low temperatures, and would catalyse the breakdown of ozone.[4] Because the resulting chlorine monoxide (ClO) enters a self-regenerating catalytic cycle, each CFC molecule could end up by destroying thousands of ozone molecules. A similar prediction was made for compounds containing bromine, a halogen-family sibling of chlorine. Simplified, the cyclical, ozone-destroying, reaction is of the form:

$$Cl^· + O_3 \rightarrow ClO + O_2$$
$$ClO + O \rightarrow Cl^· + O_2$$

Net: $\qquad\qquad\qquad\qquad O_3 + O \rightarrow O_2 + O_2$

These chlorinated or brominated halocarbons do not occur in nature. First synthesised in 1928, the CFCs are entirely human-made and have been extensively used as refrigeration fluids, spray-can propellants, blowers in foam-making and solvents. By the late 1970s, the lighter-than-air CFCs were a rapidly growing source of atmospheric pollution and scientific controversy – which was ironic since their chemical inertness (at ground-level) had been loudly acclaimed. In 1978, the US Government banned their use in spray cans. The European Community proposed a voluntary cut in their usage and, in 1983, Scandinavian countries sought a global ban on CFC aerosols. Subsequent initiatives led to the Vienna Convention of March 1985. This document, signed by 20 nations, declared good intentions rather than formal commitment. However, it set a precedent as the first multinational agreement to tackle a major environmental problem *before* its effects were scientifically demonstrable. The main basis for this precautionary policy was theoretical prediction and knowledge that ultraviolet radiation is biologically damaging, and not – as is usually the case in responding to environmental health hazards – empirical evidence of adverse effects. In early 1985 there was no published evidence that CFCs actually damage stratospheric ozone, let alone evidence of adverse biological effects due to ozone thinning.

Subsequent events moved fast. In May 1985, British scientists published clear evidence of large (40 %) losses in the Antarctic ozone layer in late

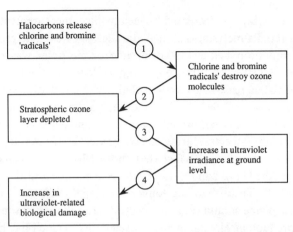

Figure 7.1. Postulated steps from release of halocarbons to increase in ultraviolet-related biological damage: piecing together the evidence of a causal sequence.

winter and early spring.[5] Here was tangible proof that something, perhaps chlorine, was actually damaging Earth's ozone mantle – and to an extent that exceeded earlier predictions. The US, Scandinavian and, less readily, the British governments moved to cut CFC usage; western European countries and Japan still resisted. During the next two years, ozone losses were reported over both poles during late winter. In 1987, thirty-six nations met in Montreal, Canada, and agreed to restrict their *release* of ozone-damaging chemicals. The Montreal Protocol sought to halve CFC emissions by the year 2000. Because chlorine-free refrigerants are expensive, less stringent controls were placed on CFC *production*, thus enabling countries such as China and India to go ahead with mass domestic refrigeration. Behind this political concession was the hope that those countries would eventually also participate in the protocol.

   The policies adopted at Vienna and Montreal, in 1985 and 1987, ran ahead of the empirical evidence. What do we now know about the presumed causal sequence, shown in Fig. 7.1, wherein CFC emissions result in risks to terrestrial health? In 1988, a striking mirror-image inverse correlation was reported between the atmospheric concentration of chlorine radicals and that of ozone at increasing latitudes between the tip of South America and the Antarctic. This was 'smoking gun'-type evidence that CFCs actually destroy ozone, as in the second step in Fig. 7.1. Corroboration came from observations that, as the atmospheric concentration of CFCs in the Antarctic had progressively increased during the 1980s (by about 50% overall), late-winter stratospheric ozone concentrations had steadily decreased.[6] Meanwhile, measurements at the McMurdo and Palmer stations in Antarctica showed that ground-level ultraviolet radiation below the springtime ozone holes was double its normal seasonal strength,[7] thus providing evidence for the third step.

A substantial increase in ground-level ultraviolet radiation (particularly UV-B) was also recorded in Melbourne, southern Australia, during December 1987.[8] This confirmed that the ozone-depleted stratosphere spreads to higher latitudes during early summer, as the Antarctic 'ozone hole' disperses. Further, the fact that those Melbourne increases were mainly within the UV-B band is significant. Since ozone absorbs UV-B and UV-C more efficiently than the longer-wavelength UV-A, depletion of stratospheric ozone should cause a 'spectral' shift that allows more of the shorter, more biologically damaging, wavelengths to reach Earth's surface – as has indeed subsequently been shown in research elsewhere.[9]

Meanwhile, what about the northern hemisphere? Over each pole, in winter and early spring, a rapidly circulating mass of extremely cold air known as the polar vortex occurs. Halocarbon molecules attach to icy stratospheric clouds and then, when the polar dawn arrives in late winter, the combined effect of freezing temperatures and solar radiation fractures the halocarbon molecules to form reactive chlorine 'radicals', as described above. Because of the geophysics of the Antarctic land mass and the associated extreme low temperature and climatic isolation of the southern polar vortex, this ozone-destroying chemistry occurs more readily over the Antarctic than over the Arctic. Even so, data gathered over the past decade show that, although there is readier replacement (from mid-latitudes) of depleted ozone over the Arctic – compared to the more intransigent depletion over the Antarctic – the average loss of ozone at mid-latitudes (60–20°) is similar in north and south. Indeed, the area affected by ozone loss may actually extend a bit further towards the equator in the northern hemisphere than in the south, perhaps reflecting geophysical factors or a greater concentration of ozone-depleting chemicals.

At mid-northern latitudes (30–50° N, including most of the USA, northern Africa, much of continental Europe and Asiatic Russia – see Fig. 7.2) the ozone layer thinned progressively by around 0.5 % per year during the 1980s and early 1990s.[10,11] Overall, during that period there was an accumulated 8 % loss in winter and a 2 % loss in summer. The summertime loss has two-fold significance: first, it suggests that bromine radicals (which are reactive at warmer temperatures than are chlorine radicals) are also important; second, it is occurring at that time of year when people are most likely to be exposing themselves to sunlight. Meanwhile, in the

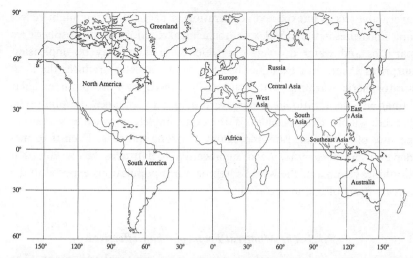

Figure 7.2. Regions of the world, by latitude (i.e. horizontal lines) and longitude.

southern hemisphere, significant cumulative ozone loss occurred during the 1980s from the pole 'down' to latitude 30° S. This includes the lower halves of Australia and of South America. Ozone depletion has continued to increase over Antarctica in the early 1990s.[12] It now appears that there are well-established trends in ozone depletion in both hemispheres, and that they are still accelerating.

From the satellite data for 1978–90, showing incremental ozone losses of several percentage points per year (with particular losses in the Antarctic spring and at latitudes 40–50° N in the late winter), UNEP estimates that the dose of ultraviolet radiation reaching the lower atmosphere has increased by 5% per decade at 30 ° N (e.g. New Orleans, Cairo, Delhi and Shanghai) and 30° S (e.g. Sydney, Buenos Aires and Durban).[13] Towards the poles, it has increased by 15% per decade at 55° S (lower tip of South America) and 40% over the Antarctic – compared to 10% over the northern polar region. However, ground-level measurements indicate that much of the predicted increase in UV-B exposure is being absorbed by anthropogenic air pollutants in the lower atmosphere over industrialised regions, particularly sulphate aerosols and (ironically) tropospheric ozone, whereas in unpolluted regions the measured and calculated increases in ground-level UV-B are in good agreement. (That poses an interesting new dilemma in relation to air pollution control!)

Of course, many uncertainties remain. Some of the recent ozone losses could have been due to natural events, such as the solar flares of 1989 or the eruptions of El Chichón in 1982 and Mount Pinatubo in 1991. Alternatively, the recent unexpected acceleration in ozone thinning might reflect non-linearities in the complex photochemical reactions on stratospheric ice particles and water droplets – in which case there may be unpleasant surprises ahead. For example, because ozone depletion reduces the local stratospheric warming that is normally caused by absorption of solar radiation, the additional temperature drop may enhance the photolytic breakdown of CFC molecules. With our incomplete knowledge, we cannot be sure of the future rate and extent of ozone thinning. Even if ozone-depleted air does not spread significantly to latitudes below 30°, that would leave most of the world's light-skinned populations exposed to the direct effects of ozone depletion.

7.1.3 The Montreal Protocol – how effective?

We cannot quickly stop the destruction of stratospheric ozone. The halocarbon genie is out of the bottle. A large volume of halocarbons has already been released into the lower atmosphere, ensuring a long-term supply of long-lived chlorine and bromine radicals when they reach the stratosphere.

The 1987 Montreal Protocol sought to minimise long-term damage to the ozone layer by reducing the emission of long-lasting CFCs. However, it was subsequently recognised that the protocol was seriously inadequate. While, by 1990, the stratospheric concentration of chlorine had increased from near-zero to around 3.5 ppb (parts per billion by volume), the original protocol would have slowed the increase such that by 2050 the concentration would have been around 7 ppb, instead of 20 ppb. Clearly, that would still have been very damaging to the ozone layer. Further, it was realised that some brominated compounds are very potent short-term destroyers of ozone. Indeed, atom-for-atom, bromine is about forty times more ozone-destructive than chlorine.[14] Thus, although the brominated halocarbons ('halons') represent only one-hundredth of all ozone-destroying gas emissions, they may account for up to one-third of the current ozone destruction.

In 1990, the Montreal Protocol was revised, with industrialised countries agreeing to stop using both CFCs and halons by the year 2000.[15] Developing countries agreed to a complete phaseout by 2010. The Protocol was extended to include carbon tetrachloride and methyl chloroform, but not nitrous oxide. However, even so, stratospheric chlorine would still have increased to 5 ppb by around 2020, before gradually declining towards the 'safe' level of 2 ppb later next century.[16] The protocol was revised again in 1992, with bans being brought forward to the mid-1990s, although methyl bromide (a widely-used fumigant of soil and produce, and recently recognised as a major source of atmospheric bromine) was excluded. The 1992 revision anticipated a peaking of stratospheric chlorine at around 4.5 ppb within 10–15 years, receding to 'safe' levels by around 2050. Thus, the world remains committed to some degree of increased ozone destruction over coming decades, even though the Montreal Protocol has already achieved a halving of the peak 1988 emission levels of CFCs. Meanwhile, October 1992 saw the biggest Antarctic ozone hole on record, along with predictions of worsening losses over the Arctic.

The ongoing loss of stratospheric ozone is one environmental problem about which the world community is having to learn and act fast. We still have far to go to redress the disruption of this vital life-support system. As ever, with disruptions of the biosphere's natural systems, there are layers of complexity: some scientists now predict that greenhouse warming, by trapping more of Earth's radiant energy in the lower atmosphere, will cause cooling in the stratosphere that will enhance the catalytic destruction of ozone. Meanwhile, in the other direction, an increase in incoming ultraviolet radiation will enhance the photochemical production of oxidant air pollutants (including ozone) in the lower, breathing-zone, atmosphere.

7.2 Direct health effects of increased ultraviolet radiation exposure

7.2.1 Introduction

Ultraviolet radiation is known to have various direct effects upon human health. Its harmful effects upon animal and plant biology are also relatively well understood and, if exposures increase significantly in future because of ozone depletion, this may indirectly affect human health by impairing the terrestrial and aquatic food chains. Indirect effects may also result from reduced air quality due to changes in tropospheric chemistry.

Before considering further the direct effects, we should set this predicted exposure increase in perspective, since terrestrial species are already exposed to 'background' levels of ultraviolet radiation. Furthermore, that exposure is naturally much greater in the equatorial regions, and at higher altitudes. Indeed, there are three-fold differences in the intensity of ground-level ultraviolet radiation between the north and south of Australia (Darwin at 12° S versus Hobart at 42° S). Therefore, projected increases of 10%, 20% or even 50% in exposure to ultraviolet radiation because of ozone layer depletion may seem unimpressive. The important point, however, is that these increases would impinge on populations of species

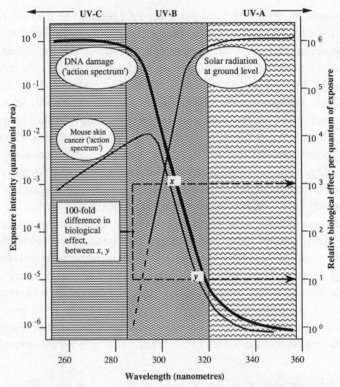

Figure 7.3. Spectral (wavelength) composition of solar ultraviolet radiation at Earth's surface and the 'action spectrum' for induction of DNA (genetic) damage and mouse skin cancer. Note the substantial change in biological 'action' with small shifts in wavelength. (Source: ref. 18)

that are biologically (and, in the case of humans, behaviourally) adapted to their local exposure levels. Hence, some scientists argue that it is not logical to argue that, because plenty of plants grow in the tropics, therefore a moderate increase in the exposure of a temperate-climate plant to ultraviolet radiation would have negligible effect.[17] After all, consider the huge increases in skin cancer that have occurred when fair-skinned humans, biologically evolved for high latitude living, have migrated to lower latitudes where previously only dark-skinned humans lived.

The UV-C band is very damaging to cellular DNA (i.e. the genetic material). However, as mentioned earlier, UV-C does not penetrate the atmosphere. UV-B, with a wavelength of 280–315 nanometres, is the most biologically damaging component that actually reaches Earth's surface, and, under experimental conditions, it induces damage to cellular DNA in direct proportion to dose.[18] Laboratory experiments show that it is many thousand times more potent than UV-A as a cause of cancer (Fig. 7.3). Much of the scientific discussion of the biological effects of ozone depletion is therefore framed in terms of estimated UV-B exposures. Meanwhile, the longer wavelength UV-A penetrates more deeply than does UV-B, reaching the underlying structural fibrous tissues – and may therefore be more relevant to skin 'ageing'.[19]

The proportion of ultraviolet radiation absorbed by the ozone layer is less at the equator and low latitudes, where the radiation passes through the ozone layer vertically, than at higher altitudes where it passes obliquely through more ozone. This accounts for natural geographic variations in ultraviolet radiation exposure at ground-level – with some supplementary local effects due to the ultraviolet-blocking effect of ozone (as photo-chemical air pollutant) in the lower troposphere.[20] As the ozone layer becomes more depleted, the ratio of UV-B to UV-A transmitted will increase, so it is a reasonable prediction that biological effects related to UV-B will become more prominent.

7.2.2 Skin damage and skin cancer

Introduction

A rise in the level of terrestrial ultraviolet irradiation will increase both the short-term and longer-term risks of harmful effects upon the skin of humans. Acute exposure to enhanced solar radiation is likely to cause more severe sunburn. Increases in cumulative exposure would cause enhanced skin thickening, wrinkling and loss of elasticity and increased risk of skin cancer. While sunburn (erythema) and skin cancer are largely due to UV-B, with action spectra that peak at around 300 nanometres, the other degenerative effects are more related to UV-A.

The risk of skin cancer is greatest in Caucasian populations and, among those, the risk is greatest in fair-skinned persons. This can be understood within an evolutionary framework. The migration of ancestral dark-skinned humans away from their place of origin (presumed to have been in hot sunlit eastern Africa) to the higher-latitude zones of Europe, China and elsewhere entailed a reduced exposure to sunlight. To maintain sufficient sunlight-induced production of vitamin D in the skin, natural selection presumably acted in favour of reduced skin pigmentation in order to let in more ultraviolet radiation.[21] Indeed, the mechanism of selection may well have been ruthlessly direct: since vitamin D deficiency causes rickets (soft, deformable bones), abnormalities of the pelvis bones in darker-skinned individual women living at higher latitudes may have directly, physically, impaired their child-bearing success – the central criterion of natural selection. Fairer-skinned 'mutants' would therefore have eventually dominated the gene pool. (Interestingly, the contemporary migration of South-Asian, African and West Indian people to European countries shows evidence of a 'replay' of this ancient problem. There have been reports of rickets in those dark-skinned migrants.[22]) This natural selection-driven lightening of skin colour in those early human populations moving to higher latitudes would also have increased their risk of skin cancer in the longer term. However, an increased risk of skin cancer in post-reproductive adulthood is of little relevance to natural selection.

Over the past two centuries, the migration of some of those lighter-skinned people to high-exposure locations such as Australia, New Zealand, South Africa and the southern USA has resulted in a substantial further increase in their risk of skin cancer. In Australia, which has the world's highest reported rates of skin cancer, recent research has shown that the rates of malignant melanoma increase more in fair-skinned migrants from the British Isles than in olive-skinned migrants from Italy and Greece.[23,24] This pattern of altered risk illustrates an important point: increased exposure of light-skinned populations to UV-B can markedly increase the occurrence of skin cancer. Africa extends above and below the equator to around 35° latitude; India lies within 30° of the equator (see Fig. 7.2). By contrast, Western Europe (excluding Spain and Portugal), Scandinavia and Russia lie predominantly within the 45–70° latitude band. The long-term inhabitants of the former, low-latitude countries are dark-skinned; those of the latter, high-latitude countries are light-skinned. However, with minor exceptions, Australia lies within the band 10–40° South and the USA (excluding Alaska) lies within the band 25–45° North. That is, both those white settler countries lie at much lower latitudes than the European countries that were the source of their light-skinned migrants – who in each case have supplanted dark-skinned native populations.

Epidemiology of skin cancer

There are three main types of cancer of the skin: squamous cell cancer (SCC – arising in the outer layer), basal cell cancer (BCC – arising in the underlying basal layer from which squamous cells form), and malignant melanoma (arising from pigmented cells in the basal layer). Overall, skin cancer is the commonest cancer in the world's light-skinned populations. However, accurate recording of the occurrence of the predominant types of skin cancer, BCC and SCC, within populations is notoriously difficult. Over the past four decades, the rates of malignant melanoma in white populations have increased by around 5% per year, and those for basal cell and squamous cell cancer also appear to have been increasing. Epidemiologists attribute these now well-established trends to successive changes in patterns of personal exposure to sunlight, occurring across successive oncoming generations. Extensive epidemiological research has shown that exposure to sunlight is the major environmental cause of these cancers, accounting for an estimated 70% of all skin cancers in the USA.[25] Since ultraviolet radiation causes characteristic damage to the DNA of actively growing skin cells (including causing linkage between adjoining pyrimidine bases on the DNA molecule), it is likely that it causes skin cancer primarily by inducing particular genetic mutations in those cells. In all three types of skin cancer UV-B may thus play a 'cancer-initiating' role, while in malignant melanoma it may also enhance the later stages of the cancer process.[26]

Basal cell carcinomas outnumber squamous cell carcinomas by about four to one. These two non-melanocytic skin cancers result from a high cumulative exposure to sunlight,[27] and therefore tend to occur in people who spend long hours in the sun. In Australia there is an increase in the incidence of these cancers with decreasing latitude, with the rate in Queensland being almost three times higher than in Tasmania. Of the two types, squamous cell cancers appear to be more strongly related to ultraviolet radiation than do basal cell cancers. These two cancers are rarely fatal; only 1–2% of all cases die from the cancer. For each type, the lag ('latency') period between first substantive exposure and the occurrence of clinical cancer is approximately three to four decades.

The least frequent, but most serious, of the three skin cancer types is malignant melanoma. This cancer arises in the pigment-producing cells, the melanocytes, within the basal layer of the skin (and, less often, within the pigmented choroid overlying the retina). If not treated early, this cancer tends to spread via the blood-stream to internal body organs, such as liver and brain, and is often fatal. The fatality rate is around 25%. The relationship of malignant melanoma to UV-B radiation appears to be more complex than for the other two skin cancers. The occurrence of malignant melanoma is not restricted to the most heavily ultraviolet-exposed parts of the body. Further, it tends to occur more readily in indoor workers (doctors, dentists, lawyers, etc.) – who presumably have occasional recreational exposure to sunlight – than in chronically exposed, tanned, outdoor workers. These apparently anomalous findings have suggested the hypothesis that the risk of malignant melanoma is increased by intermittent intense exposure to sunlight; this idea accords with most of the recent epidemiological observations.[28] The lag period between first substantive exposure and the occurrence of melanoma is around 15–25 years. The several-fold increase in incidence of malignant melanoma in developed countries over recent decades has been widely attributed to changes in styles of clothing and of recreational exposures – particularly sunbathing. In view of the lag period, it is unlikely that recent depletion of the ozone layer has yet contributed perceptibly to increased rates of malignant melanoma, or of non-melanoma skin cancers.

Estimating the increases in skin cancer

In view of the evidence that solar radiation is a major cause of skin cancer, an increase in exposure to UV-B due to ozone depletion is expected to cause an increase in skin cancers. But by how much? This question has been tackled by various groups of scientists in recent years. While a traditional epidemiologist might prefer a wait-and-see 'body-counting' approach, a socially useful answer must come from estimations made now, rather than from actual clinical observations that begin to emerge early next century.

By knowing, first, the relationship between a given reduction in stratospheric ozone and the resultant change in ground-level exposure to UV-B (called the 'radiation amplification factor'), and, second, the dose–response multiplier (the 'biological amplification factor') between increased UV-B exposure and increased risk of skin cancer, it becomes possible to estimate the future increase in risk of skin cancer associated with ozone depletion. The first relationship is being clarified by direct environmental measurements, and includes elucidating the distortions currently caused by tropospheric pollutants. The second relationship can be estimated several ways, particularly from the geographic variation in skin cancer rates in light-skinned populations in relation to variations in ambient ultraviolet radiation exposure. But we need to be careful here. How much of the observed difference in latitude-related skin cancer rates is due to differences in ambient levels of ultraviolet radiation and how much is due to associated differences in patterns of human behaviour such as occupation, recreation and clothing? Because of these entangled behavioural modifiers of average local solar exposure (well-recognised by epidemiologists as a type of 'confounding' variable, of the kind that typically besets non-experimental studies in free-living populations) the data from free-living human populations may not accurately reflect the strength of the true dose–response relationship. If, for example, low-latitude Queenslanders wear broad-brimmed hats and high-latitude

Tasmanians do not, then the true latitude-related risk of skin cancer in Australia will be underestimated by simply comparing their skin cancer rates.

In principle, we want to know what increase in skin cancer risk will occur if stratospheric ozone is depleted by $x\%$ over the coming y decades. Atmospheric scientists can estimate the stratospheric concentrations of ozone-depleting gases, present and future, and can then estimate the destructive impact of these gases upon ozone and, therefore, how much additional ultraviolet radiation will reach Earth's surface. (Indeed, both these outcomes are being directly measured by The World Climate Research Program, via a global network of reporting stations. The resultant information will enable estimation of future changes, by extrapolation.) From estimations of the increase in exposure to ultraviolet radiation, epidemiologists can then predict the increase in skin cancer.

UNEP, in 1991, estimated that for every 1% depletion of ozone the cancer-inducing dose of UV-B will increase by 1.4%, causing the incidence of BCC and SCC to increase by 2.0% and 3.5%, respectively.[13] Overall, UNEP estimates that a 1% loss of ozone would cause a 2.3% increase in non-melanocytic skin cancers. As with IPCC's estimates of global warming (chapter 6), these estimates of the radiation and biological amplification factors each lie within a band of uncertainty – approximately plus or minus one-quarter.[29] The biological amplification factor for melanoma is even less certain, and is within the region 0.5–1.0%. UNEP predicts that an average 10% loss of ozone (such as has become established at middle-to–high latitudes in recent years), if sustained globally over three–four decades, would cause at least 300,000 additional cases of non-melanocytic skin cancer worldwide each year and 4,500 extra cases of malignant melanoma – and possibly double that figure.

The effect of increased exposure to UV-B upon skin cancer incidence would be equivalent to moving the population to a lower latitude. For example, in Tasmania, Australia (around 40° S) a 15% year-round depletion in ozone could, on current trends, become established over about four decades – say, 1980–2020. This would increase the ground-level exposure to UV-B by around 20%, which would cause an increase in non-melanocytic skin cancer of about one-third. For Tasmanians, this would be equivalent to moving to live halfway up Australia's eastern coast, at about 30° S. Taking a longer view, the continued depletion of stratospheric ozone might cause 50–100% increases in skin cancer rates in light-skinned populations living at higher latitudes in both hemispheres by the middle of next century.

For the moment, all such estimates are clouded by technical and statistical uncertainties, by the unpredictable effects of adaptive changes in human behaviour (as, for example, ozone depletion reports become a routine part of our daily weather reports), and by the localised fluctuations in tropospheric air pollution. The monitoring of rates of actual skin cancer in sentinel populations around the world would provide no clear evidence of altered risk for at least several decades. To avoid such delay, the International Agency for Research on Cancer (an agency of WHO) is exploring ways of establishing an early-warning population monitoring system. Such a system might include the measurement of early cancer-related damage in skin cells, including the occurrence of specific genetic mutations, in selected populations living at different geographic locations and therefore exposed to different magnitudes of change in UV-B radiation. (See also the earlier discussion of technical options for such monitoring, in chapter 4.)

7.2.3 Effects upon the eye

To use quite the wrong metaphor, the eye is the body's Achilles' heel when it comes to natural protection against ultraviolet radiation. The one part of the body where such radiation can penetrate relatively freely is through the front of the eye. That is the unavoidable price we pay for being able to see.

The cornea (the clear external layer overlying the coloured iris and the pupil) and the light-focusing lens (which sits behind the iris) filter out the high-energy ultraviolet radiation in sunlight that would otherwise damage the light-receiving retina at the back of the eye chamber (see Fig. 7.4.). As a result, less than 1% of the ultraviolet radiation that impinges on the cornea actually reaches the retina. Nevertheless, exposure to ultraviolet radiation can gradually impair vision by damaging the cornea, the lens, and the retina.[30] Further, since the proportion of incident ultraviolet radiation that passes through the lens decreases with age, children may be particularly susceptible to effects upon the retina.

Over many decades, this protective absorption of ultraviolet radiation causes some discolouration (milky-yellowing) of otherwise clear tissues, particularly the proteinaceous crystallin within the lens. UV-B has sufficient energy to break organic peroxide molecules within the cornea and lens, thus releasing very reactive, smaller, free-radical molecules, including hydroxyl radicals. Within the metabolically active cells of the cornea, the enzyme aldehyde dehydrogenase eliminates the aldehydes produced by these reactions. Within the more inert lens material, the free

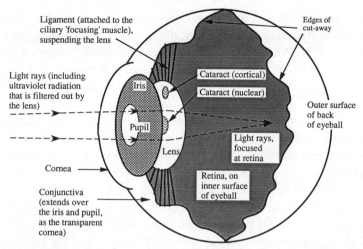

Figure 7.4. Cut-away view of inside of eyeball, viewed obliquely from front/side. Cataracts (opacities) are shown schematically in the lens, behind the light-admitting pupil.

radicals cause photo-oxidative fragmentation and cross-linking of molecules of lens protein, which leads to loss of transparency of the lens. This process is thought to be enhanced by nutritional deficiencies – both of protein and of the several vitamins (especially A, C and E) that provide antioxidant protection against the molecular hit-and-run effects of free radicals. This would help to explain the very high rates of cataract in some of the undernourished populations of Africa.

Cataracts mostly occur in old age and account for over half of the world's estimated 25–35 million cases of blindness.[31] This world total will increase markedly as life expectancy increases in developing countries. Approximately one in ten Australians over age 65 have cataracts. In the USA, over 600,000 cataract operations are done annually – it is the most frequently performed Medicare-reimbursed surgical procedure, with the total cost to society being around $2.5 billion each year. There are two main types of senile cataract: nuclear cataracts which occur centrally (behind the pupil) and cause blindness, and cortical cataracts which affect the peripheral parts of the lens and usually do not impair vision (Fig. 7.4). In developed countries, nuclear cataracts typically outnumber cortical cataracts – although often both types occur in one individual. Other types (such as posterior subcapsular cataracts) also occur.

Cataracts occur most frequently in sunny and tropical regions. However, there is rather sparse epidemiological evidence about their quantitative relationship to geographic latitude and ambient ultraviolet radiation levels. Some evidence shows that the frequency of nuclear cataract increases with decreasing latitude, including evidence from a large national survey of 10,000 Americans.[32] The prevalence of cataracts is much greater in rural Australian Aborigines than in white Australians, and, among Aborigines, there is a two-fold greater frequency of cataract in those living in the central and northwest zones compared to those living in the less UV-exposed southern zones (although some of this difference may be due to occupation rather than to ambient environment).[33] The frequency of cataract in populations around the world shows no clear relationship to average skin (or predominant iris) colour – even though one American study has shown an increased risk in persons with brown or hazel eyes.[34] If ultraviolet radiation entering via the 'open' pupil is the prime influence on lens opacity, then skin and iris colour seem unlikely to be relevant.

Higher-grade epidemiological studies have examined the relationship between estimated personal exposure to ultraviolet radiation and the occurrence of cataract. Because there are difficulties in estimating personal lifetime exposure to ultraviolet radiation, such studies have probably been a little insensitive to the true strength of the relationship. One study of 'watermen' working (primarily by fishing) in Chesapeake Bay, USA, found that a person's estimated lifetime exposure to UV-B was directly related to the occurrence of cortical (but not nuclear) cataracts; a doubling of total UV-B exposure caused a 60% increase in the probability of occurrence of cortical cataract.[35] Other such studies in the USA, Italy and India have found similar positive relationships for both cortical and nuclear cataracts,[30,36] while several studies found no relationship.[37] Cataracts have also been induced in experimental animals, particularly mice, exposed to ultraviolet radiation.[30]

From the various epidemiological studies, the US Environmental Protection Agency predicts that a 10% increase in UV-B exposure will increase the occurrence of senile cataracts by between 4% and 6%, with the increase being greater at around age 50 than at age 70.[25] That estimate lumps nuclear and cortical cataracts together, even though they may have different relationships to exposure to UV-B. Since we are uncertain about the relationship of cataracts to ultraviolet radiation and about the impending trends in ozone layer damage, estimates of future increases in cataracts can only be very approximate. The US EPA has estimated the additional numbers of cataracts that the existing American population would experience, during its lifetime, in response to six different global scenarios for CFC emissions; the estimates ranged from 10,000 to 3,239,000.[25] More recently, UNEP has predicted that a sustained 10% loss of stratospheric ozone would cause up to 1.75 million additional cases of cataract worldwide each year.[13] UNEP has noted recent evidence that long-term exposure to sunlight may impair vision by contributing to presbyopia (near-sightedness) and deformation of the anterior capsule of the lens. Presumably such effects, if confirmed, would be the result of molecular structural damage by UV-B.

There are also more superficial effects of UV-B on the conjunctiva – the clear layer extending over the 'white' of the eye and the cornea. Acute exposure to UV-B can cause photokerato-conjunctivitis (often occurring as 'snow blindness'), while sustained increases in exposure may increase the incidence of pterygium. As with the *ptero*dactyl (the winged dinosaur) a *ptery*gium is a 'wing-shaped' fleshy thickening of the conjunctiva epithelial. It is a common condition of outdoor workers in sunny climates, and is a cause of visual impairment and, sometimes, blindness. The above-mentioned study of watermen found strong positive associations between personal UV-B exposure and the occurrence of pterygium, as well as climatic droplet keratopathy (a deposition of altered proteins in the cornea, causing opacification).[30] In adult Australians, pterygium occurs in around 3% in Aborigines and 10% in non-Aborigines. It has been estimated, from rather sparse data, that a 1% increase in ultraviolet radiation would increase the incidence of pterygium by 2.5% in Australian Aborigines and by 14% in non-Aborigines.[24]

The retina, the film of optic nerve-endings at the back of the eye chamber, is sensitive to ultraviolet radiation. Although under normal circumstances virtually no ultraviolet radiation reaches the retina, a substantial loss of stratospheric ozone would increase the likelihood of exposure. The resultant photochemical damage would cause degeneration of the retina, and therefore impaired vision. Indeed, there is some, although inconsistent, evidence that this type of 'macular' degeneration is associated with cumulative exposure to sunlight.[30,38] Finally, the blackness of the retina comes from the protective choroid layer, containing melanocytes similar to those that provide pigmentation to our skin – and which, abnormally, become the starting point for malignant melanoma. Melanoma of the eye occurs only rarely (around six cases per million persons annually in Australia), and the evidence relating this cancer to the amount of ultraviolet radiation exposure received by the retina is still tenuous.

7.2.4 Effects upon the immune system

The body's immune system is our prime defence against foreign 'antigenic' material – usually proteinaceous molecules, as in living microorganisms and in inert materials such as dusts, dander and pollen grains. The immune system comprises a well-coordinated network of tissues, non-specific defender cells (macrophages and 'killer cells') and specialised, often mobile, defender cells that produce antibodies and 'patrol' the body to find and attack alien molecules. Increased exposure to ultraviolet radiation depresses facets of the body's immune system. In humans, much of this effect seems to be independent of skin pigmentation (both innate and acquired) – and therefore it potentially has worldwide implications.

This, however, is a relatively new area of research, and there is much uncertainty about it.[39] Indeed, the biological evolutionary 'purpose', if any, of any such immunosuppressive effect of increased exposure to ultraviolet radiation remains a mystery. Nor is there systematic epidemiological evidence of geographic UV-related variations in immune-influenced health disorders. Nevertheless, experiments conducted both in mice and (to a lesser extent) in humans, show that UV-B irradiation depresses the skin's contact hypersensitivity (or 'contact allergy'), reduces the number and the functions of immunologically active cells (Langerhans cells) in the skin, stimulates the production of immune-suppressing 'T suppressor' cells, and alters the profile of immunologically active lymphocytes (white cells) circulating in the blood. These disturbances in the number and function of immune-related cells persist for only a few days or weeks after cessation of the ultraviolet exposure. They are essentially *selective* effects, and they do not mimic the generalised immunosuppression that viruses and certain drugs can induce in humans.

If the immune system is severely damaged, the body cannot survive life's routine encounters with infectious micro-organisms in the environment. Thus, one possible consequence of ozone layer depletion would be a decrease in protection against cutaneous infectious and fungal diseases which are normally controlled by cell-mediated immunity in the skin.[39] The skin is a highly immunologically active tissue. The increased frequency of lesions caused by the herpes simplex virus ('cold sores') on the face during summer illustrates well the influence of ultraviolet radiation upon immune activity in the skin, and studies in mice indicate that activation and replication of the herpes simplex virus follows localised suppression of immune defenses by ultraviolet irradiation.

There is some recent evidence that ultraviolet radiation can induce a more generalised immune suppression. However, despite the potential public health importance of this effect in facilitating infectious diseases, relatively little research has yet been done in humans. Ultraviolet-irradiated mice show a reduced immune response to the tuberculosis bacterium, and a decreased ability to eliminate it from internal organs. Further, soluble chemicals ('cytokines') secreted by ultraviolet-irradiated cultured skin cells, when injected into mice, suppress the bacteria-destroying activity of macrophage cells – the first line of immunological defence – as well as the delayed type hypersensitivity response.[40] (Unwelcome corroboration of the importance of the immune system in constraining the tuberculosis bacterium has now been widely confirmed in humans infected with the immune-destroying AIDS virus. Rates of clinically active tuberculosis have risen sharply in HIV-infected persons, particularly in Africa and, more recently, in India.) This ultraviolet-induced susceptibility to infection could become important in poor countries with much diarrhoeal disease, and with a high prevalence of infectious disease problems such as tuberculosis, leprosy and cutaneous leishmaniasis – a skin disease, common in tropical and subtropical countries, that is spread by sandflies and which involves large and persistent sores that cause much morbidity and many deaths. Indeed, experimental studies in mice indicate that chronic skin infections such as leprosy and leishmaniasis may be particularly affected because of the suppression of local cell-mediated immunity within the skin.[41] Overall, these and other research reports indicate that infectious disease due to bacteria, fungi, viruses and protozoa may all be exacerbated by ultraviolet-induced systemic immune suppression.[39] Ominously, the UN Environment Programme has warned that increased ultraviolet radiation exposure may, by immune suppression, facilitate the clinical progression of AIDS.[13]

A related and potentially serious effect would be a reduced effectiveness of vaccination. To achieve good active immunity, the body must make a strong response to the antigen presented in the vaccine. For vaccinations given by inoculation into the skin (e.g. tuberculosis), the response would be impaired by ultraviolet-induced suppression of local cell-mediated response to the antigen. Although there is little evidence on this question, a recent experimental study in young adult volunteers has found that relatively small increases in ultraviolet exposure impair the skin's immune responsiveness to antigen, and that exposures sufficient to cause localised sunburn can depress responsiveness at distant, non-irradiated, sites in the body.[42] Further, both light-skinned and dark-skinned races appear to be similarly affected.[39] Although speculative, it is possible that as WHO strives to immunise the world's children against the major infectious diseases, any further such reduction in immune responsiveness of populations already immunologically weakened through malnutrition and infection could partially impede that heroic effort.

The immune system is also part of the body's defence against cancer. Strong evidence for this comes from studies of persons born with a deficiency of their immune system, of organ transplant patients who undergo immunosuppression and of AIDS patients with damaged immune systems. All have an increased likelihood of developing cancer, particularly non-melanoma cancers of the skin, lymphatic system (i.e. lymphomas) cancers and several other cancers thought to be caused by viruses. Since, in the light of new molecular biological techniques for detecting viral DNA in human cells, it is likely that viruses will be implicated in a widened range of human cancers, an ultraviolet-induced depression of immune defences could have wide-ranging consequences. The Epstein–Barr virus (EBV) is an important and interesting example, since it appears to be responsible for the lymphomas that occur in immune-depressed people. It is an ancient virus that coevolved with the human species and which is usually carried as a symptomless lifelong infection, being kept in check by the immune system's T-cells. However, depression of the immune system – whether by drug treatment in transplant patients or by the evolutionary newcomer HIV – disturbs this benign relationship and transforms the EBV into a cancer-causing virus.

Experimental exposure to UV-B depresses the immune system of mice, making them more vulnerable to cancer-inducing chemicals. Also, cancers of mouse skin transplanted to mice previously irradiated with UV-B grow more rapidly than do those transplanted to non-irradiated mice.[43] It is likely that these cancer increases are due to the ultraviolet-stimulated formation of certain white blood cells, the suppressor T-cells, which inhibit the body's normal anti-tumour defences.[44] Hence, as well as directly causing skin cancer, UV-B exposure may promote other types of cancer normally suppressed by immune system surveillance.

7.3 Indirect effects upon human populations

7.3.1 Effects on other terrestrial species

An increase in UV-B exposure will also have various deleterious effects on the world's terrestrial and aquatic biota. This is hardly surprising, since terrestrial plants and animals and aquatic surface-dwelling organisms first evolved only after the ultraviolet-shielding ozone layer had begun forming.

Increased UV-B exposure can disrupt photosynthesis in plants, thereby impairing growth. In susceptible crop plants, this would reduce agricultural productivity. From the limited evidence available, there appears to be a varied response to UV-B among different species and strains of plants.

Cells in the outer (epidermal) layers of plant leaves contain ultraviolet-absorbing phenolic substances. The inner layers of green photosynthetic cells are thus protected against the damaging effects of ultraviolet radiation, while receiving the full complement of photosynthetically active radiation to which the outer cellular layers are transparent. Ultraviolet radiation appears to impair photosynthesis via mechanisms at the whole plant, the cellular, and the molecular levels. For example, the electron-transfer reaction ('Photosystem 2'), converting light into chemical energy, may be disrupted. Likewise, the natural cellular defence mechanisms which repair the damage to important macromolecules, including DNA, may be overwhelmed by the extra damage caused by increased ultraviolet irradiation. Via other biological mechanisms, UV-B radiation may also reduce the *quality* of the crop yield. For example, the protein and/or oil content of crops such as tomatoes, potatoes, sugar beet and soybeans may fall.[45]

Of more than 200 species and crop strains screened for tolerance to ultraviolet radiation, experiments have shown about two-thirds of them to be sensitive.[45] The most sensitive plant groups include peas, beans, melons, cabbages and mustard. There is also evidence that the growth of maize, rye and sunflower seedlings is adversely affected by UV-B.[13] Experimentally, the yields of some varieties of soybeans, the world's fifth most important crop, decline by up to one-quarter when UV-B exposure is increased by 25% – although little impact seems to occur at lesser exposures. Other plant species sensitive to UV-B grow more slowly when exposure is increased. Experimental studies of soybean, wheat and rice crop growth have simulated the simultaneous increased levels of atmospheric carbon dioxide and UV-B irradiation expected by around the year 2050. Results indicate that the adverse effects of UV-B exceed the beneficial effects of the extra carbon dioxide availability.[46] These experiments suggest that UV-B impairs photosynthesis and the production of growth hormones. Further, plant pollen cells and many species of nitrogen-fixing bacteria in the soil may be endangered by increased exposure to ultraviolet radiation.[13] However, the results of such studies have not been consistent across different species of plants. One explanation may be that plants whose wild ancestors were from tropical climates have always had to cope with more intensive UV-B radiation than have ancestral plants from temperate zones.

Their biological defences are therefore likely to be better. Thus, plants originating from the Near East (around 30° N), Northern China (around 45° N), and Mesoamerica (around 20° N) are more vulnerable than those originating from mid-Africa, southeast Asia and equatorial South America. Further, some strains of plants appear to be able to respond by producing UV-B screening compounds.

For the moment, there remains great uncertainty over the extent to which increased ultraviolet radiation will reduce agricultural yield, or forest growth. Synergistic effects between ultraviolet radiation and other climatic stresses are possible. And there is also the enhancing effect that increased ultraviolet radiation has on the formation of crop-damaging photochemical oxidants (particularly ozone) from the air pollutants that form in car-congested cities. In the longer term, selective effects of UV-B upon certain plants would alter the profile of plant species within an ecosystem. Even quite small changes in competitive balance can, over time, cause large changes in ecosystem structure.[47]

7.3.2 Effects on aquatic species

In terms of photosynthetic activity, phytoplankton (micro-plants and algae) are the most important single group of primary producers in the world. They are, metaphorically, the grass of the oceans, and convert almost 100 billion tonnes of carbon to organic material annually (see Fig. 6.2). They form the base of the oceanic and coastal food webs – which supply around a quarter of all the animal protein eaten by humans.

Increased exposure to UV-B adversely affects the phytoplankton in aquatic ecosystems. There are hundreds of types of phytoplankton organisms, differing in size, photosynthetic rate, nutrient composition and sensitivity to ultraviolet radiation. Phytoplankton live near the water's surface, and generally lack defences against increased ultraviolet irradiation. For example, most cannot make compensatory changes in their position within the water column. Therefore, an increase in the amount of UV-B penetrating several metres below the ocean's surface will damage these species, primarily by impairing photosynthesis.[13]

During the early summer in polar regions, as the melting sea-ice creates a favourable dilute saltwater microenvironment, there is a dramatic surge in algal phytoplankton populations. This process lays down supplies of captured nutrients and solar energy for the marine animal food web. Ultraviolet-induced damage to phytoplankton, and to the invertebrate zooplankton (micro-animals, including krill) that feed upon the phytoplankton and which may also be directly damaged by UV-B, would cause declines in shrimp and crab larvae populations and, in turn, fish stocks. Increased UV-B exposure diminishes the length of the season that invertebrate zooplankton populations spend near the ocean surface, feeding and breeding – an effect that would diminish the usual abundance of those populations.[48]

Current estimates of the danger posed by ozone layer depletion to marine organisms in the upper layers of the sea range from insignificant to catastrophic. Adverse effects induced by solar UV-B irradiation of phytoplankton have been observed to depths of more than twenty metres in clear water and five metres in turbid water. In one study, phytoplankton populations in regions beneath the Antarctic ozone hole were observed to undergo a 6–12 % reduction in photosynthetic activity compared with phytoplankton elsewhere.[49] Because of the pervasiveness of ultraviolet radiation in the environment, many organisms in the biosphere have evolved adaptive defences appropriate to natural circumstances. In particular, many marine organisms produce UV-B absorbing 'sunscreen' substances, such as flavonoids and mycosporine-like amino acids. However, whether such mechanisms can also compensate for increased ultraviolet exposure is unknown, although some experiments indicate an increase in sunscreen production in response to increased UV-B exposure. Almost certainly, such compensation would be bought at the metabolic cost of reduced photosynthetic productivity.

Ultraviolet-induced suppression of phytoplankton activity would diminish the ocean's uptake of atmospheric carbon dioxide, because, like land plants, phytoplankton require it as a metabolic substrate. Indeed, the oceans are the largest reservoir of reactive carbon on Earth, and phytoplankton make up the crucial 'biological pump' that moves carbon from the surface waters to the deep. Ozone depletion would thus enhance the greenhouse effect by reducing the capacity of the oceans to act as a sink for carbon dioxide. UNEP estimates that a 10 % loss of marine phytoplankton would reduce the oceans' annual uptake of carbon dioxide by five giga-tonnes – an amount which equals the annual anthropogenic emissions from the combustion of fossil fuels.

There is another, more speculative, ecological disruption that ozone layer depletion might cause via the oceans. Phytoplankton release huge volumes of gaseous dimethyl sulphide into the atmosphere, at a rate commensurate with their day-to-day sunlight-driven metabolic activity. Dimethyl sulphide forms sulphate aerosol particles that act as cloud condensation nuclei. Clouds form and – in negative feedback fashion –

they curtail the ultraviolet radiation reaching the ocean's surface.[50] However, if the marine microorganism population is depleted because of ozone layer depletion, the release of dimethyl sulphide will be reduced, fewer clouds will form, and even more ultraviolet radiation will impinge on the ocean. *Positive* feedback would thus occur. As we will see many times, when it comes to perturbations of ecosystems, the plot can usually be relied upon to thicken.

7.4 Summary

The story of the slow formation and, today, the incipient erosion of stratospheric ozone has a fable-like simplicity. There is a basic irony in the sequence of events. With a little poetic licence, it goes like this.

There was once a small young planet, two-thirds covered with water. In those waters dwelled many living things, including tiny green ancestral plants which captured energy from the Sun. Now it so happened that the oxygen which was consequently given out by these same plants accumulated in the sky. Some of this oxygen was turned into ozone by the more energetic rays of the Sun. Over many years, this ozone accumulated into a layer high in the sky and shielded Earth's surface from part of these same energetic rays.

After a while the ozone shield became strong enough to allow the offspring of some of the water-dwelling plants and animals to move onto the dry land. As a result, there followed a rapid proliferation of land-dwelling plants and the animals that ate them – and that ate one another. Many years passed, and among the animals there appeared one that walked and talked, thought and fought, bred and spread. This animal also made tools, with which it came to control the other animals and plants, and the land, sea and air. These tool-making animals increased in number and knowledge, and they made more complex tools.

The use of some of these newer tools caused special gases, unknown to the natural world, to be given out. These gases accumulated in the sky and were turned into ozone-destroying chemicals by the high-energy rays of the sun. And as the ozone shield was thus weakened, so more and more of those energetic rays began again to reach the Earth's surface. This caused the plants and animals to wilt, and to curse the tool-making animal for turning the Earth's dry land back into a dangerous place on which to live.

Now, the hearer of this fable might ask: 'What about the tool-makers? Did they suffer and perish?' Well, in fact, the tool-makers saw what was about to happen, took fright, and urgently reconsidered the types of tools that should be made. But at the time this fable went to press, it had proved easier to identify the problem tools than to dispense with them – or to clean up the mess they had made. Some of the tool-makers were short-sighted; some were stubborn; and some of the less fortunate ones resented having to discard tools that they had only recently acquired. All of them found it difficult to negotiate between groupings of the more lucky and the less lucky tool-makers because of their lack of experience and wisdom in so doing.

References

1. UN Conference on Environment and Development. Protection of the Atmosphere. Chapter 9 in: *Agenda* 21. Conches, Switzerland: UNCED, 1992.
2. Lovelock argues that in preozone times the Earth had a mantle of methane that formed a protective layer in the upper atmosphere. He deduces, therefore, that early algal and bacterial life spread, film-like, over the moist surfaces of the land. See: Lovelock J. *The Ages of Gaia.* Oxford: Oxford University Press, 1987.
3. Rowland FS. Some aspects of chemistry in Springtime Antarctic stratosphere. In: Rowland FS, Isaksen ISA (eds). *The Changing Atmosphere.* Chichester: John Wiley & Sons, 1988, pp 121–40.
4. Molina MJ, Rowland FS. Stratospheric sink for chloro-fluoro-methanes: chlorine atom-catalysed destruction of ozone. *Nature* 1974; **249**: 810–14.
5. Farman JC, Gardiner BG, Shanklin JD. Large losses of total ozone in Antarctica reveal seasonal ClO_x/NO_x interaction. *Nature* 1989; **315**: 207–10.
6. Houghton JT, Jenkins GJ, Ephraums JJ. *Climate Change. The IPCC Assessment.* Cambridge: Cambridge University Press, 1990.
7. Lubin D, Frederick JE, Krueger AJ. The ultraviolet radiation environment of Antarctica, McMurdo station during September–October 1987. *Journal of Geophysical Research – Atmospheres* 1989; **94(D6)**: 8491–6.
8. Roy CR, Gies HP, Elliott G. Ozone depletion. *Nature* 1990; **347**: 235–6.
9. Blumthaler M, Ambach W. Spectral measurements of global and diffuse solar ultraviolet-B radiant exposure and ozone variation. *Photochemistry and Photobiology* 1991; **54**: 429–32.
10. Stolarski RS, Bloomfield P, McPeters RD, Herman JR. Total ozone trends deduced from Nimbus TOMS data. *Geophysical Research Data Letters* 1991; **18**: 1015–18.
11. Kerr RA. New assaults seen on ozone shield. *Science* 1992; **255**: 747–8.
12. Brasseur GP. Ozone depletion. A deepening, broadening trend. *Nature* 1991; **352**: 668–9. And: Briefings. Antarctic ozone hole hits record depth. *Science* 1991; **254**: 373.
13. UN Environment Programme. *Environmental Effects of Ozone Depletion*: 1991 *Update.* Nairobi: UNEP, 1991.
14. Because bromine is active at warmer temperatures than is chlorine it may account for some of the recent observations of ozone layer depletion during the warmer months of the year. Brominated compounds called 'halons' are widely used as fire-extinguishing gases – and they are difficult to replace in shipping, aviation and other high-risk fire settings.
15. A reprieve was granted to shorter-lived hydrochlorofluorocarbons

(HCFCs), as 'transitional' substances, on condition that they are used sparingly. Subsequently, it was found that their short-term effects are much more potent than are the early effects of the longer-lived CFCs.

16. Solomon S, Albritton DL. Time-dependent ozone depletion potentials for short- and long-term forecasts. *Nature* 1992; **357**: 33–7. Also: World Meteorological Organization. *WMO and the Ozone Issue*. Geneva: WMO, 1992.

17. Crutzen PJ. Ultraviolet on the increase. *Nature* 1992; **356**: 104–5.

18. Russell Jones R. Consequences for human health of stratospheric ozone depletion. In: Russell Jones R, Wigley T (eds). *Ozone Depletion: Health and Environmental Consequences*. London: John Wiley & Sons Ltd, 1989, pp 207–27. Also: Sterenborg HJCM, van der Leun JC. Action spectra for tumorigenesis by ultraviolet radiation. In: Passchier WF, Bosnjacovic BFM (eds). *Human Exposure to Ultraviolet Radiation: Risks and Regulations*. Amsterdam: Elsevier, 1987, pp 173–195. The action spectrum for ultraviolet-induced mutations (pyrimidine dimers) in human skin is very similar to that shown, in Fig. 7.3, for skin cancer in albino mice (Freeman SE *et al*. Wavelength dependence of pyrimidine dimer formation in DNA of human skin irradiated *in situ* with ultraviolet light. *Proceedings of the National Academy of Sciences* 1989; **86**: 5605–9).

19. Editorial. Protecting man from UV exposure. *Lancet* 1991; **337**: 1258–9.

20. This effect is enhanced by the prior effect of particulate pollutants, which scatter the incoming ultraviolet radiation, thus lengthening its average path to Earth's surface.

21. Russell WMS, Russell C. Evolutionary and social aspects of disease. *Ecology of Disease* 1983; **2**: 95–106. And: Loomis WF. Skin-pigment regulation of vitamin-D biosynthesis in man. *Science* 1967; **157**: 501–6

22. Dunnigan MG *et al*. Prevention of rickets in Asian children: assessment of the Glasgow campaign. *British Medical Journal* 1985; **291**: 239–42

23. McMichael AJ, Giles GG. Cancer in migrants to Australia: Extending the descriptive epidemiological data. *Cancer Research* 1988; **48**: 751–6.

24. NH&MRC. *Health Effects of Ozone Layer Depletion: A Report of the National Health and Medical Research Council*. Canberra: Australian Government Publishing Service, 1989.

25. Hoffman JS (ed). *Assessing the Risks of Trace Gases that Can Modify the Stratosphere*. EPA 400/1–87/001. Washington DC: Environmental Protection Agency, 1987.

26. Romerdahl CA, Donawho C, Fidler IJ, Kripke ML. Effects of ultraviolet-B radiation on the in vivo growth of murine melanoma cells. *Cancer Research* 1988; **48**: 4007–10.

27. Kricker A, Armstrong BK, English DR, Heenan PJ. Pigmentary and cutaneous risk factors for non-melanocytic skin cancer – a case-control study. *International Journal of Cancer* 1991; **48**: 650–62.

28. Armstrong BK. Sunlight and malignant melanoma: Intermittent or total accumulated exposure to the sun. *Journal of Dermatologic Surgery and Oncology* 1988; **14**: 835–49.

29. See also: Armstrong BK. Implications of increased solar UVB for cancer incidence. In: Chanin M-L (ed). *The Role of the Stratosphere in Global Change*. NATO Advanced Study Institute Series. Heidelberg: Springer Verlag, in press. There are several assumptions that necessarily underlie the attempts made to derive biological amplification factors from the existing, limited, epidemiological data.

30. Taylor HR. The biological effects of UV-B on the eye. *Photochemistry and Photobiology* 1989; **50**: 489–92.
31. Kupfer C. The conquest of cataract: a global challenge. *Transactions of the Ophthalmology Society of UK* 1984; **104**: 1–10. And: Foster A. Who will operate on Africa's 3 million curably blind people? *Lancet* 1991; **337**: 1267–9.
32. Hiller R, Sperduto R, Ederer F. Epidemiologic associations with cataract in the 1971–1972 National Health and Nutrition Examination Survey. *American Journal of Epidemiology* 1983; **118**: 239–49.
33. Hollows F, Moran D. Cataract: the ultraviolet risk factor. *Lancet* 1981; **ii**: 1249–51.
34. Collman GW *et al*. Sunlight and other risk factors for cataracts: An epidemiologic study. *American Journal of Public Health* 1988; **78**: 1459–62.
35. Taylor HR *et al*. Effect of ultraviolet radiation on cataract formation. *New England Journal of Medicine* 1988; **319**: 1429–33.
36. The Italian–American Cataract Study Group. Risk factors for age-related cortical, nuclear, and posterior subcapsular cataracts. *American Journal of Epidemiology* 1991; **133**: 541–53. And: Mohan M *et al*. (The India–US case-control study group). India–US case-control study of age related cataracts. *Archives of Opthalmology* 1991; **107**: 670–6.
37. Leske MC, Chylack LT, Suh-Yuh W. The Lens Opacities Case-Control Study. Risk factors for cataract. *Archives of Ophthalmology* 1991; **109**: 244–51. And: Dolezal JM, Perkins ES, Wallace RB. Sunlight, skin sensitivity, and cataract. *American Journal of Epidemiology* 1989; **129**: 559–68.
38. Young RW. Solar radiation and age-related macular degeneration. *Survey of Ophthalmology* 1988; **32**: 252–69.
39. Morison WL. Effects of ultraviolet radiation on the immune system in humans. *Photochemistry and Photobiology* 1989; **50**: 515–24. And: MacKie R, Rycroft MJ. Health and the ozone layer. *British Medical Journal* 1988; **297**: 369–70.
40. Jeevan A, Ullrich SE, Dizon VV, Kripke ML. Supernatants from UV-irradiated keratinocytes decrease the resistance and delayed type hypersensitivity response to *Mycobacterium bovis* BCG in mice and impair the phagocytic ability of macrophages. *Photodermatology, Photoimmunology and Photomedicine* 1992; **9**: 255–63.
41. Giannini M. Suppression of pathogenesis in cutaneous leishmaniasis by UV irradiation. *Infection and Immunity* 1986; **51**: 838–43.
42. Cooper KD *et al*. UV exposure reduces immunization rates and promotes tolerance to epicutaneous antigens in humans; relationship to dose, CD1a⁻DR⁺ epidermal macrophage induction and Langerhans cell depletion. *Proceedings of the National Academy of Sciences* 1992; **89**: 8497–501.
43. Kripke ML. Immunologic mechanisms in UV radiation carcinogenesis. *Advances in Cancer Research* 1981; **34**: 69–81.
44. Roberts LK, Samlowski WE, Daynes RA. The immunological consequences of ultraviolet radiation exposure. *Photodermatology* 1986; **3**: 284–98.
45. Worrest RC, Grant LD. Effects of ultraviolet-B radiation on terrestrial plants and marine organisms. In: Russell Jones R, Wigley T (eds). *Ozone Depletion: Health and Environmental Consequences*. Chichester: John Wiley & Sons, 1989, pp 197–206.
46. Teramura AH, Sullivan JH, Ziska LH. Interaction of elevated UV-B

radiation and CO_2 on productivity and photosynthetic characteristics in wheat, rice and soybean. *Plant Physiology* 1990; **94**: 470–5.
47. Gold WG, Caldwell MM. The effects of ultraviolet-B radiation on plant competition in terrestrial ecosystems. *Physiologia Plantarum* 1983; **58**: 435–44.
48. Damkaer DM, Dey DM, Heron GA, Prentice EF. Effects of UV-B radiation on near-surface zooplankton of Puget Sound. *Oceanologia* 1980; **44**: 149–58.
49. Smith RC *et al*. Ozone depletion: Ultraviolet radiation and phytoplankton biology in Antarctic waters. *Science* 1992; **255**: 952–9.
50. Charlson R, Lovelock J, Andrae M, Warren S. Ocean phytoplankton, atmospheric sulfur, cloud albedo and climate. *Nature* 1987; **329**: 321–3.

8

Soil and water: loaves and fishes

Of all the sectors in the world economy, it is agriculture where the contrast between the economic and environmental indicators is most obvious ... For an economist, there may be distribution problems in the world food economy, but not a production problem. To an ecologist who sees a substantial fraction of current world food output being produced on highly erodible land that will soon be abandoned, or by overpumping groundwater which cannot continue indefinitely, the prospect is far less promising.

Lester Brown, 1991[1]

8.1 Soil: source of sustenance

8.1.1 Introduction

Food, which supplies the energy and nutrients needed to build and sustain life, is the central organising principle of ecological relationships. Plants capture solar energy, convert it to chemical energy and package it within assorted organic molecules. Plants also extract inorganic elemental micronutrients (such as iron, iodine and selenium) from the soil. Herbivores obtain nutrients and energy by eating plants. We humans dine at Nature's high table, eating a mix of plant foods, herbivores and aquatic carnivores. However, in urbanised settings, we buy our food in the marketplace, eat it – and barely pause to consider its origins. Although from the high table the soil may seem a long way off, there is no such foreseeable thing as a 'post-agricultural' society. Most of our food comes from land-based agriculture.

Land accounts for 30% of the Earth's surface. Of the total 14 billion hectares (140,000 square kilometres), 3 billion are potentially arable – i.e. potential cropland – and 3 billion are pasture and rangeland. Because most of the 1.5 billion hectares of readily usable arable land is now in use (indeed most of it has been since the 1950s), bringing more land into food production is proving progressively difficult and environmentally damaging. Poor peasants in the Philippines and the Andes, struggling to cultivate fields on steep, deforested slopes, symbolise this escalating dilemma. Meanwhile, much of the world's topsoil is being overworked, degraded, built over or otherwise lost.

After ten thousand years of steady expansion in arable land, feeding today's rapidly expanding world population is straining the limits of sustainable agriculture. World food production is still edging up, after several decades of impressive gains, but the statistics are starting to falter – particularly for cereal grains, our dominant source of energy. Per-person grain production has declined since 1984. Although reasons for this are debatable, Lester Brown argues that: 'the evidence suggests that weather fluctuations and the drawdown of excessive grain stocks from the mid-eighties may have obscured the transition from an era when grain output expanded much faster than population to one in which the reverse is true.'[2] Growth in soybean production, the leading plant protein source, has slowed greatly since peaking in 1979, and per-person meat production, after almost doubling since 1950, has plateaued since 1987. Meanwhile, offshore, the total fish catch, accounting for one-quarter of the world's animal protein, appears to have reached its sustainable limit.

The basic health-related statistic is that the absolute number of frankly malnourished and starving people in the world, now around 500 million, is continuing to creep up. In addition to that hard core of developmentally stunted, health-endangered humans, another 300 million receive less than 90% of the basic daily caloric intake necessary for a fully active life and another billion or so experience lesser levels of undernourishment. These food shortages are greatest in Africa, South Asia and the Andean republics.

We should first examine the ecological background to those statistics. Through agriculture, we deliberately restructure nature to create artificial ecosystems that increase the environment's carrying capacity for the human species. By 2030, in response to massive population increase, particularly in the already-underfed Third World, we will need to double our food production.[3] That may be achievable – via some combination of expansion of arable land, better farm management and higher-yielding grains (especially wheat). However, a shortage of fertile soil, exacerbated by further losses, may well be the major limitation on feeding the human population later next century. The recent exponential growth in population has moved the world from a situation of great land abundance a half-century ago to one of impending scarcity.[4] Water for irrigation is also likely to become an increasing constraint, as aquifers are overpumped for agriculture or diverted for water-hungry urban populations. Chemical fertilisers and pesticides have achieved near-maximum effect and, anyway, further increases in their use are prohibitively expensive in many countries. We will therefore look first at agriculture's basic resource – the soil.

8.1.2 Land degradation: erosion and desertification

The world's soil is a finite and destructible resource. It is a fragile film of fertility that, over the last one-tenth of Earth's existence, has come to cover much of the planet's crust. Although strictly 'renewable', soil is produced only slowly by the oxygen-assisted weathering of rocks and the rotting of organic matter. Natural siltation processes may speed soil accumulation in some locations – as in the loess plains of northern China. In general, though, a single centimetre of soil takes anything from twenty to a thousand years to form. Yet that centimetre can be, and often is, destroyed or lost within several years. The destructive power of exploitative agriculture was vividly demonstrated by the creation of the Oklahoma dust bowl, in southern USA, early this century. Those farmers did what agriculturalists have often done; they worked the soil, did not replenish it, and then moved on to new lands. Today, in a crowded world, we have run out of easy new frontiers.

Soil is a complex mix of inert eroded rock, mineral nutrients, organic material, microbes and invertebrates – including its best known inhabitants, worms. One hectare of fertile soil in a temperate region contains hundreds of millions of invertebrates and countless trillions of bacteria, yeast cells and fungal cells. These tiny creatures break down, transport and chemically process the vital nutrients needed by plants. Healthy soil thus acquires from its inorganic, organic and living components its spongy and cohesive properties. When denuded of plant cover, overgrazed, pulverised, chemically stressed, depleted of nutrients or made salty by overirrigation, the soil dies, dries and turns to dust. Erosion by water or wind follows. Globally, 85 % of today's land degradation has been caused by erosion. Other forms of land degradation include salinisation, waterlogging, acidification and various forms of physical and chemical soil damage. Salinity occurs when the salts in irrigation water accumulate in soil as evaporation occurs. If these salts are not flushed or drained away they accumulate in the root zone. Waterlogging results from excessive irrigation, while the overworking of soil, with depletion of nutrients and organic matter, has caused widespread acidification of soil.

Prolonged land degradation can lead to 'desertification'. This human-made process differs from the natural (and often temporary) extension of preexisting deserts, such as has recently occurred with the southern Sahara. Desertification entails an essentially irreversible decline in the land's biological diversity and productivity, and is a particular problem in drylands used for grazing. Human populations have created deserts since the beginnings of settled agriculture. Today, in many of the drier parts of the world, particularly in Africa and India, human-made deserts are forming at the rate of several million hectares each year (although the available data are of uncertain quality). More generally, the degradation of grazing and agricultural land embraces every continent, particularly Asia and Africa where many of the world's poor subsistence farmers live. The eroding hillsides of Asia, the desertification in drylands adjoining the Gobi and Sahara deserts and the ravaged Brazilian savannah where rainforest once proliferated all reflect the impact of poor expanding rural populations.

Inappropriate farming methods, overgrazing and deforestation have been the main, approximately equal, causes of land degradation. Often this has resulted from replacing traditional mixed cropping and fallowing with intensive and continuous cultivation of monoculture crops. Large areas of the world's drier farmlands and grazing lands are losing their biological diversity as pressures from humans and grazing animals accelerate the loss of vegetation, including tree-cover. This causes the land to dry and harden, and the rains to run *off* and not *in*; soil erosion ensues and may lead to desertification.

The Gambia, in West Africa, shows in microcosm the problems of sub-Saharan Africa. It is a small, poor, ribbon-shaped coastal country; an ex-British colony in the midst of French West Africa. In recent decades most of the natural forest has been cleared to increase the land available for peanut crops (Gambia's main export) as well as for growing rice and for grazing goats and sheep. The large wild animals disappeared decades ago; the elephants that Gambia once displayed on its postage stamps have long gone. More recently, microclimatic change has occurred, followed by increased soil erosion. This local climate change may have been contributed to by the loss of tree cover and the alteration in the local hydrological cycle. These changes, along with regional drought, are putting increasing pressures on the sustainability of the food and water supply of this tiny country.

8.1.3 Land degradation: a worldwide problem

Land degradation is happening today on an unprecedented scale as populations expand and as agricultural practices become more intensive, more mechanised and less attuned to local ecological needs. The UN estimates that, since 1945, approximately one-fifth of all potentially arable and pastoral land has suffered moderate to extreme degradation. Erosion and desertification now threaten 40% of Africa's non-desert land, 30% of Asia's and 20% of Latin America's, while for Europe and North America the figures are 17% and 4%, respectively.[5] Desertification fuels the surge of hungry environmental refugees into the fringe settlements of Africa's big cities.

Each year, around 6 million hectares of the world's productive cropland (about the size of Ireland) is being lost through land degradation.[6] In other words, around 4% of the world's cultivated land is being destroyed or damaged each decade. In China, where land has been worked very hard and where the spread of cities is greatest in populous fertile regions, the rate of cropland loss is double that. The ecological strain of producing enough food for China's one-fifth of the world population with just one-fifteenth of the world's arable land is starting to show – exemplifying the 'stress fatigue' that typically occurs in a repeatedly assaulted ecosystem.[7] In the agriculturally mismanaged USSR a massive 20% loss of grain-land occurred between 1977 and 1991.[8] Even the grain-rich Ukrainian farm-lands have reportedly lost one-third of their valuable black humus because of erosion over the past 30 years. The vast open fields of collectivised Soviet agriculture have afforded no protection from wind and rain.

Globally, an estimated 1% (25 billion tonnes) of the world's topsoil is lost every year through wind and water erosion[5] – an amount equivalent to the entire topsoil of Australia's extensive wheat-lands. In many of the world's badly affected areas, the annual rate of soil loss exceeds the sustainable (renewal) rate by a factor of 10 or more. In just 200 years of intensive farming in the USA, the cropland has lost 25–50% of its natural nitrogen stores and one-third of its topsoil. Recently, however, soil erosion has been greatly reduced in the USA, with the deliberate resting of damaged farmland and with increasing use of minimal ploughing or no-till techniques; recent World Bank estimates suggest that the soil fertility of American cropland will decline by no more than 5% over the coming century.[3] Clearly, where resources and knowledge are available, the worst overuse of arable land can be averted. Few countries, however, have such resources – or such robust soil.

Tropical and semi-tropical soils, being geologically older, are naturally poorer, thinner and more fragile than those in temperate, northern climates.[9] Soils in Africa and Latin America are not only older than Asian soils, but Asia's alluvial or volcanic soils are naturally more fertile. Africa, with its widespread food insecurity, has naturally fragile soils that are very low in organic matter. This, along with increasingly heavy use and high-impact tropical rains, readily results in soil erosion – particularly where more fertile upland soils have encouraged dense settlement, as in parts of Kenya, Rwanda, Burundi, Guinea, Ethiopia and Nigeria. (Ethiopian farmers claim ruefully that the ancient Nile-centred Egyptian civilisation was based on the volcanic silt from Ethiopian highland farms.) The problem has become orders of magnitude more serious today, as soil erosion in the Ethiopian highlands has accelerated under the combined pressures of population growth, uncontrolled and severe deforestation and the disruptions of civil war. Ironically, much of this precious soil now ends up on the floor of the Aswan Dam doing neither Ethiopians nor Egyptians any good. In South Africa, the crowding and economic deprivation of black populations coerced into living in agriculturally marginal, racially contrived, homelands has had ecologically disastrous consequences.[10] Over several decades, the desperate daily struggle to grow food, by an impoverished and hungry population greater in number than the asthenic soil can support, has stripped much of the land bare. The limits of agricultural and fuelwood sustainability have been severely, perhaps irreparably, breached.

Although the technical know-how to arrest soil erosion and to restore soil fertility exists, the roots of the problem go much deeper. They lie in the inexorable population pressures on the land and in the poverty, power-lessness and lack of land tenure on the part of the world's rural poor. Feudal structures and the pressures of the export-oriented market economy provide no short-term incentives to good land management. Without basic social reform and population control, governments in poor countries will find it hard to preserve productive land. To compound the dilemma, greenhouse-induced climate change may cause warming and altered rainfall in temperate zones that will adversely affect some of the world's most productive lands. Unless the genetic engineers of agricultural science pull some recombinant rabbits out of hats, we will face a mounting struggle to feed ourselves.

8.1.4 Historical precursors: today's problem writ less large

Today we think of agriculture as one of the great achievements of mankind, but we forget the dereliction it has left in its train ... agriculture has only survived in places where the climate and the soil enabled the land to withstand the impact of trial and error long enough for a workable system to be established.

GW Dimbleby, 1972[11]

The early agrarian populations arose in warm, well-watered areas – particularly the Nile Valley, the Tigris–Euphrates region of the eastern Mediterranean, the Yellow River Valley in China and the Ganges–Brahmaputra region of India. Populations were small, technology was primitive and environmental damage was localised. In later millenia, as populations grew and as the proportion of urban dwellers 'parasitic' upon the peasantry increased, so the pressures on local agriculture grew, eventually subverting traditional, ecologically sustainable methods. Thus, as discussed in chapter 4, land degradation probably largely determined the fate of the ancient civilisations of Mesopotamia and the Mediterranean Fertile Crescent.[12] Over the centuries, when crops would no longer grow, animal grazing replaced agriculture. With further deterioration, sheep replaced cattle and then goats replaced sheep. Goat-herding is a last-stage ecological disaster, evident today in much of the Mediterranean region where it has prevented the recovery of desertified ancient agricultural lands. A similar sequence occurred in ancient Egypt, where the huge supplies of charcoal needed for the smelting of metals caused large wooded areas to be cleared. The newly-cleared land was then heavily exploited for the successive grazing of cattle, sheep, goats and camels.

Although largely obscured by the mists of time, farming has also transformed the European environment. Thousands of years after its early establishment elsewhere, agriculture spread into Europe from the eastern Mediterranean by cultural diffusion, migration and intermarriage with pre-agrarian communities. The felling of forests for agriculture had variable consequences, well illustrated by two examples in England. The North Yorkshire Moors and the chalk downs of southern England were long ago cleared of deciduous forests for farming land. The Yorkshire lands were farmed, increasingly intensively, for several thousand years through the Bronze and Iron Ages, after which they declined to heather-covered moor as the soil deteriorated and productivity declined. (Ironically, this is the landscape which has charmed the viewers of BBC television's 'All Creatures Great and Small'.) The southern chalk downs supported agriculture throughout the Neolithic, Bronze and Iron Ages, and then on through the Roman and mediaeval ages. Today, after recent rejuvenation, those downs afford some of Britain's best cereal-growing land.

These historical narratives show that it is impossible for agricultural societies *not* to alter the balance of ecosystems. Agriculture entails the restructuring of plant-producing ecosystems, which, depending on the mix of population pressure, local knowledge and political enlightenment, necessarily causes some level of ecological disruption. Our predecessors may have lacked theoretical understanding of ecology, but they had the option of moving on to the next patch of arable land. We, today, cannot claim ignorance – and, increasingly, we find that the next farming patch is already occupied.

8.1.5 *Natural and unnatural loss of soil micronutrients*

Lack of dietary energy (calories) is the most dramatic form of under-nutrition, frequently causing starvation and death. However, around the world there are many serious health problems caused by chronic deficiencies of essential micronutrients such as vitamin A (which causes hundreds of thousands of cases of child blindness annually) and essential elements such as iron and iodine. These elemental deficiencies often reflect ecological disruption by human populations. Indeed, one of the great but invisible forms of ecological damage being wrought by humans is the cumulative transfer of nutrients from land to sea.

Agricultural runoff, sewage disposal and organic wastes dumped at sea (directly or via rivers) all entail the loss of nutrient elements from soil, plants and animals. In nature, these nutrients would mostly be recycled on land, including via excretion of body 'wastes' and organic decay. Sewered cities, however, by eliminating offensive and hazardous wastes, sys-tematically breach this cycle. Elemental nutrients from the soil, which we obtain via plant foods, are thus diverted to the sea. Hong Kong exemplifies the problem. Each year that huge city discharges hundreds of thousands of tonnes of solid sewage 'waste' into the ocean, faecally enriched with nutrients from the soil of the New Territories (their local vegetable gardens), while it imports thousands of tonnes of energy-intensive artificial fertiliser from Europe. Recent awareness of this perversion of natural processes – compounded by mounting difficulties in disposing of burgeon-ing volumes of sewage – is inducing attempts to use sewage sludge as agricultural fertiliser. This will help restore an ecological cycle to which the sparse populations of our pre-urban ancestors contributed. Being of no fixed address, their wastes stayed while they moved on; for urban populations, the relationship is typically reversed.

Micronutrients can be lost through other environmental disturbances. The story of iodine deficiency illustrates well the ecological dimension of human dietary needs. Food grown in iodine-depleted soil is the cause of a worldwide health problem, one which in some locations is exacerbated by soil erosion. The basic problem is an ancient one, reflecting the relentless migratory spread of human populations, including into environments that were nutritionally suboptimal in their geochemistry. Regional Chinese documents from many centuries ago record that swellings in the front of the neck were common. In fourteenth century alpine Europe some of the statues in churches and cathedrals – many of which survive today – exhibited large swellings in the front of the neck, along with obvious mental handicap and physical deformity. These and other records attest to the long-standing problems of dietary iodine deficiency around the world.[13]

Elemental iodine is essential for the functioning of the thyroid gland, situated in front of the trachea (wind-pipe). If the local diet is iodine-deficient, the thyroid gland compensates by increasing in size and trapping more iodine from the passing blood supply. The swollen gland produces the externally visible 'goitre' in the front of the neck. The thyroid gland produces a hormone, thyroxine, which controls the rate of bodily metabolism. In the fetus and young infant, this hormone influences the normal development of the brain. Babies born to severely iodine-deficient mothers are likely to be mentally subnormal, and to have a squint, spasticity and poor motor coordination. Such offspring became the 'cretins' of alpine Europe. (The word 'cretin' derives from 'Chrétien', or Christian, and alludes to their perceived status as innocents incapable of evil.) Less severe iodine deficiency causes lethargy and intellectual dulling.

Iodine leaches naturally out of mountainous soils with water runoff; other ancient coastal and plains soils can also become iodine-depleted. The problem is exacerbated by soil erosion, and by disturbances of water runoff and of the natural cycles of flooding and silting.[14] Today, endemic cretinism, stunted growth, reproductive disorders and other iodine-deficiency disorders occur widely around the world.[13] WHO has recently drawn attention to the fact that areas of iodine deficiency are increasing in many regions of the world. Some 1,000 million people – almost one-fifth of the world's population – are at risk of iodine deficiency, including 400 million in China and 200 million in India. Within this global total, almost 200 million have goitre, 100 million have mental and physical impairment, 20 million have some degree of brain damage, and 3–4 million have overt cretinism. WHO has therefore recently assigned a high priority to alleviating iodine deficiency.

Iodine, like phosphorus, is an element that in nature is washed slowly but steadily from land to sea. Restitution in soil occurs via the ultralong cycles by which ocean floors rise to become dry land. Faster, lesser, cycles also operate, such as the transport of iodine from ocean, via atmosphere, to rainfall. There would be less of a problem if human settlement had not spread into marginal lands. The variable combination of population pressure, cultural practice, and loss of topsoil has increased human population exposure to this micronutrient deficiency.

8.2 Intensive agriculture

8.2.1 Introduction: techniques and issues

As food exports become major income-earners for debt-ridden poor countries, and as population pressures grow, so the scale and intensity of agriculture increases. With the increasing commercialisation of Third World agriculture, Western agrochemical companies are transforming farming techniques. The profile of crops is also changing, away from food crops towards tea, coffee, tobacco, cotton and other exportable commodities. Meanwhile, in rich countries, misdirected governmental subsidies have often exacerbated the intensification of agriculture, causing wasteful surpluses.

Crop growth is ultimately limited by the rate at which photosynthetic conversion of carbon dioxide and water to organic plant material can occur. This rate can be increased by ensuring adequate water, by supplying essential micronutrients, by reducing crop losses and by selective breeding or genetic engineering of high-yield strains. These four options – irrigation, fertilisers, pesticides and genetic selection – are the main repertoire of intensive agriculture. A fifth option is to increase the frequency and intensity of cropping. All options require sufficient fertile soil.

These intensive agriculture strategies have various ecological consequences. Agrochemicals and machinery typically cause chemicalisation and sterilisation of soil, compaction and soil erosion.[15] Artificial fertilisers supply plant-building nitrogen, phosphorus and potassium, but, unlike organic manures, do not replenish the long-term fertility of the soil. Agricultural chemical runoff pollutes waterways, sometimes causing eutrophication via an overload of phosphates and nitrates upon which blue-green algae thrive. Nitrogenous fertilisers produce emissions of nitrous oxide, a greenhouse gas. While the thirty-fold increase in global pesticide use and the nine-fold increase in chemical fertiliser use since the 1950s have undoubtedly increased food yield, they have also caused widespread chemical pollution, destruction of wildlife and disturbance of ecosystems. The clearing of land for consolidated large-scale agriculture can destroy habitat and obliterate species. Indeed, the widespread removal of hedgerows in England has caused insect infestations in various crops because the habitat of the insects' natural predators has thus been destroyed. Irrigation increases human exposure to various water-borne and vector-borne diseases. The use of genetically selected monocultures increases crop vulnerability to catastrophic disease or infestation and tends to weaken local ecosystems. Yet, as we have seen, world food production must double by 2030 just to maintain the current per-person supplies. More may be needed if living standards are to improve enough to curb fertility. However, most areas of potential cultivation lack water, are physically limited (steep slopes, poor drainage or exposure to wind and water) or have chemical constraints (alkaline soil or trace element deficiencies).[16] For those reasons, only 8% of the increase in grain productivity since 1960 has been due to increased cropland.[3]

A dramatic casualty of intensive agriculture is the former Soviet Union's Aral Sea in the southern republic of Uzbekistan. This once mighty, fish-rich, life-sustaining lake, once the world's fourth-largest lake, is likely to end as a series of salty puddles.[17] The USSR government, in 1990, finally declared it an ecological disaster 'beyond human control'. The Aral Sea is shrinking because its water sources have been diverted to irrigate vast monocultures of cotton. In the past 20 years, the volume of the sea has shrunk to half its original size. The residual water is increasingly salty, most fish species have perished, and the fishing industry is now as moribund as the countless stranded fishing boats left high and dry by the receding waters and the vanished fish. Salt from the drying and expanding sea-bed blows over the surrounding land, killing crops and vegetation. The microclimate of the region is changing because of the depletion of surface water; the lake's moderating effect on summer heat and winter cold has been much reduced. Heavy and compensatory use of pesticides and herbicides on the cotton fields has compounded this local ecological disaster. The end-result has been a dramatic reduction in the agricultural and aquatic productivity of the region. The combined economic losses now amount to an estimated US$10 billion per year, and it will require additional billions merely to prevent the blighted cotton fields becoming desert by the end of this century. There have also been reported a range of adverse effects of the Aral Sea disaster upon human health. The adjoining Kara-Kalpak population has reportedly experienced increases in liver disorders and infant mortality. Although reliable health statistics are not easily obtained from this part of the world, the *USSR Environmental Report*, a government publication, has stated that pesticide use in the region is over twenty times higher than the national average, and that rates of illness and disease in infants are about five times higher.

Pesticide use has underscored another ecological dilemma in our attempt to grow food exclusively for human consumption. As with the widening use of antibiotics in humans (see section 10.1.3), this biocidal attack by humans has caused an accelerated genetic evolution of 'pests' – insects, fungi, bacteria and viruses. Since the 1950s there has been up to a ten-fold increase in the number of pesticide-resistant species, reflected in the USA by a gradual increase in the proportion of overall crop losses.[18] Our initial response has been to use heavier spraying, more potent chemicals and targeted antibiotics to supplement broad-spectrum pesticides. Since insects and microbes have shown that they can mutate as fast as we can generate new pesticides, farmers are now moving towards integrated pest management (IPM) – i.e. the combination of manual, biological and, where necessary, chemical control of weeds and pests. In the mid-eighties, Indonesia's rice production suffered a major scare when a pesticide-resistant insect, the brown plant-hopper, evolved in the wake of massive increases in pesticide use. Further, its natural predator was killed by the same widely-used pesticide. Faced with a potential agricultural disaster, Indonesia banned fifty-seven insecticides and urgently promoted IPM. The increasing reliance upon IPM in many countries signals our enforced realisation that the long-term productivity of managed ecosystems must come from sustaining and balancing – not from destroying and un-balancing.

8.2.2 Social and health consequences of intensive agriculture

The organisation of agriculture also has profound effects upon social relations, employment and the health of agricultural workers. Recent trends to mechanised export-oriented agriculture in Africa, Asia and Latin America have had various adverse health effects, particularly among marginalised, temporary workers.[19] Large-scale investment agriculture, the concentration of land ownership and the increasing capital intensity of production have caused increased unemployment and labour insecurity. The political weakness of dispersed agricultural workers and the primacy of profit for both foreign owners and debt-ridden national governments typically cause short-changing of the social and health needs of workers.[20]

In India and Brazil, for example, the concentration of capital and land ownership has exacerbated child malnutrition and infant mortality. Plantations in these countries share common features: they have large acreages, employ wage labour (often seasonally) and produce a very limited number of crops – for marketing rather than for local consumption. Seasonal and casual labour has reduced the owners' liability for providing housing, insurance, pensions, maternity rights, injury compensation and health care benefits. In Zimbabwe, in southern Africa, the increasing dependence on foreign loans and the demands of externally-imposed 'structural economic adjustment' have disrupted the production and consumption patterns of rural communities.[20] The intensification and commercial control of agriculture, by altering the patterns of seasonal work, has adversely affected the security and health of seasonal Zimbabwean workers, especially women. Female and child labour has increased, and piecework has become more common. These insecure, non-permanent, seasonal workers have below-average health status, as reflected in higher infant mortality, lesser use of health clinics for child diarrhoea and other illnesses and lower immunisation rates. Non-permanent workers often dare not take time off work to attend to their, or their children's, health needs because of the resultant loss of pay and the possibility of losing their job. Because the tempo of rural production is now more controlled, and systematic, the opportunities for casual work occur less often each year. Family stresses and poor child health are thus increased by a system that makes no concessions to the traditional rhythm of rural family life in Africa.

8.3 Food: hunger, food insecurity and malnutrition

8.3.1 *The political profile of hunger*

The world food situation has, for decades, abounded with contradictions. Scarcity in the midst of abundance; millions of famine deaths among the poor while overfed minorities die of obstructed coronary arteries; farmers in rich countries paid by government *not* to produce food; and the commercialised export of luxury foods by poor countries while their local rural populations remain malnourished. Clearly, feeding the world's population is not simply a matter of agricultural management. Now that a global market economy has been established within the framework of Western-dominated international capitalism, the goals and methods of

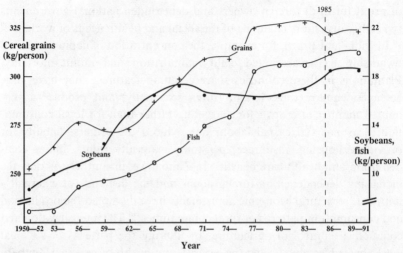

Figure 8.1. Global per-person production of cereal grains, soybeans and fish, 1950–91. (Data source: refs. 2, 42)

food production and sale have been progressively redefined. Markets, ostensibly 'open', are largely controlled by the world's rich and powerful – i.e. the developed countries, transnational companies and urban elites in poor countries. Further, in some poor countries, governments acquire cheap food from powerless peasants to placate restless urban crowds. Increasingly, food is grown for profit, and not to meet nutritional needs. Viewed from a radical perspective, food shortages are readily attributable to systematised exploitation via the reorganisation of food production and marketing by international capital, and its attendant political and economic inequities.[21] However, that interpretation – in many respects a compelling one – is not the end of the story. Even in a fair and egalitarian world, and one in which agricultural practice is attuned to local ecosystems, there are *ecological limits* to how much food can be produced. Technology and biotechnology may roll back those limits, but the basic question remains: When will we reach those limits?

The global per-person production of food has increased substantially over the past three decades, although, as we shall see below, the figures have faltered recently. Total food availability rose from an average of around 2,200 calories per person in the early 1960s to 2,700 calories in the late 1980s.[3,22] This gain incorporated a substantial increase in total food production in the developing world. The increased production was mainly due to the introduction of high-yielding cereal strains, to huge increases in the use of chemical fertilisers and pesticides, and to a doubling of the world's irrigated cropland.[15] The world's output of grain (the major food category, dominated by wheat, rice and corn) has increased faster than has population, and so the per-person grain production has risen – from around 260 kg in the 1950s to 325 kg in the 1980s (Fig 8.1).

Many poor countries and hundreds of millions of poor people have not shared in that calorific abundance. On average, people in the richest countries eat one-third more calories than they need (average 3,400 calories), while those in the poorest nations (in the range 1,600–1,900 calories) eat 10% less than the basic minimum needed for full biological function.[22] Gains in per-person food production over the past four decades have, paradoxically, been accompanied by increasing hunger and malnutrition; production is thus not the same thing as access. As agriculture in developing countries becomes integrated into the world economy, with food being increasingly commodified, so it becomes less available and less nutritionally relevant to local rural communities. As agribusiness displaces traditional rural smallholders throughout the Third World, and as populations swell, so they are less able to afford to buy food.[21] Hunger then ensues.

Although the proportion of undernourished people in the world population has almost halved since mid-century, the absolute numbers have continued to edge upwards. In Africa, around one-quarter of the population lives in a state of chronic hunger. In 1992, the Food and Agricultural Organization (FAO) estimated that there were approximately 800 million seriously undernourished people – around one-fifth of the Third World population. This figure includes 200 million children, who are therefore destined for stunted growth, vulnerability to infectious disease, and, perhaps, intellectual impairment. International agencies estimate that more than one in three Third World children below age five are malnourished.[23] In countries such as India, Pakistan, Afghanistan, Bangladesh, Burma, Kampuchea and in parts of Africa over one-quarter of babies born are underweight. Approximately 2–3 million children die each year from hunger and another 5–8 million die from hunger-assisted infectious diseases. Malnutrition impairs immunity, leading to a vicious cycle in which repeated infections weaken and cause metabolic disturbance – and further malnutrition. WHO reported, in 1992, that the prevalence of acute childhood malnutrition has been falling in Asia and Latin America, but rising in sub-Saharan Africa. Worldwide, the total number of acutely malnourished children is still increasing.[24]

The primary cause of this hunger is the systemic poverty that abounds in Third World populations. The transformation of much traditional subsistence agriculture into cash-cropping for export has exacerbated the poverty-induced hunger in many such populations. However, these basic social–political inequities as a cause of hunger are supplemented by, and must be distinguished from, ecological causes. The latter result from an overloaded environment, entailing either a degradation of agricultural resources or an overwhelming of local food supplies by population growth. Meanwhile, intense ideological debate continues to surround the issues of food production, distribution and the causes of hunger.

8.3.2 Recent trends in food production

In contrast to the longer-term trends, the world's cereal grain output plateaued in the latter half of the 1980s (Fig. 8.1). More importantly, since the peak production year in 1984, grain production *per person* has declined by approximately 1% annually.[2] The downturn in grain production appeared first in Africa around 1970, followed by downturns in Eastern Europe and the Soviet Union in the late 1970s, in Latin America during the early 1980s and in North America in the mid-1980s. More recently, some downturn has occurred in Western Europe and West Asia. Global production of root crops peaked in 1984 and has declined since, while per-person production of meat, milk and fish plateaued during 1986–91. Production of fruits, vegetables and oils has continued to edge upwards.[23] Has the bubble of indefinitely increasing food supply burst? Or is it just a temporary blip? While the overall situation hangs in the balance, the outlook for Africa is grim. For the rest of the developing world, food production prospects are uncertain.

The decline in per-person grain production has occurred because of a shortage of new cropland for cultivation, degradation of some existing cropland, reduced effectiveness (and affordability) of additional fertiliser and a slowed growth in irrigation. The plateauing in soybean production, the world's major protein crop (Fig. 8.1), reflects a lack of cropland and the fact that, as a nitrogen-fixing legume, soybeans are less responsive to chemical fertilisers. Likewise, the levelling off in meat production reflects increasing reliance on expensive grain-feeds as the availability and condition of rangelands declines.

In Africa, per-person food production has declined despite a substantial increase in cropland. This 20% decline since 1970 contrasts with Latin America and West Asia, where the figures have held constant, and East Asia where increases of 20–50% have occurred.[25] Africa became a food importer in the 1970s, reflecting the combination of rapid population growth in many newly-independent countries and widespread economic and environmental mismanagement. By the 1980s, cereal imports had increased four-fold, their costs ten-fold, and parts of Africa have since become heavily dependent on food aid. Today, Africa has the world's highest proportion of degraded agricultural land, exacerbated by rapidly increasing demand for fuelwood and grazing land. Millions now face serious malnutrition in the immediate sub-Saharan countries, particularly Ethiopia, the Sudan and Somalia. In parts of Africa, children under age five years make up one-quarter of the population but account for up to three-quarters of all deaths – the comparable statistic in Europe is 3% of all deaths. Clearly, the challenge of attaining food security in Africa is huge, and strict political and economic programmes that are attuned to ecological sustainability will be needed. Solutions to these complex social–political–ecological problems will nowhere be simple.

8.3.3 Anatomy of a multifaceted problem

Much of the growth in world food output during the 1970s and early 1980s came from the combination of energy-intensive mechanisation, increased fertiliser use, the farming of marginal land and the overpumping of slow-filling underground aquifers. By 1990, however, additional fertiliser seemed no longer to be conferring much gain in output and its use actually began to decline in the early 1990s.[26] Depletion of freshwater supplies is now becoming a widespread and serious limitation, particularly because of the ominous falls in underground water tables; beneath important grain-producing regions. These vast subterranean oceans of water fill only slowly – but can be emptied rapidly by thirsty agriculture. Agriculture also faces other problems specific to certain regions. In industrialised countries there is an increasing toll on agricultural yields from air pollution, particularly photochemical oxidants.[27] In many Third World countries, including Bangladesh, Brazil and China, flood damage due to upstream deforestation is becoming a threat to agriculture. As we have seen in chapter 6, a sea-level rise would cause loss of coastal farmland.

In the late 1980s, FAO concluded that to achieve an adequate increase in grain production in developing countries this decade would require opening up an extra 5% of arable land (83 million hectares), cropping more frequently, and increasing the use of fertiliser, tractors, and improved seed stock.[28] Critics argued that this would be squeezing a system that is already overworked, mismanaged and near to its ecological limits. Where would all that extra land be found? FAO's proposal to win new cropland from tropical forests not only overlooks the generally short-lived gains from that source (because of the intrinsically poor soils) but blithely ignores experiences such as those in Brazil and Central Asia where extension of land-use into forest areas resulted in substantially increased leishmaniasis infection of human settlers. The same FAO study also argued that increased meat production in the Third World during the 1990s would require intensification of production methods. Yet there is good evidence that sustainable animal husbandry on semi-arid lands requires *extensive* pastoralism, not radical shifts to imported *intensive* methods of husbandry appropriate for more fertile lands. Overgrazing of semi-arid lands has already caused widespread and extreme environmental problems.

Increases in meat production would also require improved control of disease vectors such as the tsetse fly (which transmits trypanosomiasis, a debilitating parasitic disease, to cattle in the warm humid regions of Africa). Previous attempts at eradication, often using saturation spraying with organochlorine pesticides, have had little impact. Even if vector control succeeded, it would lead to the clearing of vast areas of forest and savannah and their replacement with short-lived and ecologically de-structive cattle-ranching practices – such as have occurred widely in the Brazilian rainforests. Further, increased meat production would probably reduce food availability for local populations, since good quality land can yield up to ten times more protein if used to grow crops than if used to graze cattle.

8.3.4 The Green Revolution

During the 1960s, Western-dominated international aid and development agencies looked to intensified agriculture to feed and stabilise restive rural populations. Population growth rates were increasing, and the last major agricultural land frontiers had been occupied. Food-sufficient communist China was beginning to cast a long political shadow, despite the massive setbacks and famine deaths of rural collectivisation around 1960. This was also the time when many newly-independent nations were wrestling with the choice between an intensified dependence upon imported cheap grain from the perennial surpluses of the USA, the world's dominant grain exporter, and the development of more mechanised food production for local consumption and for export earnings.[29]

Newly-developed, laboratory cross-bred, 'high yield' seeds seemed to hold the key to a new agricultural revolution – echoing the impact of accidentally-discovered high-yield grass seeds upon the original agricultural revolution 10,000 years earlier. From the 1960s to the mid-1980s, these seeds brought impressive, sometimes spectacular, gains in local yield. However, the label 'high-yield' is a bit misleading, since you can't get something for nothing. Genes, after all, are merely molecular instructions. For genotype to become large-grained phenotype, these seeds required increased inputs of water and fertiliser. According to the Philippines-based International Rice Research Institute, during the latter 1980s the growth in rice yields virtually ceased.

Despite the considerable initial success, particularly in India and the Philippines, the Green Revolution has been a mixed success, and its main gains may have already been realised.[25] The newly abundant crops of the Punjab – part of India's transformation from 'a begging bowl to a bread basket' – were celebrated as the jewel in the Green Revolution crown. However, by the end of the 1980s, the success story had tarnished somewhat as the Punjab struggled with weakened soils, waterlogged deserts, pest infestations and indebted farmers.[30] The reliance on high-yielding and intensively-farmed monocultures has reduced ecological stability by eliminating the traditional mixing and rotating of crops (including nitrogen-fixing legumes), by reducing the yield of straw fodder and through increased crop vulnerability to pests and disease. Local water sources have been depleted; in parts of the Punjab the water table is declining at an estimated one-third to one-half a metre annually. In Mexico, where the Green Revolution conferred a four-fold increase in grain production, the amount of usable cropland shrank by one-tenth during the 1980s.[31]

The Green Revolution also brought social problems. Mechanisation has displaced landless labourers. Absentee landlords have taken advantage of the higher profits on offer, increasing their rents and evicting many impoverished smallholders.[21,32] In Korea, the introduction of 'miracle rice' caused farm debt to rise ten times faster than income and assets between 1975 and 1985, with disastrous social consequences.[33] Thus, much of the Green Revolution centred on the richer, larger landowners – not surprisingly, since they were better equipped and better educated for taking risks. The Green Revolution should thus be seen in relation to the political and economic forces that have shaped the modern world's food production; the necessary technical inputs of the Green Revolution have increased the control of food production by international agribusiness. A major consequence, as populations increasingly become integrated into a world market economy that sells them commodified food, is that diets have been disconnected from the local agricultural ecology and redefined through distant markets. Viewed thus, the Green Revolution looks less 'green', less ecologically authentic. However, experience in other pro-grammes has shown the gains that can be achieved by focusing on the rural masses of small landholders. Indeed, the World Bank has endorsed the view that a 'Patchwork Revolution' among small farmers could increase overall food production, while also alleviating Third World poverty and hunger.

Laboratory scientists working on genetic engineering hope that a new generation of higher-yielding crops will be created: a Gene Revolution to extend the Green Revolution. New high-yield foods suitable for Africa, such as pest-resistant, non-thirsty cowpeas, may yet usher in a green revolution there. Related developments with sweet potatoes, cassava and soybeans hold similar promise,[34] as do improved strains of triticale (a cross-breed between rye and wheat, able to grow in arid lands and poor soils).[25] The problem now is merely converting theory to practice – typically in settings of poverty, political ineptitude and civil strife!

8.4 Agriculture, food and health: other aspects

8.4.1 Patterns of consumption; dietary transition

As material standards of living rise, people replace plant foods with animal foods. Better-off urban minorities change their diets ahead of the poor majority. The rate of change has been particularly high in recent years in China, Eastern Europe and the former USSR. In rich countries, however, a plateau has now been reached, with around one-third of dietary energy coming from animal products and with substitution of sugar (refined carbohydrate) for traditional staple carbohydrates. The poorest countries continue to rely heavily on cereal grains, roots and tubers.

The increased demand for animal foods requires an increase in feed-grains, particularly when pasture and rangelands are limited. Animal feed will therefore account for about one-quarter of all cereal grains consumed in developing countries by the year 2000, compared to one-sixth in 1980.[35] It already accounts for about half the grain produced in Europe, North America and Russia. Further, as export-oriented agriculture replaces traditional agriculture in the debt-ridden Third World, large quantities of soybeans and coarse grains are sold by poor countries to rich countries as feed-grains, with various social and economic repercussions (see Fig. 8.2). The conversion of cereal grains and other food concentrates to animal products involves large losses of edible energy (the losses are greatest for cattle, intermediate for pigs and least for chickens – for whom grain is a

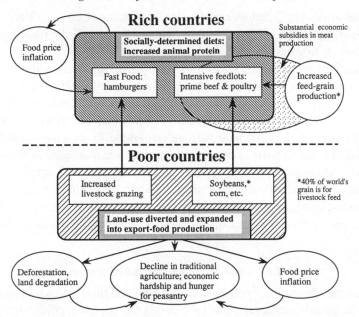

Figure 8.2. Relationship between the promotion of high-meat diets in rich countries and patterns of agriculture and land-use in both rich and poor countries. Other major social and economic consequences are indicated. (Based on a teaching diagram by P. McMichael to address the question: 'Does the North eat the South?')

natural food). Producing animal protein thus requires surplus photosynthetic energy, while also increasing the pressure on marginal grain-growing land and accelerating soil degradation. Such a resource-intensive diet would not be sustainable at a global level. A lesser meat content in the diet, particularly if it is meat from grazing and foraging animals, is more likely to be ecologically sustainable since those animals produce protein by eating food resources that are of little nutritional value to humans.

In ecological terms, it is a pity that imported grazing animals are generally preferred to indigenous foragers. For example, Caucasian Australians and Americans maintain a European preference for eating sheep and cattle – even though kangaroo and bison were the indigenous meat sources on offer. Eating meat from these ecologically benign animal sources would have lessened the damage done by the introduced edible species to forest, woodlands and ranges in both Australia and the USA. As we have seen in chapter 4, it would also lessen the dietary content of saturated fats. The beef culture is, however, culturally and commercially strong, and is embedded in ancient religious practices (e.g. the Golden Calf of the wandering Israelites and the sacrificial bulls of Classical Greece) and quasi-religious rituals (e.g. the bull fights of Spain and the rodeos of the American Midwest).

Similar environmental dissonance results from the profound influences of urbanisation and commercial advertising upon preferences for foods that are easy to prepare. In developing countries traditional staples, such as tubers, millet, maize and sorghum, are being replaced by wheat and rice – although the local climate may be less suited to growing those latter grains. Hence, in parts of eastern Africa there are unsaleable surpluses of maize, while attempts to grow wheat and rice have had mixed success, often causing environmental damage (particularly from irrigation) and requiring costly imports. As economic imperatives and 'internationalised' consumer preferences prevail, so traditional ecological wisdoms may be lost.

8.4.2 *Food safety and microbial ecology*

Food contamination by chemicals, microbes and their toxins is an age-old public health problem. New problems arise as food production, distribution and storage patterns change. These problems can thus be viewed within an ecological framework.

Food-borne microbiological diseases appear to be increasing around the world, both in poor and rich countries.[24] In poor countries, diarrhoeal diseases remain a massive problem. An estimated 70 % of these cases are due to food-borne pathogens; the rest are due to water-borne pathogens. In rich countries, steady improvements in personal hygiene, water supply, vaccination programmes and food control procedures (such as pasteurisation of milk) have hugely reduced food-borne diarrhoeal diseases. Nevertheless, many countries have recently experienced an upturn in various other food-borne diseases. In the USA, around 6.5 million cases are now reported each year, resulting in 9,000 deaths; the actual number of cases may be ten times higher than those reported.[24] In former West Germany, the annual reported number of cases of infectious enteritis (bacterial bowel infection) surged from around 5,000 in the early 1960s to 70,000 in the late 1980s.

Salmonella food poisoning has increased markedly in both North America and Europe in recent years. Poultry and meat, and in some countries eggs and dairy products, are the predominant source of this pathogen. Other relatively new food-borne pathogens now causing problems in developed countries are *Campylobacter* and *Listeria*, which contaminate meat and milk, or meat, milk and vegetables, respectively.[24]

These contemporary forms of 'food poisoning' reflect recent shifts in patterns of human food production and storage. The increasing demand for ready-to-eat foods and the vast, cost-efficient scale of commercial food production plants create new ecological opportunities for food pathogens.

Imported and commercially produced feedstuff for livestock appears to be the primary source of subclinical infection in intensively bred calves, pigs and poultry. Since World War II, animal husbandry has required enormous quantities of feed, made from diverse animal sources (blood, bone, feathers and fish) and vegetable sources. Much of it has been imported from tropical and subtropical countries and is contaminated with pathogens, including *Salmonella* and *Campylobacter*. Infected (carrier) animals readily infect other animals by shedding pathogens into feedlot soil and water – and also at the time of slaughtering and meat processing. The *Listeria* organism is able to grow at refrigeration temperature and in a wide range of pH, and has been a particular problem in cheeses undergoing extended cold storage. These recent problems with *Listeria* illustrate the continuing ability of microbes to adapt to and exploit changes in human ecology.

8.4.3 Air pollution, climate change and agriculture

Various human-made air pollutants, such as sulphur dioxide, nitrogen oxides and ammonia, impair agricultural productivity – and, via this route, present an 'ecological' risk to the wellbeing of human populations. Levels of tropospheric ozone, a major component of photochemical smog, have become high enough over the eastern USA during summer to damage crops and vegetation. Damage to crops may occur if ozone levels of at least 0.05 ppm persist for several weeks during the growing season.[36] (The exposure standard, for human respiratory health, is around 0.10 ppm.) The ambient ground-level ozone concentrations that prevailed during the 1980s caused an estimated 5–10 % reduction in crop yield in parts of the USA.[37]

Finally, in addition to the problems of feeding the world's expanding population from a limited and degradable agricultural resource base, agricultural productivity will also be affected by global climate change and by depletion of the ozone layer. These problems have been discussed in chapters 6 and 7.

8.5 Water resources

8.5.1 Introduction

Population growth and economic development, in combination, have caused the global use of freshwater to increase almost ten-fold this century.[38] Yet 98 % of the world's water is saltwater and much of the rest is locked up in polar and glacier ice masses. Most of the remainder lies underground, as freshwater aquifers. For aeons, humans have depended on the tiny portion of accessible freshwater (around 7,000 cubic kilometres – less than one five-thousandth of all of the world's water) that fills our rivers and lakes.[39] That tiny portion, plus sustainable use of aquifers, would – if evenly distributed and efficiently used – suffice for the human population's reasonable needs over the coming decades. Worldwide, those needs are: irrigation, which accounts for over two-thirds of water consumption; industry, which accounts for one-quarter; while the remainder is available for domestic, recreational and livestock use. That 'remainder' includes the basic biological requirement of humans for sufficient safe water to drink.

By early next century, our total consumption and pollution of naturally accessible freshwater will exceed supply, although the construction of dams is relieving that pressure (while creating others!).[40] Further, freshwater is *not* distributed evenly around the world, and it is often subject to seasonal and yearly fluctuations. In rich countries, people use approximately 500 litres of water per day, compared to around 50 litres in the poor urban communities of developing countries, and five litres in the rural areas of some of the very poor countries. World Bank figures indicate that in North Africa and the Middle East all the available freshwater is likely to be being consumed by the year 2000.[3] The Libyans have just completed a massive (and controversial) 1000-kilometre pipeline to bring underground water from the desert hinterland to the coastal population centres. Within a decade most of eastern Africa will face severe water shortages, as will Poland, Israel and parts of the USA. In several Middle East countries – including Bahrain, Democratic Yemen, Kuwait and Syria – the total demand for water already equals or exceeds the supply. Saudi Arabia has begun the very expensive process of distilling saltwater which, as an oil-rich giant, the Saudis can (currently) afford. Conflicts over water supply loom larger with the passage of time and the swelling of populations. Syria and Iraq fear that Turkey will reduce the flow of water down those historically famous rivers, the Tigris and Euphrates. Israel and Jordan are bristling over access to the waters of the River Jordan. Egypt worries that the Ethiopians will partially divert the life-giving flow of the Blue Nile.

Approximately one-quarter of the Third World lacks safe drinking water. Rural populations fare worse than those living in towns and cities. In the worst affected countries, three-fifths of urban dwellers can get clean water compared to only one-fifth of rural dwellers.[24] Because of lack of money and technical resources, many of the pumps, wells and latrines installed during the 1980s water-and-sanitation decade have not been maintained. In addition to malfunction, incompleted connections and losses of water are widespread. The continuing toll of disease and death caused by water-borne infections, particularly in infancy and childhood (see chapter 10), reflects the increasing pressures on supplies of clean water, a basic necessity of life. The increasing demands of family, factory and farm mean that shortage of water is becoming a worldwide problem, particularly in poorer countries. Burdened by high population growth rates, these agriculturally-based countries need more water and fertiliser for food self-sufficiency. However, the essential needs of domestic water consumption and the increasing demands of industry often combine to deprive farmers of the water they need.

There are, however, enormous possibilities for using more of the vast quantity of freshwater that falls on land and runs off to the sea. Archaeologists have uncovered small-scale methods of water and soil management from ancient cultures in arid lands in the Middle East, northern Africa, Arizona, Mexico and South America. Those ancient methods evolved to ensure sufficient productivity in arable lands that were subject to chronic water shortage – not to make the desert bloom. They suggest the potential of numerous well-sited small dams. Another major potential gain lies in finding ways of conserving and recycling water, including improving the efficiency of irrigation. Only about one-third of irrigation water actually helps crops to grow; as with agricultural pesticide use, most of it does not get to its intended target. Human communities have achieved those sorts of adaptive responses before, under pressure. We should be able to do it again, so long as next century's total human demand does not outstrip the total supply!

8.5.2 Irrigation

Along with carbon dioxide and sunlight, water is the other main requirement for plant growth. Since much potentially arable land in tropical countries has insufficient moisture, irrigation has transformed agricultural production in many parts of the developing world, particularly India and China. The amount of irrigated land has increased five-fold this century and more than doubled between 1950 and 1978 – when the world's per-person irrigated area peaked. Since then, growth of irrigation has slowed. Today, one-third of all food is grown on the one-sixth of cropland that is irrigated, including most of the world's vegetable and orchard land.

Irrigation, often a star performer in the short-term, has a number of adverse ecological effects in the longer-term. It typically causes salinisation and/or alkalinisation of agricultural land, which has fundamentally affected the fortunes of past civilisations. Around today's world, some 40 million hectares of irrigated land (i.e. about 2% of total arable land, an area about the size of unified Germany) are either waterlogged or suffer from excessive salinity.[23] The irrigation of rice fields also causes the production of large amounts of the greenhouse gas, methane.

Irrigation can alter the distribution of disease vectors, particularly the water snail that transmits schistosomiasis. Historically, in semi-arid Mesopotamia and Egypt, irrigation was well established by the third millenium BC and, not surprisingly, reference is made to schistosomiasis in ancient Mesopotamian and Chinese literature.[41] Increases in malaria, schistosomiasis, onchocerciasis (river blindness), trypanosomiasis and various arbovirus infections have been observed following major irrigation and dam projects around the world – in India, Thailand, Puerto Rico, the USA and many parts of Africa.[24] Consider three examples. In northeastern Ghana, an 'explosive' three-fold increase in schistosomiasis occurred in communities around the many newly-built irrigation dams in the early 1960s. In Asia and Latin America, the extension of irrigated agriculture has contributed to recent increases in malaria. After a large irrigation scheme was built on the Cukurova plain of Turkey in the 1970s, endemic malaria recurred because poorly drained ditches allowed increased breeding of malarial mosquitoes.

Looking beyond the immediate health effects, water is the limiting resource for agricultural productivity in the world's five most populous countries: China, India, the former USSR, the USA and Indonesia. Attempts to create new high-yielding supercrops, by genetic engineering, will count for little if water supplies are inadequate. Water is 'the most critical natural resource for future agricultural development in the Middle East, southern Europe, Egypt, the Sudan, sub-Saharan Africa, Taiwan, Pakistan, Australia, Argentina, Brazil, most of Canada and Central and South America.'[34] Yet, in many of these countries, there is an insidious depletion of water supplies, as underground aquifers are emptied much faster than they can refill.[5] These include the world's non-renewable stores of 'fossil' water, mostly laid down during the last ice age.

The US Department of Agriculture has reported that over one-fifth of all irrigated cropland in the USA is watered by depleting the water tables. The vast Ogallala aquifer, underlying eight of the American Midwest states, is falling at up to a metre per year. In some areas it has now fallen by between 100–200 metres – almost the full thickness of the aquifer – and this has contributed to stopping the irrigation of about one million hectares. In cereal-growing Tamil Nadu, southern India, the water table has fallen by an estimated 25 metres over the past decade. Bangkok's water table has plunged 25 metres over the past four decades, saltwater has penetrated its wells, and the city itself is sinking measurably. Under parts of the North China Plain the water table is sinking by 1–2 metres per year, and Beijing and many other cities are threatened by water shortages.

The other major problem with the world's drinking water is that, beyond the age-old problem of faecal contamination, there is increasing pollution by chemicals from domestic sewage, industrial effluents or agricultural runoff. Many of the world's rivers and waterways have become industrial sewers. The age-old convenience of running water to flush away industrial effluent may 'externalise' the cost of industry's wastes, but it is a growing burden on the environment. The problems in the major rivers of Europe and the USA are well known. Periodic chemical spills into the Rhine have decimated aquatic life in large stretches of the river (indeed, on one occasion the river was declared a fire hazard). During the 1980s, largely untreated effluents from industries along the Volga made up one-tenth of the river's average flow at Volgograd. Nitrate concentrations in many of Berlin's drinking water wells exceed the European standard by a factor of 10–25, and its river water contains excessively high levels of phosphate, ammonium, benzol, phenol and zinc. Industrial effluent and heavy use of chemical and natural fertilisers on farmlands around Berlin is the source of much of the problem.

In developing countries, industrial waste is often released into rivers, lakes and streams – the source of untreated water for drinking, cooking and washing. Asia's rivers are thought to be the most degraded in the world. In Shanghai, the Huangpu river, flowing through the heart of the city, receives enormous volumes of industrial effluent and has been biologically dead for over a decade. An estimated one-quarter of the length of the mighty river Ganges in India is now seriously polluted with industrial and human wastes; few industrial plants along its banks have waste treatment facilities and many of the major cities have no sewage treatment plants. Similarly, the Kalu River in Bombay receives a noxious cocktail of wastes from approximately 150 industrial plants along its banks, and its water is then used by countless residents of slums and industrial suburbs.

8.5.3 Hydrological follies

The temptation to harness major rivers for irrigation and hydroelectric power is great. Why let all that water pass unused? Currently there are grand plans for damming some of the world's great rivers in developing countries. In India, despite some earlier spectacular failures, there are controversial moves to dam the Bhagirathi River, in the western Himalayas, to provide water and power on a huge scale for at least a decade. Beyond that time, however, there is no guarantee that the usual process of silting – perhaps exacerbated by earthquakes and landslides that abound in the geologically adolescent Himalayas – will not steadily reduce the dam's yield. A similar huge scheme, for agricultural and hydroelectric purposes, is planned at the Three Gorges site on the Yangtze River in China. That massive 200-metre high dam – long fantasised about by nationalist leaders and approved in 1992 by the Chinese government – will displace a million Chinese living in a thousand villages and will generate twice as much hydroelectric power as any other dam on Earth. Its effects upon downstream irrigation and siltation remain a matter of dispute, as do its effects on local settlements and the destruction of local ecosystems.

Africa provides many examples of expensive and environmentally damaging failures. Ghana's huge Akosombo Dam, built during the 1960s, was to have transformed the country into an industrial nation with cheap electricity and extensive irrigation. Its construction displaced 80,000 people from 700 towns, but it failed to generate the wealth needed for irrigation development. Agricultural economists point to many instances when dams and irrigation schemes have adversely affected agricultural and fisheries productivity. The Aswan Dam, in Egypt, has caused decimation of the eastern Mediterranean fisheries, a substantial increase in schistosomiasis in irrigation workers, and the trapping of much of the silt that annually rejuvenated the agricultural flood-plains of the Nile delta.

There is, with hindsight, folly in many of these big water projects. Many have been undertaken to meet local political agendas, to satisfy the World Bank or to entice big foreign investors. The ambitious South Chad Project sought to turn the sub-Saharan desert of Chad, in north-central Africa, into a fertile garden. Naively, it was planned while water levels were high, and, subsequently, the lake has shrunk by 90% because of prolonged drought. Folly has not been confined to the developing countries. In Australia, research in the 1940s and 1950s indicated that the Ord River, in the remote north-west, could be harnessed to irrigate cotton crops. In 1972, the Ord River Scheme was 'opened'. By the late 1970s, it became apparent that the cotton crops would not bloom. Sorghum and rice were then tried; vast amounts of pesticide and fertiliser were applied. However, nature refused to come to heel, and the scheme was finally abandoned in the late 1980s. In the USSR, Brezhnev talked in the 1970s of reversing the flow (with the help of a few nuclear detonations) of several of Russia's northern rivers to irrigate the semi-arid southern republics. In Eastern Europe, half-completed dams along the Danube have recently been abandoned, after it transpired that there were adverse ecological and economic consequences downstream.

From decades of mixed success and environmental disruption, it is clear that the engineering response to the basic human need for water should now be tempered by ecological insight and a dash of humility. As with electricity generation, we should approach water and soil management with a greater emphasis on decentralised, small-scale techniques. In the 1950s, heroic visions were nurtured, celebrating man against nature, making the deserts bloom and harnessing the power of mighty torrents. Now the mood is changing, as threatened communities counter central authorities. The Cree Indians in Quebec, for example, are disputing the Province's proposed hydro-electric megaproject which will destroy their way of life and may poison their fisheries with newly-leached mercury. As we gain insights into these ecological complexities, so we will display less hubris in our approach to the management of the world's freshwater supplies.

8.6 Fewer fish for food?

Humans contributed to the extinction of edible land animal species 10–15 thousand years ago. This century, human hunting pressures in the world's oceans have begun to threaten many fish species with extinction. Of around 20,000 fish species, many of which are localised and relatively rare, only about 20 are taken in large commercial numbers. Those fish species are of vital importance to many human populations; they account for approximately one-quarter of the world's total consumption of animal protein – including up to half the intake in poor countries, and even more in Japan and Norway. The world's total fish catch has increased five-fold since 1950 (see also Fig. 8.1). During the 1980s, as human numbers increased rapidly, the fish catch grew to exceed the critical level of 100 million tonnes annually – estimated by FAO to be the maximum sustainable yield.[23,42]

Between 1988 and 1990, the total fish catch declined in nine of the major fishing areas, stayed stable in three and increased in five. In half of the world's ocean fisheries, the catch now equals or exceeds the estimated sustainable yield. All commercial fish stocks in the northwest Atlantic including the once-bountiful cod, haddock and herring stocks have been overharvested, and catches there are now in serious decline. The bluefin tuna population of the western Atlantic has declined by over 90 % over the past two decades – but the catch continues, buoyed up by prices of US$200 per kilogram in the Tokyo sushi market.[4] Declines in various northeast Atlantic fish stocks are following suit. Meanwhile, fishing fleets have refocused their attention on the northwest Pacific, where catches are still increasing. Fisheries elsewhere in the Pacific and throughout the Indian Ocean are now at or near their maximum sustainable yield. The extent of some of the declines has been spectacular. For example, the once huge anchovy harvest off Peru plunged sharply in the early 1970s, and has subsequently collapsed to one-hundredth of what it had previously been. A similar collapse has occurred with the Mexican Pacific-coast abalone fishery and the Californian sardine fisheries. The Australian fishing industry has peaked and declined over the past two decades, despite a steady widening of the actively-fished coastal area. The Australian government judges that the development of deep-sea fishing could not sustainably offset the coastal overfishing.

Many Third World coastal waters are now becoming depleted of staple fish stocks. Much of the depletion is due to rapidly-expanding local fishing capacity; some is due to unprincipled illegal encroachments by fishing fleets from rich foreign countries. On a visit to West Africa in 1991, I enquired about several boats on the not-too-distant horizon and was told they were probably Portuguese or Taiwanese boats illegally plying unpoliced local fisheries. These bountiful West African fisheries have already been overexploited, causing their species composition to change markedly.[23] More generally, while dollar-earning exports of fish by developing countries soared during the 1970s and early 1980s, as overfishing bites deeper the catches are plateauing.

Depletion of the world's fisheries is not just a consequence of overfishing. It is compounded by the wasteful killing of fish, mammals, turtles and seabirds by drift nets – both those under commercial control and those abandoned. (Is this to be a replay of the wasteful overhunting of the large-bodied mammals by pre-agrarian humans?) There is also the worldwide problem of toxic effluent (containing various organic chemicals and heavy metals) that kills and contaminates marine species. Nearly all of the world's catch is taken from within 200 miles of land, on the coastal shelf, and so increases in coastal pollution are becoming a significant problem. A further problem arises from the continuing destruction of coastal wetlands. Wetlands stabilise and control water cycles, and they remove toxic metals from the water (for example, various algae absorb toxins during the day and then sequester them in lower layers of bottom sediment overnight). Wetlands are breeding grounds for many fish species and they provide sustenance to adult fisheries. However, they are being systematically destroyed by land reclamation schemes that entail the filling, drainage and clearance of mangroves in order to create conventional farmland, construction sites and recreation resorts.

A further concern is that changes in the temperature and patterns of circulation of heat and water would perturb fish populations. It is well known that such fluctuations – such as that of the El Nino Southern Oscillation across the Pacific Ocean – cause changes in the distribution and size of fish populations. If warming of the oceans by several degrees occurs over the coming century, this may disturb feeding and reproduction patterns of marine life and make the waters uninhabitable for some species. Migration patterns, dependent on temperature and ocean currents, would be disrupted. An even more basic effect of ocean-warming will be to deplete, through malnutrition, the huge biomass of phytoplankton that supports the entire aquatic food chain (see also chapter 6).

International attempts to manage and safeguard fisheries have been of limited success. The Law of the Sea Convention, while still not widely ratified, seeks to bring nearly all of the world's catch under national jurisdiction via the 200-mile Exclusive Economic Zones. However, fishing fleets from rich countries continue to encroach on the waters of poor small countries. The world's fish stocks are thus under increasing pressure. Short-term solutions to the decline in edible fish stocks include switching to currently unpopular species such as octopus and squid, or, perhaps, harvesting the abundant shrimp-like krill in the Antarctic Ocean. (Krill harvesting would take food from the mouths of whales, however. As on land, the greater the harvest by humans the fewer the resources for other species.) Fish farming may ease the pressure, and already represents about one-eighth of total fish production. It has enormous implications for feeding the human population – as did the advent of agriculture 10,000 years ago! We would then have tried both loaves and fishes.

8.7 Summary

As the world moves into the 1990s, the supplies of food seem to be hanging in the balance. Much of the problem is distributional, but, increasingly, it is an ecological problem. Despite the early successes of the Green Revolution there are signs that agricultural systems are now being pushed towards their productive limits. During the mid-1980s, grain production plateaued in some of the world's most populous countries, while it continued to fall in Africa. The amount of cropland available for growing food is declining; 5–10% of the world's arable land was rendered unproductive during the 1980s, and much other intensively-farmed cropland is approaching exhaustion or depends on unsustainable use of groundwater. Some of the world's poorest countries, such as Bangladesh and many African countries, have daunting tasks ahead of them – soluble only with unprecedented international assistance. Meanwhile, we are burning the candle at both ends: while human populations continue to grow, particularly in areas where soils are relatively less fertile, so the land resource base used for farming and grazing is degrading – and the supply of freshwater, both surface and subterranean, is dwindling. Further, we have begun to overfish the seas.

There are hopes that something akin to the quantum leap forward of the Green Revolution can be repeated. But, given the current agricultural technology and the available genetic strains of plants, tempered by the limited base of usable land, this is not assured. While dramatic break-throughs in genetic bioengineering could yet buy us more time, it is hard to sidestep the argument that, as we destroy the environmental base of food production, so we inevitably reduce the world's capacity for future food production. Further, centralised large-scale approaches to agriculture in developing countries have often damaged the social fabric of local communities and the viability of small farmers, and have led to declines in community health and health care.

Increasing numbers of people are malnourished. Some of this is because of ecological disruption of the food production base, and some of it is because of the inexorable growth of populations in the poorer, food insecure, countries. For most of the well-fed and the overfed in the rich countries, these are, for the moment, little more than statistics. However, during the next century, the confluence of climate change, static agricultural yields, failing water supplies and fewer fish may well bring the statistics closer to home.

References

1. Brown LR. The New World Order. In: *State of the World* 1991. *Worldwatch Institute Report*. New York: Norton, 1991, pp 3–20.
2. Brown LR. Grain Harvest Drops. In: Brown LR, Flavin C, Kane H (eds). *Vital Signs: The Trends that are Shaping Our Future*. New York: Norton, 1992, pp 24–5.
3. World Bank. *World Development Report* 1992. *Development and the Environment*. Oxford: Oxford University Press, 1992.
4. Meadows DH, Meadows DL, Randers J. *Beyond The Limits*. London: Earthscan, 1992.
5. United Nations. *Global Outlook* 2000. *An Economic, Social and Environmental Perspective*. ST/ESA/215/Rev 1. New York: UN. 1990.
6. FAO. *Sustainable Agricultural Production: Implications for International Agricultural Research*. Research and Technology Paper 4. Rome: FAO, 1989. And: Postel S. Halting land degradation. In: *State of the World* 1989. *Worldwatch Institute Report*. New York: Norton, 1989, pp 21–40.
7. Rapport DJ. What constitutes ecosystem health? *Perspectives in Biology and Medicine*. 1989; **33**: 120–32.
8. Brown LR. Grainland shrinks. In: Brown LR, Flavin C, Kane H (eds). *Vital Signs: The Trends that are Shaping Our Future*. New York: Norton, 1992, pp 36–7.
9. Murdoch WM. *The Poverty of Nations. The Political Economy of Hunger and Population*. Baltimore: Johns Hopkins University Press, 1980, pp 114–16.
10. Durning AB. Apartheid's other injustice. In: *State of the World* 1990. *Worldwatch Institute Report*. New York: Norton, 1990, pp 11–17.
11. Dimbleby GW. The impact of early man on his environment. In: Cox PR, Peel J. (eds). *Population and Pollution*. London: Academic Press, 1972, pp 7–13.
12. Lisitsina GN. The earliest irrigation in Turkmenia. *Antiquity* 1969; **43**: 279–86.
13. Hetzel BS. *The Story of Iodine Deficiency*. Oxford: Oxford University Press. 1989.
14. Chapman JA *et al*. Endemic goitre in the Gilgit Agency, West Pakistan. *Philosophical Transactions of the Royal Society of London (Biology)* 1972; **263**: 459–90.
15. World Commission on Environment and Development. *Our Common Future* Oxford: Oxford University Press, 1987.
16. UN Environment Programme. *The State of the World Environment*, 1991. Nairobi: UNEP, 1991.
17. Micklin PP. Dessication of the Aral Sea: A water management disaster in the Soviet Union. *Science* 1988; **241**: 1170–6.
18. Weber P. A place for pesticides? *World Watch* 1992; **5(3)**: 18–25.
19. Loewenson R. Challenges to health in plantation economies: Recent trends. *Health Policy and Planning* 1989; **4**: 334–42.
20. Loewenson R. Labour insecurity and health: An epidemiological study in Zimbabwe. *Social Science and Medicine* 1988; **27**: 733–41.
21. There is a proliferating literature on this topic. Bradley PN, Carter SE. Food production and distribution – and hunger. In: Johnston RJ, Taylor PJ (eds). *A World in Crisis? Geographical Perspectives*. Second edition. Oxford: Blackwell, 1989, pp 101–24. Friedmann H. The origins of Third

World food dependence. In: Bernstein H *et al* (eds). *The Food Question. Profits Versus People?* London: Earthscan Publications, 1990, pp 13–31. McMichael P. Tensions between national and international control of the world food order: Contours of a new food regime. *Sociological Perspectives* 1992; **35**: 343–65.

22. UN Population Fund. *State of the World Population.* 1992. Geneva: UN, 1992.

23. World Resources Institute. *World Resources* 1992–93. Oxford: Oxford University Press, 1992.

24. WHO Commission on Health and Environment. *Our Planet, Our Health.* Geneva: WHO, 1992.

25. World Resources Institute. *World Resources* 1990–91. Oxford: Oxford University Press, 1990. (Also: Tolba M. *The State of the World Environment*, 1991: *Report of the Executive Director of the United Nations Environment Programme.* Nairobi: UNEP, 1991.)

26. Brown LR. Fertiliser use falls. In: Brown LR, Flavin C, Kane H (eds). *Vital Signs: The Trends That Are Shaping Our Future.* New York: Norton, 1992, pp 40–2.

27. MacKenzie JJ, El-Ashry MT. *Ill Winds: Airborne Pollution's Toll on Trees and Crops.* Washington DC: World Resources Institute, 1988.

28. Alexandratos N. (ed). *World Agriculture: Toward* 2000, *An FAO Study.* London: Belhaven Press (by arrangement with FAO), 1988.

29. One view is that the USA was able to use its Food Aid program during the 1950s and 1960s to political advantage, at a time when the world's two major political blocs were competing for hegemony within the emergent Third World. Under Congressional mandate, the USA sold its vast surpluses of grain cheaply and accepted, as 'payment', local non-convertible currency that then was required to be spent upon local development projects under the control of US 'aid' agencies. See, for example: Friedmann H, 1990 (ref. 21 above).

30. Shiva V. The Green Revolution in the Punjab. *The Ecologist* 1991; **21**: 57–60.

31. Brown LR. Re-examining the world food prospect. In: *State of the World* 1989. *Worldwatch Institute Report.* New York: Norton, 1989, pp 41–58.

32. Pearse A. *Seeds of Plenty, Seeds of Want. Social and Economic Implications of the Green Revolution.* Oxford: Oxford University Press, 1980.

33. Bello W, Rosenfeld S. *Dragons in Distress: Asias's Miracle Economies in Crisis.* San Fancisco: Institute for Food and Development Policy, 1990.

34. Wittwer SH. Food problems in the next decades. In: Botkin DB *et al.* (eds). *Changing the Global Environment. Perspectives on Human Involvement.* London: Academic Press, 1989, pp 117–34.

35. Paulino LA. *Food in the Third World: Past Trends and Projections to* 2000. Research Report No. 52. Washington DC: International Food Policy Research Institute, 1986.

36. The effects of ozone and other photochemical oxidants on vegetation *and* the effects of sulfur oxides on vegetation. Chapters 33 and 34 in: WHO (Europe). *Air Quality Guidelines for Europe.* Copenhagen: WHO (Europe), 1987, pp 386–403.

37. Haines A, Fuchs C. Potential impacts on health of atmospheric change. *Journal of Public Health Medicine* 1991; **13**: 69–80. Also: Assorted references given in chapter 33 cited in the preceding reference.

38. Meybeck M, Chapman DV, Helmer R. *Global Freshwater Quality. A First*

Assessment. Published on behalf of WHO and UNEP. Oxford: Blackwell Reference, 1990.

39. Although the proportion of global water available as freshwater at any one time is tiny, the hydrological cycle is one of the great engines of the biosphere, causing over 100,000 billion cubic metres of freshwater to fall on land as rain and snow each year – enough to cover the continents to a depth of almost one metre!

40. Brothwell D. The question of pollution in earlier and less developed societies. In: Cox PR, Peel J. (eds). *Population and Pollution.* London: Academic Press, 1972, pp 15–27.

41. Lean G, Hinrichsen D, Markham A. *Atlas of the Environment.* New York: Prentice Hall, 1990.

42. Kane H. Fish Catch Falls. In: Brown LR, Flavin C, Kane H (eds). *Vital Signs: The Trends That Are Shaping Our Future.* New York: Norton, 1992, pp 30–1.

9

Biodiversity: forests, food and pharmaceuticals

9.1 The benefits of biodiversity

In one sense, the loss of diversity is the most important process of environmental change, because it is the only process that is wholly irreversible. Its consequences are also the least predictable, because the value of the earth's biota (the fauna and flora collectively) remains largely unstudied and unappreciated.

EO Wilson, 1989[1]

9.1.1 The issues

We are living through one of the world's great mass extinctions – and it is of our own making. Yet it is not easy to see it happening on a day-by-day basis, and, anyway, the phrase 'loss of biological diversity' is a bit obscure. The world's biodiversity, the vast mosaic of living species, exists at three levels – diversity of species, genetic diversity within a species and diversity of ecosystems. In common parlance, biodiversity refers to the diversity of species. At all three levels, however, the loss of biodiversity poses threats to human population wellbeing and health.

In the light of the accelerating rate of species loss, primarily through habitat destruction, it is widely estimated that up to one-half of all species alive today could be extinct by the end of next century.[1,2] As dramatic as this tally-count statistic sounds, the loss of diversity of genetic strains *within* species is also important. Genetically distinct populations (strains) of many species exist in scattered locations, and the loss of these localised populations is a dominant element in today's extinction crisis. Not only does it reduce genetic resilience, and therefore adaptability, but it threatens the productivity and viability of ecosystems. Indeed, ecosystems can unravel because of the loss of just one or a few 'key species' that perform essential tasks, such as plant pollination by birds and insects. For example, Malaysia's durian fruit crop shrank markedly in the 1970s because of severe decline of the pollinating bat species, caused by destruction of mangrove foliage upon which the bats fed. In future, climate change, by causing certain crucial species to decline or out-migrate, may also cause ecosystems to lose vitality.

The arguments for retaining as many species on Earth as possible can be classified as either instrumental or non-instrumental.[3] At their most obvious, the instrumental arguments refer to the benefits of a wide repertoire of genetic stock for future selective breeding and bioengineering of foods, medicines and materials, and of maximising our chances of finding naturally-occurring useful substances. For example, if Madagascar's eastern rainforests are destroyed, about fifty species of wild coffee – many of them caffeine-free and not yet included in Latin America's coffee breeding genetic repertoire – would be lost. Less well understood is the instrumental argument that biodiversity makes for stable ecosystems with optimal energy capture and productivity – an essential requirement for their sustainability. There are two other non-exploitative instrumental arguments. The aesthetic argument prefers a natural world that is rich with the beauty, the exotica and the intricacies of other species. The scientific argument recognises that our knowledge about the workings of nature is enhanced by the opportunity to study all of its living facets.

The non-instrumental arguments are moral and ethical. The moral argument invokes the 'rights' of other species to coexist with us. This moral value is widely espoused among human cultures (and differs from the 'animal liberationist' argument about the rights of individual sentient animals). The related ethical argument rejects the destructive behaviour of the powerful human species towards weaker species. It also invokes the need for equity between generations of humans in their access to nature's bounty.

These general arguments for the preservation of species can, I think, be qualified by one important consideration. Some species present a major hazard to human health, while conferring no other *apparent* benefit upon us or other species. Examples are the smallpox virus and the AIDS virus (HIV). These are parasitic microbes that perform no ecological support function – unless one takes the view (framed in ecological and not teleological terms!) that they act as natural brakes on excessive human crowding and its associated social patterns. Besides, these diseases did not occur in sparse pre-agrarian human communities, but have evolved recently (from nearby animal sources) in response to varied new opportunities presented by settled human crowds. So long as we remember our species' predilection for destructive self-interested intervention in nature, there is no serious argument against the eradication of such disease-inducing species. Some 'deep ecologists' may regard that argument as an arrogant presumption by the one species that is currently endangering the fabric of life on Earth. Nevertheless, moral outrage aside, the biological reality is that humans, as products of evolution – like every other species – have self-interest software programmed into their genetic hardware. That software cannot be ignored or denied. But unless we can control it or override it, particularly in relation to population overgrowth and material overconsumption, we will continue with today's mass extinction.

9.1.2 The predicament

Pessimists – or perhaps they are the realists – predict that the ongoing extinction will rival the dramatic mass extinction of 65 million years ago at the Cretaceous–Tertiary boundary (see Fig. 1.1), when the dinosaurs disappeared. This ongoing loss of biological diversity, entailing an estimated several dozen species per day, results from a variety of human actions. These include the destruction and fragmentation of natural habitat with the spread of cities, the felling and clearing of forests and the draining and destruction of coastal wetlands;[4] the toxic effects of pollution as described by Rachel Carson;[5] overharvesting of economically useful species on land and sea;[6] the introduction of unnatural predator or competitor species via a variety of intentional and unintentional actions; or other ecological disruptions such as the various effects upon plant and animal species caused by global warming and ozone layer depletion.

As *Homo sapiens* occupies more living room, and uses more natural resources for economic development, there is less and less room, energy and nutrients left for other species. Already, on some scientists' estimation, humans have appropriated (either by consuming, preempting or destroying) one-quarter of the Earth's photosynthetic product, including 40 % of the 'net primary production' on land.[7] Simple fractions aside, the fragmentation of continental habitat by land clearance, roads, dams and cities is particularly insidious since it creates genetic and ecological 'islands'. This fragmentation leaves species vulnerable to the same fate as, historically, afflicted island populations (see chapter 1) and it obstructs evolutionary speciation.[8] Ehrlich and Wilson have estimated that the weight of human numbers and aspirations will, in the foreseeable future, require cooption of most of the remaining wilderness for farming, grazing, energy production, mining and transportation.[9] This would foreshadow an 'epochal catastrophe' for earthly life.[2]

The threatened extinction of the blue whale over the last two decades,

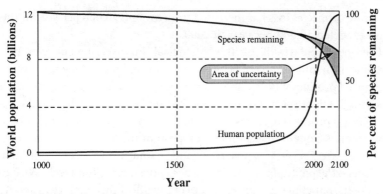

Figure 9.1. Estimated timetable of species extinction, in relation to increase in human population. (Based on a diagram in Soulé, 1991 – ref. 2)

after sixty years of profligate hunting by humans, has been a powerful symbol for a new environmental consciousness; likewise the threat to the elephant and the almost-extinct rhinoceros.[10] The point, however, is not simply that some species are disappearing, since, in nature, species always come and go. As we have seen in chapter 1, an estimated 99 % of all species that have ever lived on Earth are now extinct, many of them wiped out in mass extinctions. Climatically, geologically or cosmically induced up-heavals in the tapestry of life on Earth will occur again. Many other extinctions occur quietly, on the fringes of ecosystems, as environments gradually change and as new species arise. Today, however, the point of great concern is the *rate* at which extinctions are occurring – and their cause.

The number of living species is estimated to be of the order of 10–30 million, perhaps more. Most of these species have not yet been identified and classified. Ethics and aesthetics aside, the genetic repertoire within this biotic panoply is a potentially huge and sustainable asset that we could use to our own long-term benefit. As human numbers surge, the extinction rate continues to climb – perhaps exacerbated by the collapse of depleted ecosystems. Scientists estimate that from one-quarter to one-half of all current species on Earth will have been lost by later next century (see Fig. 9.1).[2] Currently, we are eliminating one or more species every hour – mostly small and localised plant and insect species that disappear as the world's biotically-lush tropical forests succumb to fire, axe and bulldozer. This current pace of extinction is several thousand times faster than usually occurs in (non-catastrophic) nature.[1] The World Bank estimates that the rate of loss of birds and mammals has tripled over the past 250 years and is currently about fifty times greater than 'background'.[11] The estimated total rate of species loss has increased from around 10,000 per year in 1900 to an expected 50,000 per year in 2000.

The inevitable result of this biological impoverishment will be a simplification and decline of ecosystems to lower levels of energy flux and to reduced productivity.[12] There is an important *ecological* point here. The loss of individual species is one thing, but the more generalised reduction in energy, productivity, resilience and latent genetic adaptability of ecosystems is a qualitatively different type of problem. Chapter 7 identified how increased UV-B exposure would damage the ocean's primary producers, the phytoplankton, thus eroding the marine food web. We have also noted that climate change will adversely affect the viability of species, particularly plant species. The global warming scenario predicted by the IPCC over the next century is equivalent to an altitudinal displacement of more than 400 metres and a latitudinal displacement in northern Europe of more than 300 kilometres. Therefore, all species whose habitat is currently confined to the upper 400 metres of a region (e.g. the higher slopes of a mountain range) or to the band of land within 300 kilometres of the Arctic Ocean face the threat of extinction from a warming environment. They have no escape route. Further, because of the rapidity of global warming many plant species will be unable to migrate to higher latitudes sufficiently fast, and will be 'overtaken' by an intolerable climate.[13]

From the viewpoint of our own self-interest, the weakening of ecosystems will jeopardise human health, commerce and the aesthetic pleasures that flow from nature. Yet it is the burgeoning of human numbers and economic activity that primarily endangers biodiversity. More visible is the deliberate killing of prized species and the illegal, and often decadent, trade in wildlife. The relentless demand for rhinoceros horn, tiger bones and for ivory, particularly in populous and increasingly wealthy eastern Asian markets, is a continuing sordid reminder of how we can also obliterate species by deliberate killing (see also ref. 6). Official public bonfires of seized elephant tusks in Kenya and of rhinoceros horn in Taiwan hold slight hope for changes in public attitudes – but money speaks louder than morals, particularly in the impoverished Third World hinterland where poaching persists.

This raises, finally, another dilemma. While it is easy to take a conservationist view from the comfort and security of middle-class Western society, in situations of Third World poverty and overpopulation, desperate humans will choose to feed their families rather than to preserve local species. Indeed, facile conservationism may not be attuned to local ecological realities. Recent moves by countries such as Zimbabwe and

Botswana to manage and harvest elephant populations suggest one type of model that could successfully combine ecological sustainability with employment, revenue and the social wellbeing of local human communities. However, such models would appeal more if the size of the human population was also under control. Further, human population control would not only take pressure off these high-visibility *causes célèbres* – it would also slow up the sedimentary damage that we are doing to the structures of the world's biological diversity.

9.2 Biodiversity and human health

9.2.1 Introduction

The history of human culture is replete with examples of the selective cultivation of edible plants, of the discovery of industrial and other uses of natural plant products (such as latex and cotton) and of the identification of plant substances that alleviate or cure human disease or which can be used as pesticides in agriculture or in domestic hygiene. A parallel set of examples applies to animal and animal products. The loss of these utilitarian options – particularly of those that we have not yet discovered and come to value – is one major adverse consequence of the loss of biodiversity.

It might be argued that we have already had enough time to identify the species that are immediately useful to us, and that, if utility is the sole criterion, we should therefore now know what species we need to conserve for our own uses. That argument might sound plausible for obvious and long-established uses – such as providing food, skins, fibres, fuel, spices and industrial raw materials. After all, we have already domesticated various animals and cultivated plants and learnt how to ensure their survival and propagation (often at the cost of other species). However, the argument is most certainly not true for the modern uses to which countless other species – many of them as yet unidentified – can be put, including the extraction of chemical substances or the use of wild genes for bioengineering.

9.2.2 Medicinal drugs

Humans everywhere depend heavily on nature for medicinal drugs. Traditional medicine, in many cultures, has used a wide range of naturally-occurring substances. The modern history of antibiotics is founded on the chance observation by Alexander Fleming, and its subsequent development by Howard Florey, that the natural 'penicillin' toxin given out by certain fungal moulds kills alien bacteria in the surrounding environment.

Subsequently, the lowly fungi have yielded up cyclosporin A (used to avert rejection of transplanted organs) and, more recently, gliotoxin, which may protect transplantation patients but without causing generalised suppression of their immune system.

WHO estimates that, worldwide, about three-quarters of all primary health care needs are met by traditional medicines, including almost 90 % of those needs in developing countries. The many hundreds of natural medicinal substances used by Amazonian forest-dwellers, for example, include hormonally-active substances used as contraceptives and antibiotic chemicals used in the treatment of fungal diseases. Even in the USA, where synthetic drugs predominate, an estimated half of all prescription drugs dispensed originate from natural organisms.[14] While some of these substances (most of which are plant extracts) may act by a placebo effect – i.e. the power of suggestion and belief – many have been shown to act pharmacologically. This century, Western medicine has identified and purified many of the active chemicals from these plant sources. Reserpine, for example, comes from a shrub that has been long used in traditional herbal tranquillisers, and is now used in purified form to treat high blood pressure in the West.

Overall, the higher-order plants provide developed countries with about one-quarter of all medicines currently used.[15] These include various of the clinical medical mainstays, such as digoxin and digitalis from the foxglove plant of temperate climes, a longtime standby in the treatment of congestive heart failure. The pain-relievers codeine and morphine come from poppies, while aspirin has been developed from a natural chemical occurring in willow bark. Many widely used drugs come from the tropical rainforests – e.g. quinine for malaria, and curare as a muscle relaxant in anaesthesia and in treating certain nervous system disorders. The list goes on: the antiglaucoma drug pilocarpine (from *Pilocarpus jaborandi*), the local anaesthetic cocaine (from *Erythroxylum coca*), the antispasm drug atropine (from 'deadly nightshade' – *Atropa belladonna*) and many others. A list of clinically useful drugs from tropical plants is contained in Table 9.1.

The story of quinine, the basic substance used worldwide to combat malaria, is a compelling example. The early history of European commerce and exploration in tropical countries – particularly in Africa – is riddled with deaths and disasters due to malaria. During the late seventeenth and the eighteenth centuries, European travellers discovered, via a combination of accident, luck and observation, that the extract from the bark of the Peruvian cinchona tree both prevented and treated malaria.[16] The active chemical, identified in the nineteenth century, kills the malarial parasite

Table 9.1. Some tropical plants yielding clinically useful drugs.

Species	Drug	Action/Clinical Use
Adhatoda vasica	Vasicine (Peganine)	Oxytocic (obstetric use)
Ananas comosus	Bromelain	Antiinflammatory
Ardista japonica	Bergenin	Cough suppressant
Carica papaya	Papain/Chymopapain	Proteolytic; mucolytic
Catharanthus roseus	Vinblastine,	Antitumor (leukaemia)
	Vincristine	ditto
Chondodendron	Tubocurarine	Skeletal muscle relaxant
tomentosum		
Cinchona ledgeriana	Quinidine	Antiarrhythmic
Cinnamomum camphora	Camphor	Rubefacient
Crotalaria sessiliflora	Monocrotaline	Antitumor (topical)
Datura metel	Scopolamine	Sedative
Dioscorea species	Diosgenin	Source of female
		contraceptive
Dubosia myoporoides	Atropine/Hyoscyamine	Anticholinergic
Erythroxylum coca	Cocaine	Local anaesthetic
Mucuna deeringiana	L-Dopa	Anti-Parkinsonism
Nicotiana tabacum	Nicotine	Insecticide
Physostigma venenosum	Physostigmine	Cholinesterase inhibitor
Rauvolfia canescens	Deserpidine	Antihypertensive;
		tranquilliser
Rauvolfia serpentina	Reserpine	Antihypertensive
Ricinus communis	Castor oil	Cathartic
Strychnos nux-vomica	Strychnine	CNS stimulant
Theobroma cacao	Theobromine	Diuretic

For a fuller listing see ref. 15

(plasmodium) and, as chloroquine, it has thus provided humans with a basic chemical defence. Many modern antimalarial drugs are variations on the quinine molecule's theme. However, as the humans-versus-malaria plot thickens, we may well need further assistance from Nature. In the continuing epic struggle with malaria – a struggle which humans are not yet 'winning' against this immunogenetically ingenious parasite – a continued procession of new antimalarial drugs is required to match the competitive, flexible, genetic evolution of the plasmodium. Few tropical countries are now unaffected by chloroquine-resistant strains of *Plasmodium falciparum*, which causes the most virulent form of malaria.

WHO has recently reported that drugs derived from artemisinin, extracted from a herb in China, have shown recent promise against severe drug-resistant malaria.[17] In fact this is really an old lead rediscovered. The herbal plant (qinghao) is known to Western botanists as *Artemisia annua*, and it has been in the Chinese pharmacopoiea for more than 2,000 years, principally for the treatment of fevers. Chinese scientists identified its specifically antimalarial properties in the 1970s, and its rapidity of action, including against severe, chloroquine-resistant falciparum malaria looked particularly promising. In the continuing absence of either eradication or vaccination, and with the inexorable rise of drug-resistant parasites, we rely on finding new types of antimalarial drug, from natural sources, to use against this perennial scourge of human health.

Plant species in rainforests are particularly rich in certain chemicals – such as the alkaloids which biochemists think are the plants' natural defenses against fungal diseases and insect infestation, or other chemicals which deter predators or attract pollinators.[18] Researchers at the US National Cancer Institute have already identified about 3,000 plant substances – most of them from rainforest species – that are active against cancer cells. The alkaloid chemicals vinblastine and vincristine, extracted from the rosy periwinkle (*Catharanthus roseus*) from the forests of Madagascar, have transformed the outlook for children with lymphomas (both Hodgkin's and non-Hodgkin's) and leukaemia. Where only one in five children with acute lymphocytic leukaemia underwent remission before the introduction of these drugs in the 1960s, now four in every five do so.

Nature's pharmaceutical bounty is not confined to the rainforest. Indeed, of all drugs currently used in the USA that are derived from plants, only one-fifth originate from the tropics. The most likely explanation for this is that the less-rich biota of the temperate regions has been much more accessible to Western science, and hence much better studied. Of the estimated 5,000 species of higher plants exhaustively studied as potential sources of new medicinal drugs and the tens of thousands of plants routinely screened for anticancer properties, only a minority have come from the tropical forests.[15] The Pacific yew, that grows on the west coast of the USA, is a slow-growing tree whose stocks are being rapidly depleted. The tree has no commercial value as lumber, but it is being cleared away as forests are logged. Its bark contains a chemical discovered in the 1970s and called 'Taxol' by pharmacologists. This chemical, which chemists have not yet finally managed to synthesise in the laboratory, has attracted great attention recently as an anticancer drug. It appears to be effective in the treatment of breast, ovarian and lung cancers, probably by restraining cell division. The commercial dilemma is that it takes about six yew trees to produce enough of the drug to treat one cancer patient. This dilemma actually only arises if it is assumed that forests must be logged in a no-frills clear-fell fashion according to the dictates of short-term profitability. Conservationists and cancer researchers have therefore asked the US Government to list the Pacific yew as a 'threatened' species. The matter, in 1992, hung precariously in the balance. However, good management of forests combined with conservationist logging practices would reprieve the Pacific yew. It would also accord with long-term sustainable economics.

Although the plant kingdom offers us a vast and irreplaceable pharmacopoeia, we have formally tested only about 1% of the world's plants. However, if we complemented our white-coated technical approach with ethological studies, we may well gain clues from animal behaviour in the wild. Because animals have coevolved with plants, selective survival within an animal species has probably favoured those individuals with an instinctual use of plants in times of sickness or injury. Many animal species seem to display forms of self-medication, both preventively and for active treatment.[19] For example, among Ethiopian baboons, those living in high-risk areas for schistosomiasis eat *Balanites* fruits (rich in a potent antischistosome chemical) whereas others do not. Chimpanzees in Tanzania may obtain antibiotic chemicals from chewing herbs such as *Aspilia*. Kodiak bears, in North America, appear to use the chewed *Ligusticum* root to combat skin parasites.

Animal species are also a *source* of medicinal drugs. Indeed, from within my own university one of the world's leading herpetologists, Michael Tyler, has identified several chemicals produced by frogs that have significance for the development of human medicines. Yet, as he points out with passion, frog species seem to be in decline around the world – perhaps because of the chemical pollution of their watery environments. (The cause of this apparent decline is contested. While pollution and climate change have both been proposed, other herpetologists blame natural environmental fluctuations.) In the 1970s and early 1980s Tyler participated in the discovery of the curious phenomenon of the gastric brooding frog, *Rheobatrachus silus*.[20] The adult female of this Australian species swallows her newly-fertilised eggs, which subsequently hatch in her stomach. Later she gives birth by regurgitating little frogs. Disbelieved at first, this discovery soon prompted questions about the suppression of normal stomach acids during the 'pregnancy'. How did the mother frog avoid dissolving her progeny? Subsequent research identified acid-suppressing chemicals (prostaglandins) and other stabilising chemicals which, in turn, have contributed to the further development of drug treatment for suppressing acid secretions in persons suffering from peptic ulcers. More recently, Tyler and colleagues have identified unique compounds in the skin of Australian tree frogs which may become the basis of a new generation of antibiotics, effective against a wider range of infectious agents than are the penicillin antibiotics. These naturally-occurring antibiotic chemicals are part of the frog's self-defence against the many microbes that lurk in a watery environment. They are peptides (chains of amino acids) which can be extracted from the skin of the live frog – who therefore survives to produce more. Similar frog peptides with antibiotic activity, called magainins, have been isolated from the African clawed frog. They, too, have a wide spectrum of antibiotic activity against bacteria, fungi, and protozoans such as the malarial parasite. American researchers have recently discovered a very potent painkiller secreted by a frog species, *Epipedobates tricolor*. This alkaloid chemical, which they have called 'epibatidine', is about 200 times more potent than morphine and acts via a new class of pain-related receptor cells in the brain. It is therefore ironic that, while frogs secrete chemicals that may prolong or improve human life, we humans continue to dump into the frogs' environment noxious chemicals that shorten their lives – and may hasten the extinction of some frog species.

We cannot know how many other valuable discoveries from natural sources are awaiting our discovery. We have only scratched the surface of nature's prodigious repertoire, and we can be certain that there are many such substances awaiting discovery. We do know, however, that these will be forfeited one by one as species are unwittingly or uncaringly eliminated. Likewise, important clues will be lost as the traditional wisdoms of primitive indigenous cultures are obliterated, particularly those based in tropical forests. Three-quarters of our plant-derived medicinal drugs have been first discovered by following up folk or ethnomedical practices.

Much of the publicly stated concern about biodiversity has been focused on the loss of these medicinal riches from the forests and other biological sources. Earth scientists have drawn similar attention to the need for maintenance of intact ecosystems in order to provide ecological 'services' such as the recycling of nutrients, keeping the planet's carbon, sulphur and hydrological cycles turning, and damping down climatic storms. However, perhaps our greatest concern should be for the threat that loss of biodiversity poses to the world's future food supplies.

9.2.3 Food

The conservation of species is essential for maintaining our largely unnatural system of food production. Biodiversity provides a source of alternative plant species to grow in otherwise unproductive (or, indeed, degraded) lands, and it is a source of wild genetic strains for crossbreeding or genetic hybridisation with the established food species in order to sustain viability and to enhance yield. Similar arguments apply to edible animal species. Plant genes from wild rice in the tropical forests, for example, and from wild grasses in the savannahs, have enabled us to maintain crops that are resistant to new pests, diseases and other sources of crop failure.

Historically, a total of about 5,000 plant species have fed the human population. Today, approximately 150 species meet most of our nutrient and caloric needs, and fewer than 30 species provide more than 90% of human dietary energy.[21] Just three species, wheat, rice and maize, provide half the world's food; potatoes, barley, sweet potato and cassava account for another quarter. This is a slim and vulnerable base. Yet, because of recent progress made in developing high-yield and (currently) hardy strains of plant, agricultural practice has – for reasons of short-term economics and maximising food productivity – increasingly focused on the use of these strains. Today, therefore, more than half of the Canadian prairies cultivate only the one variety of wheat (Neepawa). In the USA, in the early 1980s, nearly three-quarters of the potato crop was accounted for by just four varieties, and the whole of the pea crop by just two varieties.[22] In much of Southeast Asia, a single genetic variety of rice now accounts for as much as 60% of total rice production. The upside of this strategy has been short-term gains in productivity; the downside is that hundreds of traditional locally-adapted species have been displaced. And so the risk to future generations has been increased.

Repeated experience has shown that such crop 'monocultures' are subject to periodic pest and disease outbreaks that can be intensive and extensive – and catastrophic. The best-known historical example is the disastrous potato crop failure in Ireland and parts of Europe in the 1840s in which some two million people starved to death and another two million emigrated, primarily to the USA.[23] Between 1870 and 1890, rust virtually destroyed the vast coffee plantations of Sri Lanka (then Ceylon), and in 1942 a fungal disease in the rice fields of Bengal resulted in the death of thousands of people.[24] Because of severe and uncontrollable pest infestations, banana plantations were wholly relocated from the east to the west side of Central America, and rubber plantations have been relocated from Brazil to Asia.[25] These problems with monocultures can be particularly severe in tropical climates. The natural profusion of species in the tropics includes countless plant pests and pathogens. Further, because humid climatic conditions contribute to a luxurious 'culture medium' for pests, they can pass through many generations in a brief period (and can do so throughout the entire year), thus allowing insecticide-resistant strains to evolve relatively rapidly. In Egypt, for this reason, chemical pesticides used on cotton crop pests need to be changed very frequently.[24]

It is clear, then, that to underwrite our future food supplies we need to protect genetic diversity. Despite the cleverness of modern genetic engineering, we do not *make* genes – we *harvest* them from the wild. Some plants not yet tried may prove to be life-savers in semi-arid or marginal agricultural settings. The Marama Bean of the Kalahari desert in southern Africa and the Somalian Yeheb nut both hold promise of viable and nutritious crops in malnourished, food-insecure, Africa. The Yeheb nut was rescued from virtual extinction, and is now becoming a commercial crop in East Africa. There are a number of salt-tolerant species and strains that could be grown on soils that have become, or will soon become, too salty for other crops.

We should pause here for a moment, to be clear about the import of these strategies. At one level we are arguing in favour of preserving biodiversity, including the sustaining of ecosystems. But at another level we are extending our perennial conquer-or-be-conquered mentality. Having simplified the ecology of much of the world's land for human agriculture, we now must apply continued vigilance, and much input of energy, to sustain those unnatural systems. The challenge for us thus becomes that of anticipating and outsmarting the next 'barbarian' horde of insects or fungi. Perhaps we can learn to do so in an ecologically benign fashion – but we cannot now get off the treadmill.

Another vital need for the security of agriculture is to conserve genetic variants *within* species, to provide a defence against the vulnerability of genetic uniformity. Ever since the devastating Irish potato famine of 1845–6, we have known that the genetic uniformity of crops makes them highly susceptible to disease and pests. The blight that wiped out two successive years of the Irish potato crop was caused by a fungal microorganism that thrived in the unusually cold and wet weather conditions that occurred during those years.[26] Because of the dominant role of the potato in the Irish diet – up to three-quarters of total caloric intake in some counties – much of the population was without sufficient food when the potato crop failed. The same potato blight occurred simultaneously in other parts of northern Europe, but had much less impact since those populations had access to many other foods. Selective crossbreeding of the potato during this century has produced countless new varieties, many of which are markedly resistant to blight. Using modern genetic engineering techniques, Mexican scientists have recently developed a virus-resistant potato, expected to increase yields of this important food crop.

The same need for rescue by wild-type genes has occurred in diverse settings. Wheat and maize crops in the American Midwest have both been devastated by epidemics in recent decades. In each case, salvation came from wild strains obtained from other countries. Maize had become particularly vulnerable because of the genetic uniformity that follows years of inbreeding. However, two ancestors of the maize plant were found in a tiny and rapidly disappearing ecosystem in Mexico in the late 1970s;[27] their crossbred genes have subsequently conferred resistance to several major diseases of domesticated maize.

This intentional genetic diversification by means of crossbreeding and laboratory biotechnology is a cultural adaptation by humans that intensifies a process that occurs at a more leisurely pace in natural settings. The rise of agriculture was aided by the fortuitous occurrence of genetic variants among erstwhile small-seeded low-edibility grasses; the mutant strains were preferentially collected, eaten and resown.[28] Ten thousand years on, molecular genetics is enabling the engineering of new genetic strains of plants. Such strains may have a greater yield, an improved resistance to infestation and disease, a new independence from nitrogenous fertilisers or an increased tolerance to temperature extremes.

The recently-developed possibilities of genetic engineering, using modern biotechnology, add a new dimension of potential loss. In the eloquent words of one scientist, the extinction of species 'no longer means the simple loss of one volume from the library of nature. It means the loss of a loose-leaf book whose individual pages, were the species to survive, would remain in perpetuity for selective transfer and improvement of other species.'[29] These developments depend on the availability of genetic material from other naturally-occurring strains or species of plant that can be excised and then spliced into the genetic material of the recipient plant. Ironically, at the very time that we may need wild genes from tropical plants to genetically splice into temperate food-crop plants, to help them resist higher temperatures or increased ultraviolet exposure because of ozone layer depletion, we are depleting the tropical genetic library itself.

Already we have tens of thousands of plant strains conserved in national and international gene banks, but mostly as (space-saving) seeds. But there are problems with this horticultural Noah's Ark strategy. Seeds deteriorate, they are vulnerable to disease and their storage places the plant in suspended animation unable to keep evolving alongside a changing panoply of pests, diseases and companion plants. We therefore must also preserve wilderness and establish natural reserves in which wild plant species – especially those related to crop species – can grow and survive.

9.3 Destruction of the world's forests

An African native forest is a mysterious place. You ride into the depths of an old tapestry, in places faded and in others darkened with age, but marvellously rich in green shades. You cannot see the sky at all in there, but the sunlight plays in many strange ways, falling through the foliage. The grey fungus, like long drooping beards, on the trees, and the creepers hanging down everywhere, give a secretive, recondite air to the native forest.

Karen Blixen, 1937[30]

9.3.1 Introduction

Karen Blixen's words, written over 50 years ago, evoke the mystery, beauty, grandeur and intimacy of the ancient African equatorial forest. In such an environment, the human who watches, listens and smells will know these feelings of awe. In like vein, Alfred Russell Wallace (the naturalist, and contemporary of Charles Darwin, who independently conceived the notion of natural selection) referred to 'the sense of the past, the primeval – almost of the infinite' imparted by the Amazonian tropical forest.[31] The forest around Blixen's Kenyan coffee plantation has since been cleared and replaced by a more functional non-indigenous eucalyptus plantation. Thus do the world's forests continue to recede.

Humans have been destroying forests – particularly in temperate climates – for thousands of years. Almost half of the world's 6–7 billion hectares of forest have disappeared since the dawn of agriculture, cleared for crop cultivation (currently around 1.5 billion hectares), grazing, settlements, fuelwood and construction timber. Much of the Mediterranean region was stripped bare to build ships. Today, about 30 % of the world's land surface is forest – although less than half of that is primary undisturbed forest. The world's surviving forest comprises three main components, each of a billion or so hectares: tropical moist forest, mostly in Latin America; tropical dry forest, mostly in Africa; and temperate forest, especially in Canada and Russia. Currently, the rate of destruction is greatest, and increasing, in tropical moist forests. Around 20 million hectares (about the combined size of England, Wales and Scotland) are being cleared per year, about half of that in Brazil. Almost two-thirds of the clearance of tropical moist forests is for new agricultural settlement, in response to population pressures.

Deforestation entails loss of natural habitat and biodiversity. Forests also provide natural regulation of the hydrological cycle and of surface water movement, a livelihood and cultural integrity to around 140 million people, and a huge natural sink for atmospheric carbon. We saw in chapter 6 that deforestation adds two billion tonnes of carbon to the atmosphere annually; historically, it has contributed about half of all of the anthropogenic carbon dioxide added to the atmosphere over the past 200 years. We should now look at the biodiversity of forests in more detail.

9.3.2 Tropical forests as genetic and ecological assets

Tropical rainforests, which lie within the boundaries of the tropics of Cancer and Capricorn, have the richest ecosystems on earth. They account for about one-fifteenth of Earth's land surface, but an estimated two-thirds of all of its species. The estimated total number of species in tropical forests is in the range of 10–15 million – predominantly insect species living in the forest canopies, but also including approximately 150,000 seed plants. A single hectare of Malaysian jungle may contain more than 800 species of woody plants – about half as many as exist in the whole of North America. The Amazon alone contains one-fifth of all bird species in the world.[32] It is no wonder that the gradual but continuing destruction of the Amazonian rainforest has engendered despair among the world's environmentalists.

Sarawak, in eastern Malaysia (part of erstwhile Borneo), contains some of the richest rainforests in the world, with an untold diversity of still undiscovered species.[33] Last century, Alfred Russell Wallace, during eight years of solitary collecting in Sarawak, identified tens of thousands of new animal and insect species. Today, in those forests, there is bitter confrontation between the forest-dwelling tribes, who depend on the forest ecosystem for their survival, and the logging companies. As in many parts of the world, the basic dispute is between a minority group wishing to preserve its traditional way of life and a government which, because of financial pressures or ignorance, wants to sell off part of that ecological resource. Ironically, the short-term gain from such deforestation often becomes a long-term resource loss. We will look at the economics of this further in chapter 11.

Logging companies currently extract about 15 million cubic metres of timber from Sarawak annually. Most of the timber goes to Japan. The Japan-based International Tropical Timber Organisation (ITTO) has recently assessed that, under ideal circumstances, Sarawak's forests could yield 9 million cubic metres annually. However, says ITTO, given the prevailing forestry practices in Sarawak, the maximum sustainable yield is around 5 million cubic metres annually.[34] The ITTO, which is not generally known for its radical views, has concluded that the current logging in Sarawak is not sustainable, and that the primary forest will be logged out by around the year 2000. Indiscriminate logging of whole hillsides, in a region of tropical downpours, is ecologically very destructive of topsoil. Some years ago, our family spent a week exploring the mysterious backwaters and rainforests of Sarawak. Soon after boarding a small boat going inland on the River Rejang from Sibu, a major timber export port near the coast, my young daughters were enthralled by the sight of vast flotillas of distant crocodiles up ahead. The 'crocodiles' turned out to be logged tree-trunks, en route to Sibu. Further upstream we saw their source – extensive areas of denuded forest, sloping up bared hillsides whose topsoil would also soon end up in the Rejang River.

The fact that this destruction of rainforest will dislocate, and probably destroy, many forest-dwelling cultures has ramifications beyond the ethical and political. Many of the remaining traditional human cultures have a priceless knowledge of the identity and uses of forest species, and many – such as the Brazilian rubber-tappers – depend on the forest for their livelihood. As their forests come under threat, the plight of the Yanomami Indians of Brazil and of the Penan people of Sarawak have become well known. There are many other forest-dwellers in retreat, including now the ethnic communities in the prodigious boreal forests of Siberia, recently opened up to international entrepreneurs. Many such cultures have already disappeared; European colonisation of Brazil has eliminated about ninety forest-based cultures during this century.[18] Almost too late, Western-based bodies such as the World Wildlife Fund have begun to work with tribal cultures to salvage what they can of this remaining age-old human knowledge and folk-wisdom. Politically, this is being assisted in Latin America by the establishment of 'extractive reserves' in forests, where, in return for the prohibition of clearing and logging, the local communities are assisted in developing sustainable commercial yields of the assorted 'fruits of the forest': rubber, nuts, cocoa, berries and so on. As traditional cultures have learnt long ago, the goose, well nurtured, keeps on laying golden eggs.

9.3.3 Deforestation: the history and politics

Most of the early loss of the world's forest occurred in temperate regions.[35] As we saw in chapter 8, much of today's quintessentially European landscape is actually an artefact of human agriculture. Forests were also cleared across North Africa, North America and East Asia. Australian historical paintings from soon after European settlement reveal that the landscape around the fertile southeast fringe has been radically changed over the subsequent 150 years as farmers have cleared the forests. During that time, an estimated 30 billion trees have been cut down in Australia. Until recently, farmers in Australia received substantial tax relief for clearing land and thus extending the foundations of an agriculturally productive nation.

That much is 'spilt milk'. But the fact that most of the spilling was done by early inhabitants of today's rich countries is politically awkward, since the world community must now appraise the losses caused by the ongoing deforestation in some of the world's poorest countries. China and India, driven by increases in their massive populations, continue to clear forests. African forests, too, retreat before the drive for fuelwood, farmland and logging income. Fuelwood accounts for 80 % of energy needs in sub-Saharan Africa, and it is in very short supply. In West Africa, for example, Cote d'Ivoire has an annual forest loss of around 10 %; its remaining forests may disappear in twenty years. Indonesia, with some of the world's richest tropical forests, is embarking on widespread clear-felling for a massive expansion of its pulp-milling industry and to create room for its controversial relocation of people from densely populated Java and Bali to the outer islands.

Most of the destruction of the world's tropical forest has occurred over the past thirty years, predominantly for agricultural settlement. The WCED Report concluded that, even taking into account natural regrowth and the rudimentary reafforestation effort, little virgin tropical forest would survive beyond 2000 except in portions of Zaire, the Amazon, Guiana and New Guinea.[36] Reafforestation is expensive and its benefits are displaced into the future – whereas the clearing and logging of forest offers an immediate benefit to Third World governments hard pressed by international debt, by the restless landless poor, and by population pressure. The problem is compounded by the expanded appetite of rich countries for tropical hardwood timbers, to supplement their own supplies of oak and beech and certain long-used tropical timbers like Burmese teak, African rosewood and Honduran mahogany. In the 1950s, other versatile tropical hardwoods were discovered by industrialised countries, and the world trade in them has flourished.

Finally, history and politics aside, there are several other reasons for restraint specifically in the clearing of *tropical* forests. First, they contain many more species than do temperate forests. Second, the relatively slower cutting of temperate forests over centuries was less of an ecological disruption than is today's largely mechanised clearance and wholesale burning of tropical forests. Third, the soil under the temperate forests is generally better suited to agriculture and, once laid bare, is less subject to torrential rain and erosion. Tropical land-clearance plans usually go awry – not only because naturally productive ecosystems are lost but because, after several harvests, the thin topsoil is depleted and non-productive. Although the rich countries, for historical reasons, lack moral leverage in this debate, the ecological and economic realities of tropical forest clearance add urgency to our need to find an effective and fair solution to this erosion of biodiversity and of the forest-assisted functioning of the world's natural systems.

9.4 Summary

In the debate about the world's environmental problems, a recurrent theme is the conservation of non-renewable resources. A very special non-renewable resource is the biological diversity of all currently living species and the complex ecosystems that they make up. Species and their genes, once lost, are not retrievable. This biodiversity is a fundamental resource in various ways. Basically, we *consume* the phenotype (i.e. the mature individual, with its chemical constituents), while we *apply* the genetic information from the genotype. Not only is the panoply of species a source of many established foods, socially useful chemicals (including a high proportion of medicinal drugs) and materials, it is also a sure source of many more which have not yet even been guessed at. Genetic variation is the source of nature's basic adaptive capacity – and, for the same reason, we exploit it as the source of genetic stock, both as new species and as genetic variants of current species, for improving our food supply.

Beyond these instrumental considerations, support for the maintenance of biodiversity is also a shorthand way of saying that our world, thus maintained, will be a better and richer place in terms of the stimuli and the aesthetic enjoyment that we can get from it. Such a world will also be a better approximation of the environment in which, and for which, the human species evolved.

References

1. Wilson EO. Threats to biodiversity. *Scientific American* 1989; **261(3)**: 60–6.
2. Soulé ME. Conservation: Tactics for a constant crisis. *Science* 1991; **253**: 744–50.

3. Morowitz, asking 'How much is a species worth?', defines four categories of species – those of current commercial value, those of potential (e.g. medical) value, those of physiological uniqueness (and therefore of interest to science) and those of 'amenity value' (such as pandas, lions and redwoods). See: Morowitz HJ. Balancing species preservation and economic considerations. *Science* 1991; **253**: 752–4.
4. Wetlands are a rich source of genetic diversity. Rice plants, sago palms, oil palms and mangroves all have their origins in coastal wetlands. They are also crucial resources for much fish and bird breeding.
5. Carson R. *Silent Spring*. New York: Houghton Mifflin, 1962.
6. The world's best-known examples are the two great mammals – whales and elephants. Lesser known, pharmacologically dubious, examples are the continuing slaughter of rhinoceros, bears and tigers to supply traditional 'medicinal' products to Asian markets. The Chinese market for tiger-bone medicines for arthritis and nervous disorders and pickled tiger-penis for failing virility has caused an estimated 93 % loss of the Asian tiger this century.
7. Vitousek PM, Ehrlich PR, Ehrlich AH, Matson PA. Human appropriation of the products of photosynthesis. *Bioscience* 1986; **36**: 368–73.
8. Erwin TL. An evolutionary basis for conservation strategies. *Science* 1991; **253**: 750–2.
9. Ehrlich P, Wilson EO. Biodiversity studies – science and policy. *Science* 1991; **253**: 758–61.
10. Douglas Adams, of *Hitchiker's Guide to the Galaxy* fame, has written poignantly and engagingly about his pilgrimage to see the surviving remnants of various almost-extinct animal species, including the white rhinoceros in northern Africa, the kakapo parrot in New Zealand, and the Yangtze freshwater dolphin. (Adams D, Carwardine M. *Last Chance to See*. London: Heinemann, 1990.)
11. World Bank. *World Development Report* 1992. *Development and the Environment*. Oxford: Oxford University Press, 1992.
12. In nature, when widespread extinctions have occurred, there has not been the long-term destruction or alienation of environmental resources that humans are now increasingly perpetrating. Rather, the slate is wiped partially clean, a 'new game' is declared and profuse speciation takes advantage of the myriad ecological niches. If humans were to go into decline – or extinction – then that natural sequence of events would presumably ensue.
13. Roberts L. How fast can trees migrate? *Science* 1989; **243**: 735–7.
14. Farnsworth NR, Soejarto DD. Potential consequences of plant extinction in the United States on the current and future availability of prescription drugs. *Economic Botany* 1985; **39**: 231–40.
15. Soejarto DD, Farnsworth NR. Tropical rain forests: Potential source of new drugs? *Perspectives in Biology and Medicine* 1989; **32**: 244–56
16. Cartwright FF. *Disease and History*. London: Hart-Davis, 1972.
17. Editorial. Rediscovering wormwood: qinghaosu for malaria. *Lancet* 1992; **339**: 649–50.
18. Bird C. Medicines from the rainforest. *New Scientist* 1991; **1783**: 32–7.
19. Newton P, Wolfe N. Can animals teach us medicine? *British Medical Journal* 1992; **305**: 1517–18.
20. Corben CJ, Ingram GJ, Tyler MJ. Gastric brooding: Unique form of parental care in an Australian frog. *Science* 1974; **186**: 946–7

21. Wittwer SH. Food problems in the next decades. In: Botkin DB, Caswell MF, Estes JE, Orio AA (eds). *Changing the Global Environment. Perspectives on Human Involvement.* London: Academic Press, 1989, pp 117–34.

22. Organization for Economic Cooperation and Development (OECD). *Economic and Ecological Interdependence.* Paris: OECD, 1982.

23. Carefoot GL, Sprott ER. *Famine on the Wind: Man's Battle Against Plant Disease.* Chicago: Rand McNally, 1976.

24. Biswas AK. Climate and development. In: Biswas AK (ed). *Climate and Development.* Dublin: Tycooly International, 1984, pp 1–25.

25. National Research Council, Committee on Selected Biological Problems in the Humid Tropics. *Ecological Aspects of Development in the Humid Tropics.* Washington: National Academy Press, 1982.

26. Boyden S. *Western Civilisation in Biological Perspective. Patterns in Biohistory.* Oxford: Oxford University Press, 1987. Of further ecological interest, the fungus is thought to have been brought from America by the ever-faster trans-Atlantic steamships.

27. Tatum LA. The Southern Corn Leaf Blight epidemic. *Science* 1971; **171**: 1113–16.

28. There is speculation that higher-yielding cereal crops arose from accidental crossbreeding between small-seeded wild wheat and goat grass, causing a tripling of the chromosome count and therefore a bigger and more edible seed – a seed that was therefore preferentially used. Humans and the big-seeded species would both have benefitted, in coevolutionary fashion. Bronowski J. *The Ascent of Man.* Boston: Little Brown & Co., 1973.

29. Eisner T. Chemicals, genes, and the loss of species. *Nature Conservancy News* 1983; **33**: 23–4.

30. Blixen K. *Out of Africa.* Harmondsworth, England: Penguin, 1954 (first published 1937).

31. Wallace AR. *My Life.* New York: Dodd Mead, 1905.

32. Editorial. Costing the Earth. *The Economist* September 2, 1989: 16.

33. Hanbury-Tenison R. No surrender in Sarawak. *New Scientist* 1990; **1745**: 12–13.

34. Cross M. Logging agreement fails to protect Sarawak. *New Scientist* 1990; **1745**: 7.

35. Miller KP, Reid WV, Barber CV. Deforestation and species loss. Responding to the crisis. In: Mathews JT (ed) *Preserving the Global Environment. The Challenge of Shared Leadership.* New York: Norton, 1991, pp 78–111.

36. World Commission on Environment and Development. *Our Common Future.* Oxford: Oxford University Press, 1987.

10

The growth of cities

10.1 Urbanisation and social change

The world ... is faced both by the massive degradation of the natural environment and by the accelerating decline in the quality of life of many of those who live in the built environment of cities. The two crises are related. The consequences of urbanization make a major contribution to the global environmental changes that threaten the very existence of life in the future, while changes in the biosphere increasingly affect health and social conditions in the cities.

WHO, 1991[1]

10.1.1 Introduction

Cities reflect the 'fact' that we, like ants and bees, are an intensely social species, given to building structured environments in which thousands of occupationally differentiated individuals live. Earlier chapters have considered the health consequences of various forms of global environmental change. Why now include a chapter on the health impact of urbanisation? After all, the most obvious impact of cities arises from their character as a 'built environment' – not as a disrupted natural environment.

The growth of cities, however, has great ecological significance for at least three reasons. First, in poor countries, the crowded makeshift fringes of cities testify to the population pressures on overloaded rural economies. Second, the artificial ecological setting of high-density city life alters the profile of health hazards – e.g. more 'crowd infections' and physical (industrial and traffic) injuries. Third, the existence of cities has adverse impacts on ecosystems, both directly by pollution and by encroaching on farmland, wetland and coast, and indirectly by distancing and disengaging humans from the rhythms of nature. This separation, this psychological detachment of humans from unpaved nature, helps confirm those social values and priorities that we in the First World have acquired and

259

amplified since early in the industrial revolution. Urbanisation is thus both a consequence and a cause of global environmental change.

10.1.2 The rise of cities

Cities have long been the star performers in human culture. Historically, towns and then cities formed when settled farming communities generated enough surplus food to support an urban population. Archaeologists have traced the first remnants of this evolution from peasant village to trading town in the Palestine–Syria–Anatolia region, dating back to around 9,000–8,000 BC. Subsequently, as wealth and power accrued at the centre, cities evolved. Aristocrats, academics, artisans and artists live in cities, which have thus become the seat of learning, religion, scientific discovery, culture, commerce and government. While cities have poured forth a stream of literature, music and fine arts, there are also darker sides – poverty, squalor, high crime rates, and exposure to infectious diseases, accidents and antisocial behaviours.

Cities attest to the unique capacity of humans to superimpose the complexities of culture upon basic biological destiny. Living in cities distances us from nature; this reduces some risks to health, but creates others. As cities grow, and crowds multiply, so many of these risks are amplified. Historically, early towns and cities lived in some sort of balanced reciprocity with their surrounding countryside. Skills and money were exchanged for food. While urban populations have generally had the upper hand, and have tended to be parasitic upon rural populations,[2] there have been redeeming reciprocal benefits. In today's world, however, as populations surge, as economies become internationalised, and as rural production in the Third World becomes hostage to international debt, so the historic localised urban–rural relationship is breaking down. So too is the fabric of much of urban life around the world. The implications for human health are great.

An increasing proportion of the world's burgeoning population lives in cities – that is urban settlements (towns or cities) of over 10,000 people. In 1800, only 5% of the world population lived in cities – i.e. around 40 million people. In 1900, the figure was around 250 million, or 15%. By 2000, half of the world's six billion people will be urban-dwellers, and, by early next century, *Homo sapiens* will have become a predominantly urban species. (At the same time, the proportion of the world's population aged over 65 years will have increased by around 40% between now and 2020.) While this dramatic shift in where people live has occurred in both rich and poor countries, the proportional rate of growth is now several times greater in less developed than in more developed countries. Although the level of urbanisation is low in sub-Saharan Africa, the *rate* of urbanisation in many African cities is very high (up to 10% per year), as the displaced rural poor migrate to the cities. (The word 'urbanisation', as used by demographers, refers to an increase in the proportion of the national population living in cities.) Rapid urbanisation represents a profound transformation of human ecology – a transformation that is generally outstripping social and political responses.

The most dramatic evidence of accelerating urbanisation has been in the growth of very large cities, with populations greater than five million. In 1950 there were seven such cities, by 1970 there were twenty, and by the year 2000 there will be around sixty – three-quarters of them in developing countries. The four largest cities, in 1990, were Tokyo-Yokohama (29 million) and Mexico City, New York and Sao Paulo (each of 16–19 million).[3] On current trends, the population of greater Mexico City will exceed 25 million by the end of this century, while Sao Paulo, Dacca, Bombay, Calcutta, Delhi, Shanghai, Cairo and others will reach 15–20 million. The urban population of the developing world will then be double that of the developed countries; by 2025 the differential could be as great as four-fold. Environmental refugees, already numbered in millions, will add further to these bulging numbers, and will gather precariously at the socially fragile fringes of urban populations.[4] Nearly all the world's rapidly-growing cities are in poor countries, and approximately half of this Third World urban population lives in conditions of extreme deprivation.[5] In some cities, like Ibadan in Nigeria and Calcutta in India, around 70% of the population live in degraded shanty towns or slums; in Addis Ababa in starving, war-exhausted Ethiopia the figure is 80–90%.[6] WHO warns: 'Nations must face the devastating facts that this deprived sector of the urban population is not only at greatest risk of disease, injury and starvation in the city environment, but by the end of this century the extreme poor in the cities of the world could represent a quarter of the global population.'[7] One-half of these urban poor will be in Asia, while Latin America and Africa, with their smaller populations, will each account for around one-quarter.

The current surging growth of many large Third World cities resembles that of industrialising Europe last century – but on a much larger scale. Within many poor countries, displaced rural workers are moving to cities, supplementing the already-high fertility within those impoverished urban-fringe communities. Their move reflects a mix of influences, including the need for paid work, the search for food and shelter, and displacement from rural life by environmental deterioration or overcrowding. Much of this enforced ('push') migration differs from the dominant prosperity-driven 'pull' of economic and lifestyle attraction to city life that, this century, has underlain the prolonged growth of cities in rich countries. It is misleading, however, to envisage the urban poor as politically or economically marginal. Many squatter settlements are stable, with spontaneous forms of social organisation and support networks and with long-term dwellers who are well integrated as cheap labour in industry or as participants in the 'informal sector' of the economy.[8] This largely self-employed 'informal' sector is, in many ways, complementary and functional in relation to the formal economy. In sub-Saharan Africa the informal sector – called the *jua kali* ('hot sun') economy in Kenya – accounts for as much economic activity as the formal economy and, in the estimation of the UN's International Labor Organization, employs 59% of the urban workforce.

Urban migration is also occurring *between* countries. Most of the people who migrate between countries move to larger cities in more affluent countries. Many of these migrants are poor or displaced, seeking a new life in a new country. Consequently, the new urban underclass of cities like London, Paris, Berlin, Stockholm and New York is dominated by immigrants from hot poor countries – such as Pakistan, Bangladesh, India, northern Africa, Turkey and Puerto Rico.

Not all of the world's population movement, however, is in the direction of the city. In many rich countries, much residential and industrial development is now occurring outside, but near, major cities. These developments draw on the high-quality infrastructure and services available from them: piped water, sewers, telephones, television, garbage collection, paved roads and public transport. Indeed, this 'suburbanisation' has been evident around various of Europe's larger cities throughout this century.[8] In recent decades a more clearcut 'counter-urbanisation' has developed, with development reorienting to smaller towns and rural areas. This has been facilitated by extended and improved roads and by workers having faster cars and higher incomes. Meanwhile, many of the once dominant industrial cities are going into decline as the wealthiest countries move towards post-industrialism; examples include Cleveland, Detroit and Pittsburgh in the USA, Liverpool and Belfast in the UK, Lille and St. Etienne in France, Dortmund, Essen and Duisburg in Germany, Rotterdam in the Netherlands, Bilbao in Spain and Genoa in Italy.

Although urbanisation strains social and political infrastructures, physical facilities and local ecosystems, it often confers benefits on national economies. The World Bank estimates that, in recent years, the one-third of the population of developing countries living in cities has produced nearly two-thirds of their national 'wealth'. While, in conventional economic terms, that may vindicate urbanisation, this calculus ignores the externalised costs of urban-industrial activity to the environment and to the health of the workforce. Besides, urban economic performance is very vulnerable to receding economic tides. Poor urban populations therefore have insecure prospects, but, for many, it beats being landless, hungry and exploited in the countryside.

10.1.3 Urban versus rural life

The formation of early human settlements in villages reflected a need for security against human enemies, wild animals, hunger and natural disasters. It also reflected human sociability.[9] Long-term stable settlement required a steady local food supply, through agriculture and animal husbandry. Subsequently, non-peasant elites established control over rural workers and provided them with armed protection and skilled services in return for food, fibre and animal skins. This led to the radical innovation of densely-settled fortified towns, populated with princes, priests, soldiers, traders and bureaucratic shufflers of clay tablets. As civilisations flourished, and as governments grew stronger and populations grew larger, so towns grew into cities.

But living in a settled urban crowd entails new types of risks to health. By far the main problem is that of infectious diseases, particularly those transmitted by person-to-person spread. Crowded urban populations represent a rich ecological niche for microbes and have, presumably, stimulated the evolution of crowd-dependent bacteria and viruses. Let us examine this more fully.

The human species is, typically, a recent entrant to localised ecosystems in which a long-established tapestry of relationships between infective agents and animal species already exists. Our hunter-gatherer and nomadic ancestors established an uneasy equilibrium with the particular microbes that had been a long-time natural part of their local ecosystem. To a lesser extent, primitive agrarian humans may have done likewise. These ancient parasites therefore evolved alongside early humans and their hominid ancestors and, over time, would have tended to have lessening effects upon human health. Often, they are transmitted between humans via an intermediate 'vector' species – commonly an insect such as the mosquito – and the long-term coevolution of host, vector and parasite tends to result in a satisfactory level of survival in each (see also chapter 2). Sometimes the human host is unaffected; sometimes there is mild, albeit chronic, debilitation; sometimes there is more serious disease.[10]

Once crowds of humans settle more densely in cities, the dynamics of infection change. New bacteria and viruses – those that cannot survive for long outside the human host, and that have not been around long enough to have 'learnt' how to hitch rides on (or lurk in) a go-between vector species – now get an ecological opportunity. Transmission can occur via contaminated drinking water, faecally contaminated hand-to-mouth passage, coughing and sneezing, physical contact and multiple-partner sexual transmission. Therefore, in this early urban ecological setting the bacterial and viral 'crowd' infections first appeared. Most such infectious agents were transferred to human populations from animal herds (see also section 4.3.2).[2] Their descendants today include measles, chicken pox, mumps, whooping cough, diphtheria, smallpox, cholera, typhoid, tuberculosis, leprosy, gonorrhoea and syphilis. Vector-borne diseases can also flourish because of more intensive management of livestock around human settlements. The stock may amplify the parasite, particularly arboviruses, as happens with the amplification of the Japanese encephalitis virus within domestic pig populations. In Africa, it is likely that livestock herds amplify the populations of mosquitoes and tsetse flies, while livestock in eastern Asia help to sustain the schistosomiasis parasite's life cycle by acting as an alternative host.

Whatever else might be said of the impact of cities upon health – and much of the impact has been beneficial – it is unarguable that city life entails hugely increased risks of infection. These infectious organisms are mostly contagious rather than vector-borne. Because many of these infective agents are, in evolutionary terms, new arrivals within the human species, and because there is negligible selection pressure in favour of subtlety of effect when there is a copious supply of uninfected persons, many of these microbes have severe or lethal impact. With the exception of sexually transmitted diseases, which need to linger to be passed on, most crowd infections are hit-and-run events because the microbe cannot withstand the onslaught of the host's antibody response. Measles, for example, can only infect each person briefly and once – not because those persons die, but because they acquire lifelong immunity. However, in an urban population of over half a million people, measles can continue to circulate indefinitely, infecting each new oncoming generation of children.

The saga of human plagues is essentially the saga of decimations of urban populations – the plagues of Pharaonic Egypt, the Justinian Plague of latter-stage Rome, the Black Death, syphilis in sixteenth-century seafaring Europe, the panic-inducing epidemics of cholera in nineteenth-century Europe, tuberculosis, the post-World War I influenza pandemic and, today, HIV (AIDS). This century, as large cities have sprawled mightily and huge urban conglomerations have formed, age-old infectious disease problems have proliferated among the crowded impoverished Third World slums – particularly childhood diarrhoeal disease, respiratory infections, tuberculosis (especially in Africans coinfected with HIV) and sexually transmitted diseases. Meanwhile, in First World cities an ominous trend has arisen over the past decade; drug-resistant forms of many familiar infectious diseases have emerged – tuberculosis, streptococcal and staphylococcal infections and several types of gastroenteritis.[11,12] Tuberculosis has recently increased by around 60% in the USA, half the increase being due to the 'liberating' effect of HIV infection and half to drug-resistant strains of the bacterium. This drug resistance has profound ecological significance; it suggests a third stage in the urban struggle with infections. Initially, city crowds enabled microbes to proliferate; then, with sanitation, vaccination and antibiotics, humans achieved dominance; now the microbes, in response to the sustained selection pressure of antibiotic use (and the special opportunities of the hospital environment) are genetically adapting to the changing human environment. The inexorable processes of biological evolution are at work – and, in microbes, changes can occur fast. (This, ironically, may be one of the few areas in which human intervention is actually increasing biodiversity.)

Various non-infectious health problems are also associated with this massive social experiment of living in large cities. Many of those problems arise from, or are compounded by, population size, crowding, breakdown in social order and intractable poverty. Traditional rural or nomadic life, for all its hardships, has the social and emotional benefits of community-type living. So too does life in villages and other small settlements. However, life in big cities is different. To move into the city is to make a transition from rural community-type society to urban association-type society. The former is conservative, hierarchical and stable; it entails relationships with, and social support from, an extended kinship and close friends. Association-type life in the cities entails fluidity, more equality and more liberty – but the sense of community is often reduced or gone. Urban immigrants living in slums and shanty towns may retain important elements of their 'community', or may establish new interdependent networks. But there are strong undercurrents tending to pull those fragile communities apart. These include long and inflexible working-hours for many of the men, the demoralisation and financial stress of unemployment for others, insanitary and crowded living conditions, the exhaustion of overworked and underfed mothering, physical dangers from makeshift housing, unguarded machinery and uncontrolled traffic and variable access to other basic needs and services (fresh food, medical care, schooling).

This, then, is the demographic and ecological outline of urban living in today's world. What are the effects upon population health?

10.2 Urbanisation and human health

10.2.1 Introduction

The urban environment differs greatly from the natural environment in which, and for which, human biology evolved. Many aspects of urban life are conducive to good health, including material and educational resources, social support, physical security, and access to organised medical and hospital care. In much of the world, the vital statistics of urban populations are better than those of rural populations; unlike ninteenth-century industrialising Europe, Third World child mortality in the late twentieth century has actually been lower overall in urban than in rural populations. As we shall see, however, this comparison is fraught with information gaps and various biases, and generalisations are risky. On the debit side of city life, since the days of Jericho (the world's earliest known city) there have been health hazards associated with city-living: overcrowding, contagion, the accumulation of human excrement and household waste, occupational hazards, the discharge of industrial effluent and various forms of social disorder.

Environmental pollution by noxious chemicals is typically at its most extreme within urban populations. Indeed, streetside oxygen booths are proposed for Mexico City to alleviate the adverse health effects of that city's rampant air pollution. (There has even been talk of installing huge fans atop the surrounding hills, to disperse the intermittent heavy smog caused by 3 million cars and 35,000 factories.) Lurid accounts of environmental pollution and blighted population health have recently emerged from the industrial cities of Eastern Europe and the former USSR. The steel manufacturing and metal smelting centres of Russia are an ecological disaster (see also section 4.6). However, in this chapter we will look beyond the well-known problems of noxious pollutants, to a broader, ecologically-oriented review of the impact of the urban environment upon health.

Colonial influences last century caused many cities of the southern hemisphere to be modelled on those of northern metropolitan powers. Today there is a divergence in styles and quality of Third World urban life, reflecting a reassertion of local culture and the harsh realities of economic need. Urban-industrial development has added new dimensions of environmental pollution, and has often been abetted by trans-national companies relocating production centres to sites of cheap labour. Meanwhile, the impoverished outskirts of many East African cities are becoming 'ruralised' as slum dwellers recreate pockets of subsistence agriculture.

The rising tide of migration from poor to rich countries – which, increasingly, reflects environmental decline and economic hardship – brings many poor migrants seeking work. They change the cultural profile of the host city and introduce new foods and cuisine. While many succeed through self-denial and hard work, many fail, and the ensuing poverty, unemployment and social-cultural dislocation breeds illegal economic activity and crime. Drug trading, prostitution and other organised crime then become desperate sources of income and survival. Both those that succeed, such as the eastern Asian migrants to California, and those that fail, such as many Hispanics in New York and North Africans in western Europe, are liable to be resented and attacked by alienated segments of the host population.

In cities nearly everywhere, the rate of urban expansion is outstripping the planning and resource capacity of governments. Coping with the resultant problems in New York City is difficult enough. However, it is an order of magnitude more difficult in the Third World, where the fragile infrastructures of most of the big cities cannot accommodate the growth in physical and social demands. Consequently, poverty, deprivation, environmental hazards and poor health proliferate. In Egypt, the population of 12 million in Cairo relies on a water and sanitation system designed for a population of 2 million. Similarly, Karachi, in Pakistan, makes do with a central sewer system largely unchanged in three decades. The lethal explosions in 1992 within the decrepit sewer system of Guadalajara, Mexico's second city, were symptomatic of the same widening chasm between urban population pressure and public resources.

10.2.2 Sources of health hazards

Viewed ecologically, today's large cities represent an extraordinary conglomeration of individuals living at population densities which in most mammalian species would evoke adaptive (perhaps aggressive or self-destructive) behaviour. They also live remote from the sources of food, domestic fuel and, often, water. Nevertheless, extraordinarily – and it is a tribute to human culture – our cities have, by and large, not been the health hazard that ecological first-principles might predict. Via social and technological adaptation, the supplies of food, water and energy have been sustained, much infectious disease has been controlled, communications have been maintained, and the adverse consequences of crowding have been ameliorated. But there are ominous signs that we now face a heightened constellation of problems in the world's hypertrophied cities. As we have begun to overload Earth's natural systems, so we are overloading urban systems.

Urbanisation in the Third World has been largely uncontrolled and is becoming an increasingly serious public health problem. WHO has identified the following six major hazards:[1]

1. Rapid and massive urban population growth.
2. Increased use of land in previously unsettled ecosystems, and the occupation of urban land prone to landslides, floods and other natural hazards.
3. Extreme increases in population density.
4. The increased prevalence of extreme poverty – especially among women and children.
5. Increasing biological, chemical and physical pollution of air, water and land because of industrialisation, transportation, energy production and improper management of commercial and domestic wastes.
6. Inadequacy of the physical (especially housing, sanitation and roads), social and health-care infrastructure.

The major urban health problems that result from these circumstances are infectious and gastrointestinal diseases (the 'diseases of poverty'), chronic degenerative diseases associated with poor living and working conditions, and disorders due to the stress of social isolation, insecurity, family dissolution and cultural conflict. The Third World urban poor are at the interface between underdevelopment and industrialisation, and their disease pattern reflects this worst-of-both-worlds. They have a heavy burden of infectious disease and malnutrition. They are also vulnerable to various chronic diseases (such as respiratory disorders from air pollution and the effects of exposure to heavy metals in the ambient environment), along with disorders that relate to social disruption. Their rank poverty causes further pressures on the urban environment's absorptive capacities – and further degrades the quality of the land, water and air with which they have day-to-day contact.[13]

In crowded Third World slums, individuals often have less than one square metre of space each. Person-to-person transmission of tuberculosis, influenza and meningitis therefore occurs easily. For example, in the shanty towns of industrial Porto Alegre, in Brazil, pneumonia is the leading cause of infant deaths – the rates are six times higher than in other areas of the city.[14] New settlers, often malnourished and sometimes from refugee camps, bring a steady top-up of infectious disease organisms into this crowded human 'culture medium'. Dilapidated housing allows the proliferation of many infectious-disease vectors, particularly fleas, bedbugs, ticks, triatomine bugs, flies and mosquitoes. For example, Chagas' disease (affecting around 20 million people in urban and rural Latin America) is transmitted by insects that thrive in cracks in the walls and floors of mud or wooden houses.[15] The insect vector (the blood-sucking triatomid bug), which naturally transmits the infectious agent (*Trypanasoma cruzi*) to rodents and marsupials, has adapted to human encroachment on its natural habitat by colonizing the human domestic indoors habitat. The high prevalence of Chagas' disease, for long affecting 5–10% of the population, makes it the second most important vector-borne disease (after malaria) in Latin America. It causes long-term, essentially untreatable, debilitation and, by affecting the heart muscle, it is a major cause of chronic heart disease – and of sudden death in apparently healthy young persons. Meanwhile, outdoors, various of the 'clearwater' mosquitoes adapt quickly to breeding in dirty pools and puddles, underscoring their capacity for continuing ecological adaptation as the human-made environment changes.

Cities in rich countries also have health problems arising from deficiencies in the urban environment. Many of the conurbations of rich countries are now undergoing a crisis of post-industrial living, in which the urban elites are moving from the inner cities, leaving behind an underfunded social infrastructure, a demoralised civic administration, and a socially disadvantaged population residue beset by unemployment, family breakdown, loneliness, violence, drug abuse and various forms of criminality. This flight from the inner city has created what the Americans call 'donut cities', comprising an empty, decrepit, impoverished city centre surrounded by a sprawling suburbia replete with have-a-nice-day shopping malls. In some of these cities, urban renewal programmes are now seeking to reverse this process, making inner city living attractive again. There is a risk that, unless done within an overall and socially equitable urban planning context, 'urban renewal' may simply reverse the residential relationship of rich and poor, and consign the marginal workforce to the remote outer suburbs, perhaps unserviced by public transport.

Many social scientists wonder if we have departed too far from a human scale, a sense of comprehensible community, and opportunities for meaningful and creative work and leisure. Emile Durkheim, a pioneer of modern sociology, late last century described a trend towards 'anomie' within urban populations of the industrialised world. This condition, entailing various forms of social disorganisation, arises from the decline of social norms. Today, the various signs of social stress – such as chronic behavioural disorders, violence, drug abuse, alienation and hopelessness – seem to be increasing in many of our cities. The violent racial and class hatreds of the Los Angeles riots in 1992 revealed potent tensions not far below the surface of modern urban America. René Dubos (see chapter 3) would have said that these are all problems of disordered human ecology, arising within that part of the environment that we ourselves have made – the city.

The intransigence of the worldwide problem of 'hard drugs' to orthodox law enforcement is another reminder that the fortunes of urban populations are not insulated from those of rural populations. Lavish law enforcement measures against drug trafficking continue to fail because they cannot defy economic gravity. Financially desperate rural populations in poor countries, often abetted by debt-ridden governments, grow the crops that sell best in the international markets, both legal and illegal. Between 1986 and 1990, world opium production doubled and coca production went up by a quarter while, simultaneously, the world market price for orthodox commodities continued to fall. The case of Colombia, designated by the USA as the villain of the cocaine piece in recent years, is instructive. In 1990, just as President Bush committed an extra US$8 billion to the war on drugs, the USA caused coffee prices to plummet by withdrawing from an international coffee agreement. Now, since coffee is Colombia's largest legal export earner, if impoverished Colombian peasants cannot sell coffee profitably to rich American consumers then they will grow cocaine as an alternative cash-crop. (Policy inconsistency is one thing; hypocrisy is another. A former US Surgeon-General has said of

US trade pressures on Third World countries to import US-made cigarettes: 'at a time when we are pleading with foreign governments to stop the export of cocaine, it is the height of hypocrisy for the United States to export tobacco'.[16])

Finally, there is another important way in which urbanisation adversely affects our environment and therefore our health. The spread of cities, the growth of industry and its effluent, and the concentration of human, domestic and commercial wastes puts many pressures on local ecosystems. Wetlands and mangrove swamps get filled in and built over. Freshwater supplies get diverted by local agricultural needs. Coastal and estuarine fisheries get polluted with chemicals, vast quantities of nutrients (including those in sewage) get dumped in waterways and oceans (often causing algal blooms), and aquifers become polluted with toxic chemicals from landfill dumps. Air pollution (particularly photochemical smog) damages surrounding vegetation and crops. There is also, in a much amplified form, the age-old problem of overworking and eventually exhausting the local agricultural system in order to feed a growing urban population. These sorts of effects are unavoidable at some level if humans are going to live together in cities. However, it is the sheer scale of these effects and the rate at which environmental damage occurs that is the ominous hallmark of urbanisation in the world today. Latin American specialists point out that most of the areas in Central and South America that are undergoing rapid population growth are within the more fragile eco-zones, where sustainable use of natural resources is already difficult.[17] Many of our cities are thus becoming ecological malignancies that may exceed, and exhaust, the carrying capacity of their local environment.

10.2.3 *The urban health profile*

Many health indices, particularly in rich countries, suggest that the *average* level of health is better in large cities than in the countryside. These statistics often mislead, however, because measures of urban health are heavily weighted by the superior health experience of the affluent segment of city-dwellers. In the Netherlands, for example, while the life expectancy for the urban poor is five years less than for the well-to-do, the overall figure is close to that of the latter majority group.[1] In reality, the poorer segments of urban communities often have much worse health status than the rural poor.[18] However, the poorest of the urban poor are often statistically elusive: they may be illegal immigrants, they may be transitory, and their health and vital statistics are usually not recorded. In Third

World countries, further bias results from the selective tendency for the more severely disabled and sick to stay with their rural community and not migrate into the non-supportive urban milieu.

WHO estimates that 600 million urban-dwellers in developing countries live or work in 'life and health threatening environments'. For many poor urban residents, the main daily struggle is for food. Their typical intake is around 60% of the city average, while the intake of various key micronutrients such as vitamin A is usually proportionately much less. Severe malnutrition is common in slum children, and the health and nutritional status of their mothers – poorly educated, struggling to breastfeed and care for a succession of young children and subjected to discriminatory social practices – is particularly vulnerable. In many urban slums, a malnourished child, lacking safe water and sanitation, is 40–50 times more likely to die before reaching five years of age than is a child in a rich country.[19] The dominant causes of illness and death in young children are diarrhoeal diseases, respiratory infections, vaccine-preventable diseases (especially measles) and nutritional deficiencies; all are strongly associated with poverty, overcrowding and poor environmental conditions.

Many epidemiological studies have described gradients in health, particularly child health, between urban and rural populations and between different segments of urban populations.[1,20,21] The following are examples of the former comparison. The rates of leprosy are lower in urban than in rural populations in many African countries; in the UK, lung disorders are more prevalent in cities than in countryside; and in Spain, there is a higher prevalence of mental retardation in rural areas than in cities. Studies *within* cities show consistent health deficits among the poor. The infant death rates from diarrhoea are correlated with lack of piped water in urban populations of southern Brazil. In Quito, the capital of Ecuador, the infant death rate has been 20–30 times higher in squatter settlements than in upper-class districts.[21] In the slums of Haiti's capital, Port-au-Prince, the infant death rate is three times that of the rural area and many times greater than in the richer areas of the city, while in Manila, capital of the Philippines, the infant death rate in squatter communities has been three times higher than for the rest of the city.[19] Various studies have shown consistently higher levels of intestinal parasitic infestations (especially *Ascaris* and *Trichuris*) in the poorer areas of the cities and a variety of nutritional deficiencies among those socioeconomically disadvantaged groups.[8] It is clear from these and a great many other studies that there is a strong link between poverty and generally poor health – although some care is needed in the interpretation of such statistics since their quality may vary between the compared subpopulations.

Some of these studies have sought also to identify the actual environmental factors or circumstances that are the immediate cause of poor health. For example, a study of infant diarrhoeal deaths in Pelotas, Brazil, showed that households sharing taps or using public stand pipes were at higher risk than those with in-house piped water.[22] Readers familiar with the history of epidemiological research will hear echoes of John Snow's seminal studies in cholera-afflicted London 140 years ago. Presumably we will go on documenting this deadly relationship for so long as there is persistence of the underlying poverty that leaves communities without safe drinking water. Lack of clean water and adequate sanitation is widespread in Third World cities and remains a major killer. The previous Director-General of WHO, Halfdan Mahler, claimed that the number of water taps per thousand persons is a better indicator of health than is the number of hospital beds.

Today, about 80 % of the world's city-dwellers are supplied with safe water, either by in-house connections or public stand-pipes, and about 65 % have sanitation via sewer connections, septic tanks or latrines. The corresponding figures are lower in rural populations. Most of those without are in the Third World (see Fig. 10.1), where the general lack of maintenance further erodes these statistics. Up to one-half of the piped water is lost or unavailable because of leakages, broken water mains, or malfunctioning stand-pipes. Some of the official statistics are hardly credible anyway – for example, Nigeria and Liberia reported in the mid-1980s that 100 % of their urban populations had access to safe water. (Presumably the word 'access' encourages creative statistics.)

Poor sanitation is an even more widespread problem. In many large cities, particularly in Africa and Asia, there is no sewerage system, and so much of the human excreta and household wastes end up, entirely untreated, in rivers, stream, canals, gullies and ditches.[23] People largely depend on backyard pit toilets, while, in the worst slum areas, open defaecation in lanes and alleyways helps to spread hookworm, other intestinal worms, typhoid, dysentery and diarrhoeal diseases.[24] One-seventh of the world's population is infested with hookworm, resulting in 50,000 deaths annually This parasite enters via the skin, usually of the feet, and makes its way via the blood stream to the intestine. There it attaches to the intestinal lining, where it grows (thus depriving the host individual of scarce nutrients), and sheds eggs. These are transferred, via faecally contaminated groundwater and soil, to the next pair of unsuspecting

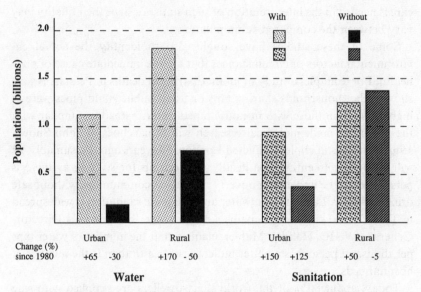

Figure 10.1 Numbers of people with and without access to safe drinking water and adequate sanitation in the Third World, 1990. (Source: World Bank, 1992[41]).

passing feet. Other widespread intestinal worms, such as round-worms, whipworms and others, have similar life cycles. They all spread readily in urban environments where poor sanitation creates special ecological opportunities. In Mexico City, for example, cysticercosis (a tapeworm that is usually transmitted in contaminated pork) is now spread more widely via sewage-contaminated vegetable gardens.

In addition to unsafe water and lack of sanitation, urban populations in poor countries are prone to other causes of disease and death – including malnutrition, inadequate housing, lack of facilities for solid waste disposal, narrow and dangerous roadways and overcrowding. Some of these health hazards are clearcut, and their effects are hardly contentious. For example, the accumulation of solid wastes in, or next to, urban slums encourages the spread of parasitic diseases such as leptospirosis, primarily because of the garbage-associated proliferation of insect and rodent vectors. Physical injuries frequently result from unsafe buildings, which often collapse, and from children playing on dangerous unplanned roads. Cities also pose social–environmental health problems, especially to young people. In Third World countries, approximately half of all urban children under age fifteen live in conditions of extreme poverty. About one-fifth of these children are 'street children' with minimal or no family support. They work at odd jobs, scavenge or beg for food and often have to seek out shelter. They are exposed to great physical, emotional and sexual hazards. The problem is severe in Latin America, where there are an estimated 20 million homeless 'street children'.

There is a more obviously ecological aspect of urban population health, reflecting the interface between cities and surrounding ecosystems. As the expanding urban fringe encroaches on adjoining agricultural land, waterways or forest, so opportunities arise for endemic infectious disease vectors (e.g. certain species of mosquitoes) to adapt and thereby include the urban population in its habitat. Various water-related parasitic diseases, previously primarily diseases of rural populations, are now becoming endemic in many cities of the developing countries. These include schistosomiasis and the mosquito-borne infections malaria, dengue, filariasis and, in some regions, yellow fever. For example, irrigation ditches around urban fringes nurture the clearwater mosquito, *Aedes aegypti*, resulting in heightened epidemics of dengue fever that can be a major cause of child mortality. In a dengue epidemic in Thailand in 1987, 1,000 deaths occurred, predominantly in children.[23] Meanwhile, in the contaminated water of back-street ditches and pit latrines of eastern Africa and parts of Asia the *Culex quinque fasciatus* mosquito is thriving and causing increases in filariasis. In Brazil, the transformation of schistosomiasis into an urban disease has been associated with migration from rural areas in the north-east of the country where schistosomiasis is endemic.[23] Thus, as ecological boundaries are breached, naturally confined pockets of infection become generalised urban health problems.

10.2.4 Where the poor (must) live

In cities and towns around the world, the poor live in the most run-down and crowded housing, often in degraded environmental settings. In developing countries they often live close to industrial plants, since poverty and lack of transport oblige many urban migrants, as low-cost factory workers, to settle nearby on land that nobody else wants. Environmental pollution is a serious health hazard for those living near to industrial sites, power plants, dumps and busy motorways. The burning of fossil fuel causes local air pollution and, from Mexico City to Shanghai to the industrial cities of Eastern Europe, this has blighted the respiratory health of local communities. In lead smelter towns, nearby housing accumulates lead in the house-dust, yard and garden – jeopardising the mental development of children. In many Third World cities, in addition to the usual slum-dwellers' problem of water-borne infection, contamination of water by industrial wastes occurs – as is reflected in increases in childhood health disorders due to heavy metal poisoning.[19]

As environmental standards become restrictive in developed countries, large companies may relocate hazardous factories to poor countries with minimal environmental controls. In the 1970s, for example, many companies that manufactured asbestos products transferred production from the USA to Latin America, particularly Mexico and Brazil. Some US car manufacturers built factories in Mexico that have seriously polluted the air and water of the urban environment. When recipient countries, keen (or desperate) for industrialisation and employment, create special-ised export-processing zones with a high concentration of industry, then health risks can escalate. South Korea has followed this path, and the hugely expanded population living around the Ulsan/Onsan industrial complex has suffered increases in respiratory and nervous system disorders, while local agriculture and fisheries have also been damaged by pollution.[23] There may also be a heightened danger from accidents. The Bhopal tragedy in India, when explosive dispersion of methyl isocyanate at the US-owned Union Carbide plant killed around 2,000 persons and injured many tens of thousands, underscored dramatically the vulnerability of poor urban-dwellers to industrial hazards.

In most Third World cities, there are spontaneous settlements on illegally occupied land. These have varied names – in Argentina, *villa miserias*; in Brazil, *favelas*; in India (Delhi), *jhuggi-jhompris*, and so on – and they typically exist on land ill-suited to housing. These settlements have no basic plan and little room for thoroughfare. Two-thirds of the slum population of Sao Paulo, Brazil, live in areas prone to flooding and landslides; most have no drinking water or public lighting and 98 % have no access to sewers or septic tanks. On the lower slopes of the denuded hillsides around Rio de Janeiro, 3 million squatters have built make-shift shelters. In February, 1988, torrential rains sent avalanches of hillside mud into these slums, killing several hundred people and leaving 20,000 without any shelter. In Banjul, capital of The Gambia in West Africa, I saw such a settlement taking root – a group of dispossessed rural immigrants from neighbouring Senegal were building a rudimentary settlement with salvaged tin, cardboard and plastic bags, at the edge of the refuse dumping area, next to a mosquito-infested swamp. Located similarly, the hillside squatters of Rio de Janeiro and Sao Paulo have experienced many outbreaks of the rat-borne infectious disease, leptospirosis, when flooding has displaced rats and stray cats and dogs from the adjoining rubbish dumps.

Twenty years ago, the physical clearance of illegal squatter settlements, often coupled with enforced resettlement in low-cost public housing, was a crude but accepted approach to solving urban housing problems in developing countries. However, while this eliminated some of the health problems arising from the cramped, squalid and unserviced squatter settlements, it simultaneously created new health problems.[25] Slum clearance schemes have generally fallen out of favour, as governments now try more supportive housing policies, including public construction programmes and the provision of social services and basic physical services. While the latter approach is more enlightened, it is adaptive rather than preventive. Curitiba, in Brazil, shows what can be achieved in the short term with such an approach – but, ironically, such success lures hordes of hopeful rural poor. So long as rapid population growth and entrenched poverty persist, so will the problems of urban deprivation, environmental degradation and poor health.

10.2.5 Other urban health problems

Despite what the philosopher Jean-Jacques Rousseau might have implied about city life 250 years ago when he blamed civilisation for mental illness, urban life is not obviously more hazardous to mental health than is rural life. Nevertheless, among city-dwellers, the poor, the socially marginalised and the isolated are particularly vulnerable to mental disorders; and this vulnerability is often compounded by lack of social support and a feeling of entrapment. Many studies in North America have shown that the distribution of mental disorders in cities closely follows the distribution of poverty in inner city areas, and that a lack of social and cultural cohesion particularly endangers mental health. There is some evidence that schizophrenia, one of the most common serious mental disorders, occurs more often in lifelong city-dwellers than in rural populations.[26]

Various other mental health problems are related to aspects of the urban environment.[25,27] On the one hand, crowding and lack of privacy and of control over one's living space may cause disordered social relationships, aggressive behaviour, child and spouse abuse, or drug and alcohol abuse. On the other hand, isolation and loneliness, without family or social support, may cause depression, suicide or drug and alcohol abuse. Poor quality housing, insecure tenure, unemployment and marital breakdown all contribute to psychosocial health problems in socioeconomically disadvantaged urban-dwellers. Among adolescents, the frustrations and demoralisation caused by such deprivations or setbacks, and from the lack of facilities for social and physical recreation that result from poverty, urban decay or bad town-planning, can provoke vandalism, drug-taking and indiscriminate graffiti. Research has consistently shown, however, that the existence of strong social networks and a sense of local community mitigate these mental health hazards.

In recent decades, many cities in rich countries have undergone a commercialisation of their central areas, with soulless glass-box office towers and commercial enterprises replacing much of the inner city housing. In place of stable residential communities, impoverished minority groups, alcoholics, drug addicts and behaviourally disordered people have moved into the residual housing. Where this demographic shift has been followed by economic downturn and de-industrialisation, then endemic poverty, urban decay, and deviant and criminal behaviour have typically ensued. Attempts to eradicate the problems by building high-rise public housing have often compounded this inner-city malaise, as has the thrusting of concrete freeways through inner-city neighbourhoods. Now, in many countries, a wave of realisation is breaking over city planners, leading to attempts to recreate the cohesive and attractive high-density inner city living that occurred rather more spontaneously in much of the early life of European cities.

Alongside the problems due to urban design and disordered social relations are various major problems of urban environmental pollution, in cities around the world. One obvious health problem arises from the high levels of ambient air pollution, particularly from the burning of fossil fuels. These problems have been evident in the cities of developed countries for many decades and, around mid-century, periodically caused acute surges in deaths in various coal-burning metropolises. The most dramatic episodes were in Liege, Belgium, in 1930, in Donora, Pennsylvania, in 1948 and in London in the famous 'pea-souper' winter smog of December 1952. While much of that particulates-and-sulphur air pollution has been eliminated in developed countries, many developing countries and Eastern European countries – particularly coal-burning China and Poland – now have this type of air pollution. It has been reported that in Weimar, in eastern Germany, wintertime concentrations of sulphur dioxide in recent years have exceeded the infamous 1952 levels in London.[28] In heavily industrialised southern Poland the wintertime levels of local air pollution rival those seen in some of Poland's most heavily exposed occupational environments. Today, there is a rapidly growing worldwide problem of photochemical smog in car-congested cities, while many cities in developing countries are also blighted by the 'traditional' air pollution caused by industrial and domestic coal-burning.

Another air pollutant, lead, derives from car exhausts (leaded fuel) as well as from industrial emissions and is attracting wide attention as an urban health problem. In Bangkok, Mexico City and Djakarta, it is largely a problem of automotive exhausts; in inner-city Chicago, Washington and Sydney it is as much a consequence of flaking leaded housepaint in dilapidated housing areas. Unlike modern synthetic chemicals or photo-chemical smog, lead is not a new health hazard. Indeed, it may have been the first metal smelted by humans.[29] The Greeks and Romans wrote about lead poisoning and, in one historian's view, lead poisoning of the upper classes may have contributed to the decline of the Roman Empire.[30] The combination of lead plumbing, lead-based pottery glazes for food and wine storage vessels, and the addition of lead arsenate to wine as a sweetener would have ensured a substantial daily dose of lead for well-to-do Romans – whose fertility and intelligence might then have suffered. (Those who subscribe to a cyclical view of human history would savour a recent report, from England, of high levels of lead in sherry stored in expensive high-lead crystal decanters, such as are found on the sideboards of stately homes.) The advent of motor vehicles this century has increased the use of lead, in fuel and batteries. It has also been used in house paint and in solder in food cans. Community exposures have thus increased, and concern now focuses on the adverse effects of environmental lead exposure (associated with blood lead concentrations of less than around 20–25 micrograms per decilitre) upon the developing brain of young children. Blood lead concentrations typically peak at around age two years, reflecting the hand-to-mouth lifestyle of early childhood. Recent studies in Australia and the USA have demonstrated the occurrence of adverse, although small, effects due to such environmental lead exposure.[31,32] From this research, it is estimated that children in heavily lead-polluted urban environments may have lost up to 3–4% of their measured IQ, by comparison with children in relatively unpolluted urban environments.

Unlike synthetic pesticides and photochemical smog, lead is a natural component of our environment, present in trace amounts. It is the scale and form of human economic activity that creates the public health problem – we have hugely increased the 'flux' of lead through the environment. Cities are part of that 'scale and form' of expanded economic activity that has caused health-endangering exposures to lead, as, yet again, we tip the scales in the direction of environmental overload.

10.2.6 Cholera: a reminder of urban vulnerability

Cholera, an acute infectious disease, is a recurring and ancient scourge of urban slums and poverty. The cholera bacterium, *Vibrio cholerae*, is spread by contaminated water and food. It produces a toxin that causes damage to the lining of the gut, diarrhoea, and a rapid, often fatal, dehydration. Cholera is currently on the move again, as a pandemic, having extended into the equatorial countries of Latin America in 1991.

In the early and middle decades of last century cholera terrified the rapidly-urbanising populations of western Europe, as successive waves of the dread disease rolled in from eastern Europe and beyond. With only brief delays, these epidemics breached the Atlantic buffer and swept through urban North America. The cause of cholera was unknown; scapegoats were sought. The nature of infectious disease was not yet understood, and officialdom believed that contaminated air-borne miasmas were the source. In mid-century London, John Snow (who was Queen Victoria's anaesthetist at her two confinements) concluded from meticulous epidemiological research and clear-headed biomedical reasoning – and against the weight of learned medical opinion – that cholera was water-borne. This, he said, explained why it occurred most frequently in households drinking water from sewage-contaminated sections of the Thames. Although Snow's view was rejected by rear-guardists in the medical scientific establishment for several decades, as sanitation and hygiene progressively improved in Europe so cholera receded.

Meanwhile, epidemic outbreaks of cholera occurred periodically around much of the world. There have been seven recorded pandemics since early last century. The most recent pandemic began in Indonesia in 1961, then, assisted by human migration, spread through Southeast Asia, and westwards via the USSR and West Asia. Eventually it reached previously uninfected West Africa where it became endemic, particularly in crowded coastal areas. During the late 1980s, Africa accounted for the great majority of reported cases of cholera.[33] In January 1991 it reached South America, creating an epidemic in that continent for the first time since the 1880s.

The epidemic spread fast in Peru, infecting nearly 200,000 people and killing over a thousand in the first three months of 1991. It then spilled over into Ecuador and Colombia, and subsequently hundreds of thousands of cases have occurred in many South American countries. The most rapid spread occurred in the slums and shanty towns, largely bereft of sanitation. These urban fringe-dwelling populations, like those of early cholera-prone industrialising Europe, have formed as a result of rural migrants settling in cities faster than new infrastructure can be provided – or afforded. Impoverished Peru has sought foreign loans to build sewage treatment plants, but its international credit rating has been deemed too low. Consequently, 40 % of people living in Lima, the capital of Peru, have no clean drinkable water. More generally, the Andean South American population is very vulnerable to water-borne infection because of a widespread lack of clean drinking water and sewage disposal. In 1992, the disease established itself in inland western Brazil, underscoring the spectre of possible spread into the huge, warm, unsanitary slums of Brazil's overcrowded coastal cities. Once established within poor urban populations, infectious diseases such as cholera can become endemic, recurring year after year – as has long occurred in India and Bangladesh.

The cholera organism – unlike typhoid – does not persist in human carriers. However, by establishing permanent environmental reservoirs in slightly saline environments, in which it proliferates, cholera becomes endemic. In Bangladesh, for example, where cholera is endemic, there is an annual cholera epidemic that coincides with the blooming of the blue-green algae (*Anabaena*) in semi-saline waterways along the coast. Likewise, in Africa, it appears that the anchoring of endemic cholera has occurred in aquatic ecosystems, not just in brackish water but also in freshwater. The cholera vibrio is now thought to 'hibernate' beneath the mucous covering of algae and phytoplankton. Plankton sampled in the coastal waters of Peru in 1991 were found to harbour the vibrio.[34] Predictions of future algal proliferation, in response to global warming, carbon dioxide 'fertilisation' and nutrient runoff from urban populations, therefore have ominous implications for the environmental spread of cholera.

The solution to health problems such as cholera in large, expanding and impoverished cities will require ecological insight. Further, while WHO estimates that it would cost US$5 billion a year for the next decade to provide safe water and sanitation for Latin America, poor countries cannot afford a fraction of that sort of expenditure. Cholera is a disease of poverty and urban crowding, facilitated by poor education, lack of sanitation and environmental deterioration. Money spent on cholera vaccination (if and when a cheap and effective vaccine becomes available) would prevent cholera in the vaccinated population. However, money spent on safe water, sanitation, fertility control and other social infrastructure needs would prevent not only cholera, but other water-borne infectious diseases and poverty-related health problems.

10.2.7 Sexually transmitted diseases

By the middle of this century, in rich countries, the long decline in most of the age-old infectious diseases – especially those that killed young children – had finally been capped by the advent of modern antibiotics and vaccinations. Infectious disease seemed no longer to be a serious public health problem in those countries. However, the 1980s have reminded us of the ever-present hazards of infectious diseases: microbes, the tiniest members of our ecosystems, are always with us. Their genetic profile changes restlessly as, in effect, they probe for new ecological niches in which to multiply, or for defences against human anti-microbial factors (such as antibiotic drugs).

During the past decade there has been a widespread outbreak of sexually transmitted diseases (STD) – 'wide' both in the number of infectious agents and in the range of human populations affected. The attack has come predominantly from four bacteria: those responsible for syphilis, gonorrhoea, chancroid and chlamydia; and from four viruses: the herpes simplex virus, human papilloma virus, hepatitis B virus (HBV) and human immunodeficiency virus (HIV – i.e. the AIDS virus). These sexually transmitted diseases are particularly associated with urban poverty and with the disintegration of traditional forms of social and family organisation. In developed countries, and particularly in the USA, the rates of most STDs are highest among the urban poor. These are primarily diseases of teenage years and young adulthood, and the age profile of much of the world's urban poor renders those populations susceptible. Not only is the age-profile loaded towards risk of infection from sexual activity, but, in the case of HBV and HIV, intravenous drug abuse compounds the risk.

AIDS was first identified in 1981, in the USA, and has subsequently been widely reported from around the world. The circumstances in which the virus first entered the human species remain uncertain and controversial; there may have been undetected cases of AIDS in the 1960s, or even earlier. Speculation about its origins has encompassed the possibilities of contamination of early polio vaccine grown on monkey kidney tissue, transfer of infected primate blood to volunteers during laboratory studies of malaria transmission, and localised African ritual contact with monkey blood. Validity aside, these possibilities remind us again that changes in human demography, culture and technological practice create ecological opportunities for microbes. In late 1992, WHO estimated that 10–15 million men, women and children had been infected worldwide. Of these, 1.5 million had progressed to clinical AIDS. The HIV pandemic continues to grow and has now also become widely established in the big cities of India and Southeast Asia (particularly Thailand) and Latin America.

By 2000, an estimated 40–50 million people will have become infected with HIV – perhaps more according to less conservative estimates. If so, around one-tenth of the world's annual tally of deaths would be attributable to AIDS by early next century. Most new HIV infections during the 1990s will occur in poor countries, particularly in urban populations, and including 5–10 million children. In Africa, the infection is approximately equally distributed between men and women and is being passed on to an increasing proportion of new-born children. By the late 1990s, there will be around 10 million uninfected African children orphaned by the death of both parents from AIDS. Some demographers now think that AIDS will cause the worst-affected parts of sub-Saharan Africa to experience a reversal in population growth. Meanwhile, in developed countries, HIV infection has been much more closely associated with male homosexuality and intravenous drug abuse than in Africa or Asia. Heterosexual transmission, while still only a minor contributor in developed countries, appears to be gradually increasing.

Behind these statistics lies the ecological reality that social disruption and change due to the breakdown of traditional agricultural and employment patterns, uncontrolled urbanisation, persistent poverty and continued population growth all potentiate the spread of STDs. The 'silent epidemic' of STDs and infertility among women of the rural Third World reflects the combination of non-traditional rural–urban–rural mobility of men with the traditional oppression of women (which often precludes the use of barrier contraception and access to medical care).[35] Through our sexual values and behaviours, we continue to create new ecological opportunities for the ever-alert microbes with whom we share the biosphere.

10.3 The future: designing cities for healthy living

10.3.1 Introduction

The goal of designing healthy cities has figured repeatedly in the history of Western society. Plato, wrestling with the ideals of the Greek city-state, gave it serious attention. The nineteenth-century enthusiasm for Hygiea, the City of Health – a reaction against the urban blight and scramble that was unleashed by early industrialisation – has been discussed in chapter 3. Today, the European Office of WHO is mounting a Healthy Cities Project;

countries from around the world have joined in this project. Its proponents argue that cities are built by people; so they should be *for* people, not for commerce, cars, or concrete manufacturers. Decision-making should therefore involve the communities and groups concerned – the urban environment should be conceived and planned on a 'human' scale, to promote community identity and autonomy and to foster social contact and good health.

Historically, in today's developed world, communities and their policy-makers have usually planned only for the foreseeable future, and so their cities – like elderly European cathedrals – have growed like Topsy, bit by bit. Today, under the growing pressures of migration, hard-core poverty and social disintegration, many of these cities are faltering under demographic and social strain. In Third World countries, poverty and a lack of planning persist as the dominant influence on the shape of tomorrow's megacities – undermining the prospects for better health in those cities next century. WHO has recently argued that sustainable, health-promoting urban development requires policies, plans and programmes based on an understanding of the urban system as an organic whole.[23] Urban development should not exceed the capacity of local ecosystems, and should support rural and regional health and prosperity. The attainment of those ideals is, of course, impeded by population growth and poverty. Therefore, at least as a holding operation, until widespread urban poverty and its worst effects are alleviated, we need broad physical guidelines and social strategies to guide urban growth and land use.

10.3.2 The car

Cars have become a special problem for cities. During the last two-thirds of this century, in developed countries, the car has progressed from being servant to master; from being a novelty for the idle hamper-packing rich to a consumer necessity for many. It provides mobility, emancipation from the suburbs, recreation – and social status. But cars also cause congestion, air pollution, trauma, noise, greenhouse gas emissions and social fragmentation. Discarded, they linger in piles on the urban fringes, the ultimate non-biodegradable junk heap.

The voracious needs of the automobile dominate city landscapes throughout the developed world. Two-thirds of the land area of inner Los Angeles is devoted to roads and parking lots. Main streets in the cities of developed countries are blighted by used-car saleyards. Many residential communities have been surgically divided by new highways and freeways.

Corner stores and neighbourhood shops, once within easy walking or cycling distance, are superseded by soulless suburban shopping centres surrounded by asphalted acres of parking space. In the words of Peter Newman, an Australian urban-environmental scientist: 'The standard approach to [building] roads has been to see them as a conduit for traffic where little was allowed to get in the way of a smooth flow. . . . This is reductionist engineering that forces cities into a mould of concrete and bitumen, denying the importance of streets as meeting places, recreation areas, flora and fauna havens and the aesthetic glue that holds the city together.'[36]

Ironically, the dual proliferation in car ownership and population steadily reduces the benefits to owners. In many cities, peak-time traffic moves slowly, if at all. At the centre of London, in 1905, the average speed of horse-drawn traffic was 8 mph, and, in 1990, that of motorised traffic was 11 mph.[37] This congestion is commonplace in cities like London, Rome, New York and Los Angeles, where executive car-phones are used as an antidote to waiting in stalled traffic. If congestion-relieving motorways are built, then more people are enticed to drive, and public transport continues to decline. Such congestion is increasingly becoming a feature of big cities in poorer countries – and in Mexico City and Seoul, at peak periods, it is quicker to walk or bicycle. In India, for the moment, the ubiquitous unhurried urban cows maintain a commendable level of passive resistance to the excitable urgency of horn-tooting motorists.

In England, between 1952 and 1987, the number of passenger kilometres travelled per year by car increased almost seven-fold, while that of bus travel halved and that of cycling fell by four-fifths.[38] Yet, it is widely known that public transport is safer, more energy-efficient, uses less space and produces fewer emissions per-person than does private car use. A car carrying one person uses 5–10 times more energy per passenger-kilometre than a bus carrying 45 people. Bicycles are, of course, even more energy-efficient; further, they are quiet, emission-free and need little parking space. In pedestrian-friendly and bicycle-friendly cities such as Copenhagen or Amsterdam they are also relatively safe. Globally, they have become the dominant form of transport, as moderate gains in income have occurred in the developing world, particularly Asia.[39] In 1990 over twice as many bicycles as cars were built. As the world's bicycle fleet approaches one billion this decade, that of cars approaches half a billion. Four-fifths of these cars and light trucks are in the developed world, while two-thirds of the bicycles are in Asia (including Japan, where they are widely used for commuting).

As well as causing over 300,000 traumatic deaths around the world each year, along with much injury and property damage, cars contribute mightily to urban air pollution. These air pollutants, particularly carbon monoxide, nitrogen oxides and ozone, have various adverse health effects, particularly respiratory disorders such as asthma and bronchitis.[40] Various components of exhaust gases, both from conventional and diesel engines, are suspected of contributing to human lung cancer. The health effects of car exhaust gases, particularly those of yellow-brown summertime photo-chemical smog, have been widely documented – and, in California, have finally impelled strong legislative responses. From later this decade, a progressive increase will be required in the proportion of zero-emission cars sold. Such responses are much less easy in poor countries, where basic catalytic converters are unaffordable. This, and the increase in vehicle numbers, points to a four-fold increase in urban exhaust emissions by 2030 in developing countries in the absence of strong abatement measures.[41] Cars are also major contributors to greenhouse-enhancing carbon dioxide emissions. Cars and light trucks account for around one-fifth of all anthropogenic carbon dioxide emissions. (For readers who prefer a tangible statistic, using a tank of fuel in a four-cylinder car produces an amount of carbon dioxide that weighs about the same as the driver.)

Despite these negative impacts of cars upon the urban environment, there are many incentives to use them. For example, governments hugely subsidise road construction, keep fuel prices vote-winningly low and accept cars as a non-taxable part of every business executive's salary package. However, inner-city populations in developed countries are starting to object. Cities such as Athens, Milan, Copenhagen and Los Angeles have recently introduced restrictions on car usage to reduce congestion and pollution, and to allow pedestrians the unmolested pleasures of walking, talking, sitting and browsing. Amsterdam has been constraining the downtown use of cars, hoping that its canals will then be rehabilitated for transport. In a remarkable turn-around, car-infested Los Angeles has embarked on a huge smog-reducing programme of light-rail construction. In Palo Alto, California, the city government pays its employees to cycle to work. Government action in both the Netherlands and Denmark, including the provision of extensive tracks and parking facilities for bicycles, has greatly boosted cycling. Meanwhile, in China, spacious fenced-off cycle lanes on main roads, copious open-air parking space (with official, and usually very elderly, guards in attendance) and subsidised prices all ensure that tides of cyclists still dominate the roads.

Restrictions on the urban use of the car, and a switch from fossil fuel to solar or hydrogen power, may yet produce a socially and ecologically sustainable version of this culturally-powerful icon. So, what urban planning options will reduce our actual use of cars? From studies in cities around the world, three policy needs stand out: promotion of rail-based transit; various methods of traffic calming to liberate cyclists and pedestrians; and the development of high-density urban villages integrated with rail transit.[42] This reference to 'urban villages' raises questions about the optimal size and form of cities. Read on ...

10.3.3 Optimal size and form of cities?

Animal behaviourists have written much about the optimal size of animal populations in the wild, and how various physiological and behavioural mechanisms act to stabilise population size. Many such populations only come together at special times, such as at annual breeding or migration. However, since only a few humans now live 'in the wild', and since, unlike most other species, nomadic humans live in a great variety of very different habitats around the world, it is not possible to specify an optimal population size that would generally apply to pre-agrarian humans. Besides, our prime interest today is in the optimal size and form of human *settlements*, as the foundation for good health, happiness and material wellbeing.

In the early 1970s, *A Blueprint for Survival* proposed the decentralisation of urban populations into small village-like settlements of about 500 people, aggregated into larger communities of around 50,000.[43] In this way, 'human-scale' communities would replace the social and ecological problem of the modern, ever-expanding, megalopolis. They would have their own internal sense of identity, with a sense of shared responsibility and cooperation. Critics argued that such small and self-contained settlements would breed social pressures and moral coercion, thus infringing the freedom of action and the opportunity for independence and anonymity of city life. They predicted the recreation of the pettiness, rigidity and tedium of life in small villages. Others objected to the intellectual and cultural constraints that would apply, arguing that modern massive cities are, like mediaeval cathedrals, an expression of human creativity, spirituality and striving. Clearly, in the debate about the optimal size of human settlements, there are romantics on both sides of the fence.

Related to size, but distinguishable from it, is the consideration of patterns of social contact. Peter Wilmott, an English researcher, argues that even superficial interactions between local residents, if regular, encourage some sense of attachment to the locality. Regular residents, he says, 'get to know each other by sight. ... Recognising and being recognised by others creates a sense of belonging.'[44] However, the creation of small and cohesive 'communities of attachment' – whether in the post-industrial rich world where people will live, work and play predominantly in and around their homes, or in the urbanised populations of poorer countries where, in coming decades, activities will be simpler and more traditional – needs to be balanced against considerations of efficiencies of scale for the provision of services.

Other commentators argue that humans have a basic need for contact with nature and its spontaneous forms. Modern Japanese techno-cities, with their over-designed and chaos-free environments, along with social isolation of non-working spouses, have acquired a reputation for depressive illness and high suicide rates – the so-called 'Tsukuba Syndrome'.[45] The colour green, say psychologists, is the most soothing to the human psyche. In some of the heavily concreted high-rise housing estates of big cities, the only blades of grass visible are those that penetrate the cracks in the concourse. Cities that sprawl unrelentingly to the horizon, that lack parks and gardens, that do not plant street trees, and that provide no ready public transport for the poor are environments in which nature is neither much known about nor experienced. In such environments children think that apples grow in supermarkets. For such reasons, there is increasing advocacy for planting 'edible landscapes' in cities. This has been done in the city of Davis, California, where market gardens and vineyards within housing developments and fruit trees within the adjoining public spaces now produce more food than when those areas were plain farmland. In many large African cities, the slums and shanty towns are being 'greened' as people recreate, on slivers of land, microcosms of village agriculture. Around the world, the ancient urge to garden, to grow food and to get dirt under the finger-nails still finds many and varied outlets for city-dwellers. That urge would be (partially) met by restoring greenery and localised food-production to city life.

Urban villages

There is a growing consensus among city planners that an ecologically sound approach to urban planning will require greater population density, clearer community identity, green areas, and better transit, biking and walking areas. Car dependence increases sharply as population density decreases below around 30 people per hectare of urban land. There are other positive reasons for increasing the density in urban areas, including the fact that sprawl is very costly and that, as an area ages, its older people need options for smaller, closer housing.

One very productive way of achieving density increases is by building semi-detached 'urban villages', each with integrated high-density mixed land use. This enhances sustainability because it facilitates greater local self-sufficiency, and thus shorter trips with more walking and biking. Research in Australia has shown that if urban villages became the basis for city development then at least a 20% per-person reduction in greenhouse gases could be achieved by 2021, saving $5.5 billion in infrastructure costs for the government, with extensive light-rail and traffic calming paid for out of reallocated road funds. Additional savings would be achieved in reduced road accidents, noise and smog costs.[46]

The urban village concept has so far largely been confined to European city-planning, but it is now increasingly on the social agenda in car-based cities in North America and Australia. Existing examples may not yet be very radical, but they provide something that is more integrated, less car-dependent and more oriented to the 'urban commons' than standard low-density, privatised suburbia. Says Newman: 'They reach back to the roots of how cities first began with a mutual co-operation and sharing that enables diversity to be achieved in human activity. They provide a model for us in cities that have lost their way ecologically and socially because of a total orientation towards optimising life for private needs and neglecting the public good or urban commons.'[36]

10.4 Summary

Villages, towns and cities are a natural expression of human sociability, skills, hopes and fears. Over thousands of years, they have played the pivotal political and cultural role in agrarian and industrial populations. However, cities have become the locus of a new generation of ecological, demographic and political problems. An increasing proportion of the world's expanding population, particularly in poor countries, is living in cities. Fertility rates in the Third World urban poor are high. Much of the in-migration reflects the decline of rural life; ecological disruption and environmental disasters are likely to increase the move to the cities. By the year 2000, half of the world's population will live in cities. By early next century, various large cities in India, Latin America and elsewhere will contain 20–30 million people.

Three-quarters of the cities that contain more than 5 million people are in developing countries. In these cities there are an estimated 100 million homeless adults and perhaps as many as 80 million homeless and, in many cases, abandoned children. Although the *average* level of health has usually been better in cities than in rural areas, there are stark contrasts in health within those expanding urban populations, many of which are now seriously outstripping the urban infrastructural capacity. Overall, one in four people in these Third World cities lack safe water and over one-third of city-dwellers lack access to sanitation. The figures are much higher in the slums, where diarrhoeal disease is consequently rife – and continues to kill vast numbers of young children. In many of these cities, air pollution has now become a much graver public health problem than it is in rich countries.

Very large cities are a recent, unplanned experiment in human ecology. The sheer scale, rapidity of growth, and the dimensions of poverty, misery and poor health now evident in many Third World cities raise basic questions about the sustainable healthy forms of urban social organisation. The fact that infectious diseases thrive on urban populations has been underscored by recent surges in cholera, tuberculosis (the world's major killer infectious disease), AIDS and other sexually transmitted diseases. In addition to their immediate physical influences upon population health, cities distance urban-dwellers from the rhythms of nature, potentiate industrial expansion and encroach on and violate local ecosystems. Humans have never before tried living in cities with populations of several tens of millions of people. Such cities are rapidly becoming a spectacular, but formidable, social and ecological experiment.

References

1. WHO. Environmental Health in Urban Development. WHO Technical Report Series, No. 807. Geneva: WHO, 1991.
2. McNeill WH. *Plagues and Peoples*. Anchor Doubleday: New York, 1976.
3. UN Department of International Economic and Social Affairs. *World Urbanization Prospects* 1990. New York: United Nations, 1991.
4. Despite the dramatic megacity statistics mentioned above, most of the urban-dwellers of Africa, Latin America and Asia live in smaller cities – many of them less than 100,000 persons. (Large cities require a powerful economy to sustain them, and in some poor countries, lacking export income, there has been little urbanisation.) Meanwhile, the most common type of human settlement is the rural village, with between 10 and 10,000 people. In India, in 1980, 98 % of all settlements were villages with less than 5,000 inhabitants.
5. WHO. Urbanization and health in developing countries. The urban crisis. *World Health Statistics Quarterly* 1991; **44**: 189–97.
6. Jensen L. Developing countries begin grappling with urban woes. *World Development* 1988; **1(4)**: 4–8.

7. Report of Urbanization Panel, WHO Commission on Health and Environment. WCHE/URB/2/8. Geneva: WHO, 1992.
8. Harpham T, Stephens C. Urbanization and health in developing countries. *World Health Statistics Quarterly* 1991; **44**: 62–9.
9. Recent biological–anthropological studies have suggested that the brain:body ratio of humans, relative to that of other apes, would predispose to stable groupings of around 100–150 individuals. For chimpanzees the figure is around 60, while for gorillas it approximates to 8. (See: Dunbar R. Why gossip is good for you. *New Scientist* 21 November, 1992: 28–31.)
10. Humans living in relatively undisturbed ecosystems must sometimes have become incidentally infected by a vector-borne parasite whose primary host was another species. For example, humans have acquired the yellow fever virus (monkeys are the primary, unaffected, host and mosquitoes are the vector) and the sleeping sickness trypanosome (ungulates are the primary, unaffected, host and the tsetse fly is the vector). The *Plasmodium* parasite, which causes malaria in humans, infects many other species of birds, reptiles and forest mammals. (See: Knell AJ (ed). *Malaria*. Oxford: Oxford University Press, 1991.) Where four species of malarial parasite infect and affect humans, more than twenty infect, but do not noticeably affect, monkeys. Although the evolutionary pathway of malaria is uncertain, human participation as 'host' for the parasite is probably a recent accidental intrusion on a long-established benign three-way relationship between the parasite, mosquito and monkey. The advent of a new potential host species, *Homo sapiens*, opened up new ecological opportunities for these parasites. Accordingly, there has been less time for this newer relationship with humans to evolve into benign relationships that optimise survival and reproduction for both parasite and host. (In the presence of abundant urban populations, however, there is less 'pressure' on the parasite to reduce its virulence.) For malaria, the heterozygote human phenotype for both sickle-cell anaemia and thalassaemia, entailing abnormalities of the haemoglobin upon which the plasmodium feeds, may be the beginnings of adaptive evolution by humans. In the presence of endemic malaria, heterozygotes survive better than unaffected persons.
11. Krause RM. The origin of plagues: Old and new. *Science* 1992; **257**: 1073–8.
12. Bloom BR, Murray CJL. Tuberculosis: Commentary on a reemergent killer. *Science* 1992; **257**: 1055–64; and: Cohen ML. Epidemiology of drug resistance: Implications for a post-antimicrobial era. *Science* 1992; **257**: 1051–55.
13. World Commission on Environment and Development. *Our Common Future*. Oxford: Oxford University Press, 1987.
14. Lima Guimataes JJ, Fischmann A. Infant mortality in Porto Alegre. *Pan American Health Organization (PAHO) Bulletin* 1985; **19**: 235–51.
15. Starr MD *et al.* Chagas' disease: risk factors for house infestation by *Triatoma dimidiata*, the major vector of *Trypanosoma cruzi* in Costa Rica. *American Journal of Epidemiology* 1991; **133**: 740–7.
16. Barry M. The influence of the US tobacco industry on the health, economy, and environment of developing countries. *New England Journal of Medicine* 1991; **324**: 917–20.
17. WHO. Urbanization and health in developing countries. Urbanization and the urban environment. *World Health Statistics Quarterly* 1991; **44**: 198–203.

18. Tabibzadeh I, Rossi-Espagnet A, Maxwell R. *Spotlight on the Cities*: *Improving Urban Health in Developing Countries*. Geneva: WHO, 1989.
19. WHO. *Urbanization and its Implications for Child Health*: *Potential for Action*. Geneva: WHO, 1988.
20. Williams BT. Assessing the health impact of urbanization.*World Health Statistics Quarterly*. 1990; **43**: 145–52.
21. Harpham T, Lusty T, Vaughan P (eds). *In the Shadow of the City*: *Community Health and the Urban Poor*. Oxford: Oxford University Press, 1988.
22. Victoria CG *et al*. Water supply, sanitation and housing in relation to the risk of infant mortality from diarrhoea. *International Journal of Epidemiology* 1988; **17**: 651–4.
23. WHO Commission on Health and Environment. *Our Planet, Our Health*. Geneva: WHO, 1992.
24. In 1991 I visited Banjul, the capital of The Gambia in West Africa, on the edge of the Sahel. Ten years earlier, at the beginning of the UN's Water and Sanitation Decade (see chapter 8), the government had torn up the old sealed roads in order to put in a sewer system. The installation is still proceeding, at a very West African pace. Meanwhile, the pit latrines and open drains dominate the olefactory landscape – and the dry red dust, liberated from beneath the old roadway bitumen, blows everywhere.
25. Mayo SK. Shelter strategies for the urban poor in developing countries. *Research Observer* 1986; **1**: 183–203.
26. Lewis G, David A, Andreasson S, Allebeck P. Schizophrenia and city life. *Lancet* 1992; **340**: 137–40.
27. WHO. Urbanization and health in developing countries. Selected health problems. *World Health Statistics Quarterly* 1991; **44**: 208–33.
28. Godlee F. Air pollution: I – From a pea souper to photochemical smog. *British Medical Journal* 1991; **303**: 1459–61.
29. Krysko WW. *Lead in History and Art*. Stuttgart: Springer Verlag, 1979.
30. Gilfillan SC. Lead poisoning and the fall of Rome. *Journal of Occupational Medicine* 1965; **7**: 53–60.
31. McMichael AJ *et al*. Port Pirie Cohort Study: Childhood blood lead history and neuropsychological development at age four years. *New England Journal of Medicine* 1988; **319**: 468–75. And: Baghurst PA *et al*. Life-long exposure to environmental lead and children's intelligence at age seven: The Port Pirie Cohort Study. *New England Journal of Medicine* 1992, **327**: 1279–84.
32. Needleman H, Bellinger D. The health effects of low level exposure to lead. *Annual Review of Public Health* 1991; **12**: 111–40.
33. Glass RI *et al*. Cholera in Africa: lessons on transmission and control for Latin America. *Lancet* 1991; **338**: 791–5.
34. Tamplin ML, Parodi CC. Environmental spread of *Vibrio cholerae* in Peru. *Lancet* 1991; **338**: 1216–17. Also: Epstein P. Cholera and the environment (Letter) *Lancet* 1992; **339**: 1167–8.
35. Jacobson JL. The other epidemic. *World Watch* 1992; **5(3)**: 10–17.
36. Newman PWG. *Sustainable cities*: *International and Australian progress*. Proceedings of EcoCity Conference, Adelaide, April 1992.
37. Hass-Klau C. Public transport and integrated policies in large metropolitan areas of Europe. *The Planner* 1990; **76(20)**: 13–20.
38. Godlee F. Transport: a public health issue. *British Medical Journal* 1992; **304**: 48–50.

39. Lowe M. Bicycle production outpaces autos. In: Brown LR, Flavin C, Kane H (eds). *Vital Indicators. The Trends That Are Shaping Our Future.* New York: Norton, 1992, pp 72–4.
40. Mage D, Zali O. *Motor Vehicle Air Pollution. Public Health Impact and Control Measures.* Geneva: WHO, 1992.
41. World Bank. *World Development Report 1992. Development and the Environment.* Oxford: Oxford University Press, 1992.
42. Newman P, Kenworthy J. *Cities and Automobile Dependence – An International Sourcebook.* Aldershot: Gower, 1989.
43. Goldsmith E *et al.* (eds). *A Blueprint for Survival.* Boston: Houghton Mifflin, 1972.
44. Wilmott P. *Community Initiatives, Patterns and Prospects.* London: Policy Studies Institute, 1989.
45. Sonoda K. *Health and Illness in Changing Japanese Society.* Tokyo: University of Tokyo Press, 1988.
46. McGlynn G, Newman PWG, Kenworthy JR. *Towards Better Cities: Reurbanisation and Transport Energy Scenarios.* Melbourne: Australian Commission for the Future, 1991.

11

Impediments 1: conceptual blocks

11.1 Introduction

For over three billion years the biological health and reproductive success of each evolving species has been the currency of natural selection. The narrative has not been smooth; over the aeons there have been major environmental dislocations to the otherwise more graduated, orderly processes of biological evolution. However, in circumstances of relative stability, the capacity of a well-adapted species to thrive within the prevailing, fluctuating environment is what ensures its continued survival. Hence, we would expect, logically, that *rapid* degradation of that environment would impair the thriving and surviving. My central argument has been that erosion of the natural systems within which the biological evolution of *Homo sapiens* occurred, and upon which we therefore remain dependent, will necessarily impair human population health.

Seeking solutions to these macroenvironmental problems requires us to ask unsettling questions. What do we really mean by 'standard of living'? How should we value material consumption against other things (such as good health) that contribute to the quality of life? Should we repudiate conventional short-term economic accounting, with its 'externalising' of social and environmental costs and its discounting of the future? Can we achieve a programme of international aid and concessionary developmental timetables that redresses past inequalities in opportunities between rich and poor countries? Are stronger political and economic incentives needed to ensure an immediate decline in fertility in poor countries?

These final three chapters explore some of the underlying impediments, and some of the ways of helping to overcome them. They seek to promote dialogue, to help define a framework for discussion – a framework that takes explicit account of the implications for human health and survival. Our capacity to act on behalf of unseen distant populations and unborn future generations will be the acid test for *Homo sapiens*, that unusual evolutionary experiment of nature. Lynton Caldwell, reflecting on the transition that human society must now make, writes: 'Today we can believe in the possibility of a world that may be better – but not its certainty. Whether humans in sufficient numbers and in time will make the choices required for a sustainable and sanative future remains to be seen. The odds are unknown, but a reason for hope lies in the demonstrated capacity of humans to learn and, when necessary, to learn quickly.'[1]

While there are many impediments, only a few are explored in these chapters. Conventional economics, with its assumptions and limitations, is one basic impediment to change. So too is the recurring exploitative relationship between rich and poor, as also is our misuse of the Global Commons. Other impediments include our innate insensitivity to distant, uncertain and intangible threats, and – relatedly – our disregard of risks to population health as a primary purpose for seeking ecological sustainability. Since these last two characteristics may be simpler to analyse, I will examine them first.

11.2 A lack of distance vision

11.2.1 Distinguishing trends from events

Ecological disruption, measured against the puny timescale of a human life, is a slow-motion process. Those processes, which are dramatic when measured against geological timescales, are trends, not events – and they therefore register only faintly on our alarm system. As with other species, biological evolution has programmed us to react to current events rather than to future threats.[2] The struggle for survival is conducted in the present tense; the physiological processes of 'fight or flight' are for dealing with imminent danger. Whereas we are readily drawn to, and distracted by, the more immediate and localised aspects of environmental degradation (such as oil spills and the Chernobyl disaster), we are much less able to address the bigger or longer-term aspects. In *The Plague*, Albert Camus wrote: 'A pestilence isn't a thing made to man's measure; therefore we tell ourselves that pestilence is a mere bogy of the mind, a bad dream that will pass away.' Camus' argument is that the confined space-time framework in which we humans usually operate – and for which we have been equipped by our own biological evolution – makes it difficult for us to address larger-scale threats and crises.

Nevertheless, as global environmental changes become increasingly evident, our gaze is being forcibly lifted to more distant horizons. For example, the US National Academy of Sciences, having concluded that the world may soon warm despite any concerted international actions, has recommended adaptive preparations: agricultural research should seek plant varieties adapted to warmer climates, irrigation systems should be planned for different rainfall patterns, buildings and bridges should be proofed against storms and floods, and – by a horticultural analogue of Noah's Ark (i.e. storing plant seeds so that ecosystems can be recreated in new locations) – biological diversity should be protected against the adverse effects of climate change.[3]

These circumstances pose a new test to our cerebral capacity. Can we muster the understanding, foresight and collective will to respond to dangers on, or beyond, these more distant horizons? The fact that the environmental change that now tests the human species is self-inflicted makes no difference to the logic of the situation. If we cannot change our current pattern of behaviour – not just by adapting to the current level of environmental deterioration, but by preventing further deterioration – then we may well fail the test.

11.2.2 *Failure to perceive the health hazards*

Human health, the environment and the economy are inextricably linked ... we exist not simply as individuals, but also as highly active parts of an ecosystem that is itself alive and finite. The health of that environment determines the health and safety not only of ourselves, but of our children, and of theirs, and of theirs to follow.

Canada's Green Plan, 1990[4]

You might expect that a national government's Green Plan that contains the above statement would contain a substantive discussion of the importance of environmental management to the health of human populations. After this eloquent nod in the direction of health, however, *Canada's Green Plan* (1990) then reverts to a conventional litany of other environmentalist preoccupations: Respect for Nature, Preservation of the Economy-Environment Relationship, Ensuring the Efficient Use of Resources, and so on.[4] The few passing references to health are at the familiar level of the need to control toxic pollutants. The plan displays little awareness of the ecological dimension of human health.

During 1990–1, WHO's Commission on Environment and Development conducted a broad review of the impact of environment upon health. The review focused on four major categories of human activity: urbanisation, industrialisation, agriculture and energy use. Its report, published in 1992, documented a wide range of adverse effects of environmental abuse and decline upon the health of human populations.[5] The report discussed the health consequences of rampant urbanisation and intractable poverty in the Third World. It also recognised, albeit briefly, the potentially serious repercussions for human health of climatic disruption, ozone layer depletion, acid rain, overfishing and degradation of farmland. But it did not elucidate the essentially ecological complexion of those problems. Rather, it implied that today's environmental problems are direct extensions of previously-documented problems, particularly those due to air and water pollution in industrialised populations and the health hazards of Third World urban slums.

The IPCC drew specific attention to various possible adverse effects of global warming and ozone layer depletion on human population health. Other international and national bodies are starting to explore the public health consequences of ecological disruption. In Australia, the National Health and Medical Research Council has published substantive reports on the health implications of ozone layer depletion (of particular relevance to sun-loving southern Australians) and of climate change.[6,7] Nevertheless, in a subsequent comprehensive review of 'ecologically sustainable development' Australian policy-advisors have remained generally opaque to the notion that concern for population health should be a *primary* criterion for sustaining ecosystems. Similarly, the political manoeuvring and declamations at the 1992 UN Earth Summit made clear that the main motivating issues were not human health and wellbeing, but economic considerations, national autonomy and the rigid defence of ideological (or, in the case of the Vatican, theological) positions.

Why has little attention been given to the human impact dimension of global environmental change? I suggested in the Introduction that we have not yet developed an understanding of the intimate relationship between the biosphere's natural systems and our biological health. Today's dominant human cultures convey little awareness of the nature of ecosystems and their internal fabric of interdependency and life-supporting mechanisms. We now understand well the notions of infectious and toxic agents, with their direct and demonstrable effects upon the health of living organisms. However, it requires a *qualitatively* different understanding to appreciate why and how ecosystem disruption will endanger health and survival. Indeed, it is easy to be misled by the facile observation that, in the short term, consumption-driven economic growth has been accompanied by improved population health (especially increased life expectancy). The fact that long-term unconstrained growth must eventually overload the ecological underpinnings of health is beyond our field of vision.

Another part of the problem is our tendency to perceive health at the personal and day-to-day level rather than at the population level. Once it becomes clear that the health of whole populations will be affected by such things as climatic instability, increased spread of infectious diseases, food shortages and increased ultraviolet irradiation then the topic becomes overtly political – and more dramatic. The consequences of global environmental change are likely to become a (perhaps *the*) major focal point of public health over the next few decades.

11.3 The limitations of neoclassical economics

11.3.1 Introduction

Much of the environment debate is couched in terms of economics. Indeed, the central theme of the 1992 UN Earth Summit was the tension between the pursuit of economic growth and the maintenance of ecologically intact environments. We should therefore examine the assumptions and values that mainstream economists bring to that discussion. It is not that economists are necessarily the bêtes-noir of environmentalism. Rather, as John Maynard Keynes pointed out, the ideas of economists and political philosophers, both when they are right and when they are wrong, are more powerful than is commonly understood.

It may be useful, and fair, to distinguish here between economics and what is sometimes called 'economism'. Economism presumes the primacy of economic forces and values in human society – and that those things that are important to society are measurable in monetary terms. It is economism, particularly in the hands of crude political practitioners, which is antithetical to environmental values. On the other hand, neoclassical economic theory in Western industrialised countries is regarded as one of the major social institutions of this century. Grounded in the belief that the free market provides the best-available means of allocating resources and as having its own intrinsic wisdom in assigning a money-value to tradeable goods and services, it provides a framework for market transactions, summary measures of what transpires in the marketplace, and guidelines for intervention in the marketplace to achieve socially-desired outcomes.

Neoclassical economics does not actually proscribe the inclusion of demands made upon environmental assets (e.g. forests) or environmental services (e.g. biodegradation of chemical wastes) in its price-setting calculus. Indeed, there is a peripheral body of economic theory that has addressed this very issue. However, our conventional economic models have not yet included those less tangible and generally unpropertied items in society's accounting. We 'own', and can therefore sell, our labour, our house and our car. (We therefore have an additional, direct, incentive to look after these things.) We do *not* 'own' the air, the river water, the ocean, or the stocks of wild animals – and so in market transactions we do not charge for the 'costs' incurred against these environmental assets.

Mainstream economics thus tells only part of society's story, albeit the part in which modern governments and their voters have most immediate interest. Society's economic activity is measured in terms of its total marketable (value-added) production – i.e. the net value of goods produced and services rendered. For nations, it is the Gross National Product (GNP) that summarises this economic activity. For individuals, the average 'standard of living' is measured as per-person consumption of goods and services. While accepting that the GNP and related aggregate indices are significant social inventions, the point to note here, again, is that only those things that are assigned monetary value in the market-place figure in the arithmetic of those conventional indices. Externalised social and environmental costs are excluded.

This means that reliance upon the GNP, as an economic index, can lead to ecologically bizarre accounting. Consider the example of the Exxon Valdez oil-spill which greatly boosted the GNP of Alaska because of the increased economic activity associated with cleaning up the fouled marine environment. It can also distort society's priorities. An American economist, Rathjen, offers a second example. In considering the consequences of greenhouse warming, he argues that since agriculture only accounts for 2% of the USA's GNP, therefore a substantial fall in agricultural productivity would have little economic consequence.[8] One wonders what he and fellow economists eat. Terminating the nation's agricultural production may well reduce GNP by only 2%, but it would make for a lot of disgruntled Americans rattling empty trolleys in supermarkets – not to mention the disaster for a hungry world that still partly depends on food surpluses from the naturally fertile lands of North America. Economic indices should be the means to achieving social ends; they are not ends in themselves.

11.3.2 *Historical background*

Classical economic theory, as formulated by Adam Smith over 200 years ago, postulated that the opportunity to trade goods and services in the open marketplace enables individuals to optimise their own interests. A self-correcting balance is achieved between supply and demand. Smith argued that the 'invisible hand' of the free market is beneficial to society because it achieves maximal social good, measured as the sum of individual welfare. Historically, this central, liberal idea of classical economics helped to free individuals from traditional hierarchical authority, while also stimulating the production and flow of goods and services. But markets have never worked as simply or as well as in the classical model. Nor have they worked to everyone's advantage. Competing post-classical economic theories have therefore evolved, with different approaches to the management of the market-place.

The prevailing economic system in today's democratic states derives from Keynesian macroeconomic theories that evolved in Western society earlier this century, and which proposed a major role for government in modulating society's economic activity. Communist countries have drawn on the ideas of Karl Marx, as gilded by Lenin, Mao Tse Tung and others, and have employed a much heavier-handed centralised control of production, distribution and exchange. More recently, economic rationalism, popularised by Thatcher, Reagan and lesser acolytes, has sought to diminish state responsibility and to champion the imputed wisdoms of the deregulated, self-correcting marketplace. Nevertheless, all these models have used the same basic game-board and equipment. None has yet recognised the extent to which the biogeochemical processes of the biosphere are the main source of our wealth and sustenance.

Classical economists recognised three sources of wealth: land, human resources (labour) and invested capital. All goods and services acquired their value from these three sources, and could then be traded in a morally neutral marketplace. In his critique of early capitalism, Marx wrote about the struggle between the latter two sources of wealth – labour and capital. He paid little attention to natural resources as a source of value, and he saw society's historical calling as being to conquer nature, rather than to be subject to it. Marx referred to nature as 'a matter of utility' for mankind and condoned the exploitation and degradation of nature because of its 'enabling function' for industrial development.[9,10] This century, Keynes propounded his economic theories amidst the gloom of the 1930s Depression, when commodity prices were rock-bottom and when natural resource scarcities were the least of society's worries. Keynesian economics therefore largely ignored natural resources as a source of national income and did not allow for their depletion and depreciation. As part of this legacy of partial cost accounting, today's neoclassical economics practitioners – who would never exempt an industrial plant from depreciation allowance to cover future costs of renovation or replacement – nevertheless regard natural resources as 'free'. The depletion of natural-resource stocks and the filling of natural sinks with pollutants are simply not included in the balance sheets.

Overall, it seems fair to conclude that conventional economic theories of all ideological hues have overlooked the run-down in our life-support systems that results from much of our economic activity. That may have mattered little when the engines of modern economies were first being cranked up and the scale of environmental impact was so much less. But we now need an urgent modification of our economic theories to achieve full-cost accounting. Indeed, various economists (Daly, Pearce, Repetto and others) have begun exploring alternative formulations, and a network of research has emerged around the ideas of 'ecological economics'.[11]

11.3.3 'Costing' the environment

The impairment of human health by environmental pollution or by ecological disruption is a direct 'cost' to the exposed generation. Likewise, the depletion of non-renewable resources and the long-term impairment of natural systems (such as UV-shielding properties of the stratosphere) is a 'cost' to be borne by current and future generations. Conventionally, however, we ignore these costs; in polite economic parlance, we 'externalise' them. Purists might argue that it is not possible to assign monetary 'costs' unless we can directly measure the 'opportunity cost' occasioned by blighted health, depleted resource or impaired functioning of the biosphere. In practice, when pressed, economists can find ingenious ways to assign money values to many – but not all – things.[12]

The aim of national accounting should be to provide a comprehensive and empirical framework that operates in the long-term interest of society at large. Lester Brown asks: 'How can basic biological indicators be so bearish and economic indicators so bullish at the same time? The answer is that the economic indicators are flawed in a fundamental way: they do not distinguish between resource uses that sustain progress and those that undermine it.'[13] He cites, as an example, countries that overcut their forests – and which thus do better in the short term, on conventional accounting, than do those that manage their forests on a sustained-yield basis. Logged trees, directly saleable, are counted as income; meanwhile, no subtraction is made on the balance-sheet for the depletion of the forests from which those trees come. However, the success of this delusionary,

market-driven, accounting is short-lived because overcutting eventually
destroys the resource base. Robert Repetto, of the World Resources
Institute, has shown how the apparent economic growth of countries such
as Indonesia and Costa Rica is substantially reduced when account is taken
of loss of forests, topsoil and oceanic fish stocks.[14] The recent average
annual increase in Indonesia's agricultural productivity, he argues, is
almost exactly equalled by the annual loss in soil fertility – around 4% in
each case. Yet, under the UN's prevailing system of national accounting,
the latter is ignored. Recognising this fundamental accounting flaw, the
WCED concluded that: 'In all countries, rich or poor, economic
development must take full account, in its measurements of growth, of the
improvement or deterioration in the stock of natural resources.'[15] This
raises an acutely political question: Can this be done within the existing
economic framework?

The free-market model, rooted in capitalism and modern democracy,
has clearly outperformed centrally-planned attempts to solve day-to-day
social–economic problems. The dramatic recent collapse of Eastern
European and Russian communism reveals the inadequacy of centralised
planning, at least in the hands of doctrinaire, autocratic and self-serving
bureaucracies. The market-based economic model, responsive to socially-
agreed values, has the *potential* to make adjustments towards full-cost
environmental accounting. However – and this is an important condition
– without an infusion of a new set of values that replace environmental
exploitability with sustainability, supported by internationally-coordina-
ted governmental prescription, the free market will not go far or fast in this
direction. Yet, that is what is needed. Therefore, the menu of economic
models from which society chooses must be extended beyond the old
ideological warriors, since none have been able to cater to the wider needs
of ecological economics.

Neoclassical economic doctrine has not only discounted impacts upon
the environment; it has explicitly encouraged excessive extraction, harvesting, consumption and waste – all in the exalted cause of expanding the
GNP. The need for a recasting of economic theory has been colourfully
summarised by Herman Daly (now an economist with the World Bank)
and John Cobb:

Further growth beyond the present scale is overwhelmingly likely to increase costs
more rapidly than it increases benefits, thus ushering in a new era of 'uneconomic
growth' that impoverishes rather than enriches. This is the fundamental wild fact
that so far has not found expression in words sufficiently feral to assault successfully
the civil stupor of economic discourse.[16]

This mounting critique of conventional economic practice is causing a rethink within the august chambers of institutions such as the World Bank. The idea that there are limits to society's growth has actually been around, in European culture, for two centuries. Indeed that is one reason why economics is referred to as the 'dismal science'. The physiocrats of eighteenth-century Enlightenment France could see no possibility of new wealth being created by industrialisation; for them agriculture was the only true source of wealth. Malthus, Ricardo and the nineteenth-century English economists who foresaw the depletion of coal stocks all predicted situations in which supply would not meet demand. (In the event – at least in the short term – they were proven wrong by unforeseen technological innovation.) It is only more recently that expectations of unlimited economic growth have arisen, inspired by a heady infusion of optimism from the industrial boom decades that followed the Second World War. Taken at face value, the conventional measures of economic performance in industrialised countries this century could easily justify that conclusion. However, it is the mounting evidence of collateral environmental damage that is now impelling us to look beyond the facial cosmetics of economic performance indicators.

Green parties argue for replacing the existing 'throughput' economy (featuring high input in raw materials and energy, and high output of emissions and waste products) with a circular 'conserver' economy that maximises recycling and use of renewable sources of materials and energy.[17] The GNP, they point out, was never intended as an indicator of human wellbeing. An ecologically sustainable economy should be evident from 'real-life indicators' that incorporate measures of the sustainability of the use of energy and non-renewable resources, population health, quality of air, water and food, maintenance of biodiversity and – as 'soft' as it might seem to hard-nosed economists and laboratory scientists – life-satisfaction and neighbourhood conviviality.

To this same end, Daly and Cobb have proposed an Index of Sustainable Economic Welfare based on a comprehensive set of measures of consumption, distribution and environmental degradation.[16] Whereas the GNP has doubled in the USA during 1950–90, this index has hardly increased at all. Similarly, the United Nations has recently proposed a Human Development Index, which combines population measures of longevity, knowledge and average purchasing power.[18] By measuring longevity and literacy this UN index reflects the social equity of within-population distribution of health-related knowledge, access to health care and adequacy of food, clothing and housing. Consequently, countries with great social inequality such as Brazil and the US rank well below their per-person income 'peers' on this index. Likewise, various of the oil-rich Gulf states, with relatively poor profiles of education (especially of females) and child health, score low on this UN index. Meanwhile, the public mood seems to be changing in rich countries – and, to a surprising extent, among disparate groups in poor countries.[19] Despite recent buffeting from the chill winds of economic recession, these shifts in public opinion appear quite strong. Public opinion polls in the USA during the 1980s saw an increase from around 45 % to 75 % in the proportion of citizens who *said* they wanted environmental protection included in the nation's house-keeping expenditures irrespective of the cost.

11.3.4 Reconciling economics with ecology

There is contention over whether the current economic framework needs overhaul or replacement. Many economists argue that the goals of environmental protection and economic progress are not innately in-compatible and that it is merely a matter of society specifying what it values – and therefore wishes to include in the national accounts. Others argue that market-based economics makes too many assumptions about the (self-interested) 'nature' of human behaviour and the 'socially normal' form of transactions and relationships. While that debate is evolving, what are the possibilities for making running repairs?

In the USA, several government-funded studies have concluded that the country would not be economically disadvantaged by stabilising carbon dioxide emissions at current levels by the year 2000, and that energy efficiency and renewable energy technologies would actually save many billions of dollars annually. Similar assessments are emerging from many other countries, suggesting that energy efficiency measures could save money and make exports more competitive. (As US energy expert Amory Lovins has reportedly said: 'Not just a free lunch, but one we could be paid to eat!') Relatedly, the widespread fear that saving forests or reducing greenhouse emissions means losing jobs may also be exaggerated. During the recent era of stricter pollution control in the USA, very few closures of manufacturing plants have occurred because of new environmental standards. Nevertheless, a move towards greater energy efficiency and less carbon dioxide emissions will cause some plants to close down. Coal-miners around the world will lose jobs over coming decades, causing hardship and social dislocation. But, as ever with evolving technology, new jobs will arise and, in a fair society, the gains on new economic fronts would be used to compensate losses on old, contracting fronts.

Meanwhile, there are other profound influences at work. Late-industrial Western society appears to be beset by 'structural unemployment', in which many jobs are disappearing because marketable surplus value comes increasingly from energy-intensive technology and more efficient extraction of raw materials – not from long lines of blue-collared workers. Since 1950, the total manufacturing output in the USA has increased more than three-fold, energy use has tripled and capital expenditure has quadrupled, but employment has only increased by one-third. As industrial production is thus being transformed, and as transnational corporations redefine the world's labour markets to minimise their production costs, so unemployment grows in the traditional industrial sectors of rich countries.

By opting for environment-sparing technologies, however, within a full-cost accounting context, the employment outlook could well benefit. Technologies that use less energy and more recycling of materials require more human work. The Worldwatch Institute estimates that the per-unit production of electricity with renewable energy installations (solar thermal and wind), using current and reasonably foreseeable technology, would employ as many workers as in nuclear and coal-fired power plants.[20] In the Great Lakes region of North America, it has been estimated that a 50% increase in the use of biomass energy (alcohol fuel, wood, municipal waste, etc.) instead of fossil fuels would result in a substantial net increase in employment. There is a counterpart to this in the agricultural sectors of poor countries. Mechanised energy-intensive agriculture, geared to short-term, export-based profitability and hence often destructive to the land, has displaced hordes of traditional rural workers. A reversion to ecologically sustainable farming, incorporating a blend of traditional methods and human-scale technology, more equitable land-ownership and supplemented by education and social support for rural communities, would create new jobs.

Despite our tendency to reify the economy, it is *not* a thing apart; it must operate within the bounds of a global ecosystem that has finite capacities to produce freshwater, form new topsoil, absorb pollution and furnish materials. As a subset of the biosphere, human economic activity simply cannot outgrow the world's biophysical limits. The economy is an expression of human social behaviour within that greater, real-world framework. Interestingly, the two words 'ecology' and 'economy' share a common etymological root. The classical Greek word 'oikos' means 'home': oikologos refers to maintaining the household in a liveable state, while oikonomikos refers to balancing the household budget. To bring these seemingly distant relations into compatible cohabitation again, neoclassical economists and ecologists need to bridge a communication gap that entails a major conceptual difference. Economics is largely based on incrementalist assumptions, wherein social circumstances evolve gradually and smoothly, and for which fine-tuning is therefore the appropriate response. This entails watching the marginal changes, discounting the distant future, deferring intervention until the costs become compelling – while throughout remaining unaware of the wider ecological limitations. However, ecologists do not presume a smooth linearity; ecosystems, pushed beyond limits, can change rapidly as thresholds are exceeded and non-linear responses are triggered.

11.3.5 *Risks and cost-benefit analyses*

Some earlier economists, such as Pigou in the 1920s, recognised the problem of 'externalities'. From them arose the notion that economics needs a way of valuing all those things that people care about and which provide the resource base for sustained human social existence. Such valuation, however, may be easier said than done. Consider the example given earlier: although the timber value of a logged tree can be readily determined, its value as part of a productive ecosystem, wildlife habitat or recreational amenity is much harder to estimate. Setting money values entails 'operationalising' subjective judgements about such things as a human life, a healthy year of life, a lake, a rhinoceros or an unblemished sunset. Economists use various, sometimes ingenious, ways to assign money values to such extra-market commodities. For example, indirect market-based estimates of the social cost of poor health can be made by totting up health-care costs and lost income (productivity); likewise, the cost of fully remedying the pollution of a lake can be used to estimate its 'value' in a pristine state. Alternatively, one estimates people's 'willingness to pay' – for example, finding out how much travel expenditure people will/would incur to go and look at a particular pristine lake, a sunset or a free-range rhinoceros.

Once values have thus been estimated, then conventional cost-benefit analysis can determine the balance of the costs (including the cost of avoiding damage to environmental assets and the cost of impaired health due to environmental damage) and the benefits.[21] The unit cost of preserving an environmental asset increases as that asset dwindles – and the implied unit value rises once it becomes apparent that the asset is dwindling. Invoking the economist's tool of marginal analysis, we may be willing to pay more to save one of the last ten northern white rhinoceroses in Africa (that's about what is now left) than to save the thousandth-last rhinoceros. But there are limits. Americans have sanctioned their government paying US$15 million up-front to save that B-52 of bird-life, the Californian condor; in future, however, they might well baulk at paying US$15 billion. The removal of lead from urban-industrial environments, where young children experience a blunting of intellectual development approximately in proportion to the amount of lead exposure, becomes more expensive as the lead concentration in soil and air is lowered. Thus, in money terms, a law of diminishing marginal returns applies to environmental lead abatement; each additional unit of health gain for the community costs more units of money. Social judgements must therefore be made, and it is usually easier to do so when the costs and benefits of alternative policies can be compared in common currency.

Such analyses often discount the future; that is, greater weight is attached to present costs and benefits than to those to be borne (perhaps by others!) in the future. This 'discounting' occurs for varied reasons: because the distant future seems less important; because there are uncertainties about the shape and circumstances of the future (what new technologies, or social values, might then exist?); or because it may seem wiser to invest money in known ways of creating new wealth than in possibly premature precautionary investments – which would incur opportunity costs. This question of how to value the future, enmeshed with the moral dilemma of intergenerational equity, is a complex and controversial issue. Some forms of ecological disruption will do damage that will last for centuries or millenia. In the case of the mass extinction of species, it is hard to even put a sensible number on it.

In conventional economics, the balancing of costs, risks and benefits seeks to maximise the efficiency of allocation of 'resources' in circumstances of limited supply. The resources can include clean air, condors and soil fertility. However, this classical, at-the-margins accounting is vulnerable to changes in perceptions, priorities and presumptions about the needs (and values) of future generations. It implies more completeness and precision than exists, and more environmental wisdom in the marketplace than is possible. Some economists therefore advocate a more qualitative approach to decisions about economic activities; one that would draw a prudent dividing line between what is ecologically 'safe' and what is simply too risky because the ecological consequences may be either irreversible (species extinction being the obvious example) or too costly for this or future generations to repair.[22] This categorical approach moves us towards a more socially-driven, rather than market-driven, view of 'ecological sustainability' and intergenerational equity.

11.3.6 'Sustainable development'

Sustainable development is...improving the quality of human life while living within the carrying capacity of the supporting ecosystems.

IUCN, UNEP, WWF, 1991[23]

The report of the WCED, in 1987, highlighted the central concept of 'sustainable development'. Subsequently, the word 'sustainable' has become widely used, often in contradictory and ambiguous fashion. Put simply, something that is sustainable can continue forever. Societal development and the economy are complex, evolving entities, and so the notions of sustainable development and a sustainable economy require further consideration. On the other hand, the notion of 'sustainable growth' is an ecological nonsense – nothing physical can grow indefinitely. Nevertheless, as we have seen, expansion of the GNP has become the exalted touchstone of social progress in the modern world. Indeed, there are widespread fears that an obligatory move to sustainable environmental management will disrupt economic growth and slow the creation of wealth.

We should clarify these words 'development' and 'growth'. Where growth refers to quantitative change, development refers to qualitative change. In talking of our children, for example, we refer to physical *growth* and to mental *development*. The former refers to body size, the latter to brain maturation. If, by sustainable economic 'development' we actually mean continuing, material, economic 'growth', then we should look again at the arithmetic in the WCED report, showing that attempts to raise the material standard of living of all of the world's poor to the hypertrophied and wasteful level of rich societies, using current or foreseeable technologies, would simply not be ecologically sustainable. If, however, 'development' refers to social equity, the protection of ecosystems and increased life-satisfaction from resource-sparing activities, then it can be ecologically sustainable.

The word 'sustainable' also needs a little more attention. Some economists and politicians assume, or hope, that it is a nice way of saying that, with a little fine-tuning, a little nipping and tucking at the edges, we can continue with business-as-usual. However, the missing word, which is implicit but which should be made explicit, is 'ecologically'. *Ecologically sustainable development* means that we are seeking social and material progress within the constraints of sustainable resource use and environmental management. Thus: renewable resources (plants, trees, animals and soil) will be used no faster than they are regenerated; non-renewable resources (such as fossil fuels and metals) will be used no faster than acceptable substitutes can be found; and pollutants will be generated no faster than can be absorbed and neutralised by the environment. Further, the vitality of ecosystems will be maintained. Thus, even if a forest is managed so as to supply timber on a sustainable commercial basis, it should not be deemed to be 'healthy' if its capacity to provide habitat for natural diverse fauna is degraded. Sustainable development, so defined, will require a fundamental change in the way we use the environment in our pursuit of economic growth. It will require us to deal not just in *market* prices but in *social* values (the economist's 'shadow prices') – i.e. the values that people place on things irrespective of whether they are, or could be, sold in the conventional market-place.[24]

The notion of 'sustainability' has introduced a *dynamic* ecologically-based dimension into the debate about development and environment.[23] The WCED envisaged that the needs (not wants) of the present generation could be satisfied 'without compromising the needs of future generations'. This would require us to live off the environment's sustainable yield – the interest, not the capital. The Commission also argued that, through equitable sharing and careful environmental resource management, the aspirations of the world's poor could be met – even though this would necessitate a substantial increase in global economic activity by the middle of next century. However, since the total economic activity of today's unevenly 'developed' world already depends on massive use of non-renewable energy sources and on humans commandeering almost half of the planet's 'net primary production', radically new environment-sparing technologies will be needed if ecologically sustainable development is to be attained.

11.4 Summary

There are various obstacles on the path of reorienting human society to ecologically sustainable development – i.e. development which sustains indefinitely the good health of human populations. Our 1400 cc brain, with its convoluted frontal lobes dedicated to abstract thought, is 3–4 times larger and much more complex in its 'higher-level' cerebral wiring than are the brains of early hominoids. Yet it is still a brain that has evolved for dealing with the here-and-now of individual and group survival. This preoccupation with the present distracts us from contemplating future adverse health consequences, particularly those that differ from today's familiar types of environmental health problems. That blind spot is compounded by our meagre insights into the role of intact ecosystems in maintaining the biological health of species. We have seldom had to worry about that before. Over the past 50,000 years, and particularly during our agrarian and industrial existence, we have disturbed and disrupted many local ecosystems with our increasingly heavy-handed tenure of much of the world's surface. But we have not previously been confronted by the prospect of regional or global breakdown in the processes of the biosphere.

We have systematically discounted our environmental assets within a marketplace oriented to commercial immediacies. We buy and sell goods and services according to the perceived value of the ingredients and to the measured human input – while turning a blind eye to the run-down of nature's stock, the exhaustion of nature's sinks and any adverse impact on the health of exposed working or residential populations. The priorities of the marketplace have allowed a neglect of the Global Commons. While the business-as-usual economic rationalists argue 'No economy, no environment', ecologically aware economists are now retorting: 'No environment, no economy'. Ecologists and public health scientists argue: 'No ecosystem, no population.' And any simpleton can see that: 'No population, no economy.' As a species we need to think more imaginatively and more providently about the world's – and our – future.

References

1. Caldwell LK. *Between Two Worlds. Science, the Environmental Movement and Policy Choice.* Cambridge: Cambridge University Press, 1990.
2. Ornstein J, Ehrlich P. *New World, New Mind: Changing the Way We Think to Save Our Future.* London: Methuen, 1989.
3. National Academy of Sciences. *Policy Implications of Greenhouse Warming: Report of the Adaptation Panel.* Washington DC: National Academy Press, 1991.
4. Government of Canada. *Canada's Green Plan.* Ottawa: Ministry of Supply and Services, 1990.
5. WHO Commission on Health and Environment. *Our Planet, Our Health.* Geneva: WHO, 1992.
6. National Health and Medical Research Council. *Health Effects of Ozone Layer Depletion: A Report of the National Health and Medical Research Council.* Canberra: Australian Government Publishing Service, 1989.
7. National Health and Medical Research Council. *Health Implications of Long Term Climatic Change.* Canberra: Australian Government Publishing Service, 1991.
8. Rathjen GW. Energy and climate change. In: Mathews JT (ed). *Preserving the Global Environment. The Challenge of Shared Leadership.* New York: Norton, 1991, pp 154–86.

9. Redclift M. *Sustainable Development: Exploring the Contradictions.* London: Methuen, 1984.

10. Some modern apologists seek to exonerate Marxism with the convoluted argument that certain of Marx's theories can still provide useful approaches to dealing with ecological problems. (See: Grundmann R. *Marxism and Ecology.* Oxford: Oxford University Press, 1991.) However, the ecological consequences of Marxist economics, at least as recently practised in the USSR and Eastern Europe, stretch the credibility of this scholastic argument.

11. Costanza R (ed). *Ecological Economics. The Science and Management of Sustainability.* New York: Columbia University Press, 1991.

12. In doing so they may well change (and perhaps degrade) the resource, and thus turn a piece of pristine nature – e.g. migrating whales, or a sunset over a Pacific island beach – into a viewable and commercially garnered commodity. Via the Disneyland ethos we can 'add value' to nature everywhere.

13. Brown LR. The New World Order. In: *State of the World* 1991. *Worldwatch Institute Report.* New York: Norton, 1991, pp 4–20.

14. Repetto R. Accounting for environmental assets. *Scientific American.* 1992; **266(6)**: 64–70.

15. WCED. *Our Common Future.* Oxford: Oxford University Press, 1987.

16. Daly HE, Cobb JB. *For the Common Good: Redirecting the Economy Toward Community, the Environment, and a Sustainable Future.* Boston: Beacon Press, 1989.

17. Jacobs M. *The Green Economy.* London: Pluto Press, 1991.

18. UN Development Program (UNDP). *Human Development Report, 1990.* New York: Oxford University Press, 1990.

19. Brown LR, Flavin C, Postel S. *Saving the Planet.* New York: Norton, 1991.

20. Flavin C, Lenssen N. Designing a sustainable energy system. In: *State of the World* 1991. *Worldwatch Institute Report.* New York: Norton, 1991, pp 21–38.

21. The following equation estimates the Net Present Value (NPV) to society of continuing, from time t_1 to time t_n, with an industrial activity which is income-earning but which is also both environmentally-damaging and a risk to human health:

$$\mathrm{NPV} = \sum_{t=1}^{n} [(B_t - C_t - aR_t)/(1+i)^t]$$

Σ indicates summation of the sequence of calculated values of the square-bracketted expression for each unit of time from t_1 to t_n. In each time interval, B_t is the income-generating benefit of continuing the activity; C_t is the cost of controlling the pollution; R_t is the risk to the population's health; a is the coefficient which converts that risk (which may be measured in such units as number of deaths averted or number of healthy life-years gained) to financial units; t refers to each unit of time within the specified time period t_1 to t_n; and i is the discount rate applying to health risks across each unit of time (such that the further into the future the greater the cumulative discount).

22. Toman MA. The difficulty in defining sustainability. In: *Global Development and the Environment. Perspectives on Sustainability.* Washington, DC: Resources for the Future, 1992, pp 15–23.

23. IUCN, UNEP, WWF. *Caring For The Earth. A Strategy For Sustainable Living.* Gland, Switzerland: Earthscan, 1991.
24. Pearce D. Sustainable futures: Some economic issues. In: Botkin DB, Caswell MF, Estes JE, Orio AA (eds.) *Changing The Global Environment. Perspectives on Human Involvement.* London: Academic Press, 1989, pp 311–23.

12

Impediments 2: relationships

12.1 Relations between rich and poor countries

12.1.1 Historical context

The world has become increasingly internationalised during this century. Fond notions of the world as a patchwork quilt of traditional colourful cultures recede, as the integrating influences of the global market-based economy advance, supplemented by distance-shrinking electronic communications. Those influences bear the stamp of the historically rich and powerful Western countries – and, increasingly, Japan. Whatever the historical or geographic reasons, those countries enjoy most of the world's comfort, privilege and political control. In this rapid internationalising of world affairs, however, lies the seed of hope of new forms of enlightened multilateral world governance, on behalf of the planet's overloaded biosphere. Ranged against that hope is the worry that the international economic engine, moving into higher gear, will accelerate and compound the damage.

The legacy of history does not augur well for a quick change to international fairmindedness and cooperation. Last century, the competitive scramble to acquire a piece of Africa reflected the explicit needs of major European powers to maintain economic and political parity with one another. As Cecil Rhodes, the founder of Rhodesia (now Zambia and Zimbabwe), said: 'We must find new lands from which we can easily obtain raw materials and at the same time exploit the cheap slave labour that is available from the natives of the colonies. The colonies would also provide a dumping ground for the surplus goods produced in our factories.'[1] No words minced there. Colonies provided the metropolitan powers with cheap food, raw materials and labour, new markets for manufactured goods and new investment opportunities. In today's post-

313

colonial age the words are less explicit, but the economic relationships are of similar general form. Through international trade agreements, through the yoke of escalating international debt and through the sheer size and power of transnational companies, traditional imperialism has been replaced by economic colonialism. The Opium War fought by Britain last century in order to secure the right for its maritime traders to sell opium to China has a modern counterpart in the actions of multinational cigarette companies, with the trade-policy back-up of their home governments, to gain access to the vast markets of the Third World.[2] The US Government has played a particularly pernicious role in this matter, threatening trade sanctions against Southeast Asian governments that restrict the access of US-based cigarette companies to their domestic markets.

After the 1930s Depression and World War II, the industrial countries set up a new world order to cushion themselves against such economic buffeting in future. A prime aim was to integrate the current and former colonies more closely with the emerging global market economy, dominated by the Western industrial system. Such integration would provide a continuously expanding market outlet for manufacturers and a continuing supply of raw materials. This would require two developments in these countries at the periphery:[3] an increased purchasing power and an enhanced capacity to supply the commodities. Three key international institutions were duly established which facilitated these developments – the International Monetary Fund (IMF), the World Bank and the General Agreement on Tariffs and Trade (GATT). Over the past forty years, these three agencies, generally dominated by US interests, have essentially controlled the world economy.[4] More recently, with the shifting centre of gravity of the world's economic power, Japan's influence upon these bodies is increasing as US influence wanes. (This may mean that policies will become more pragmatic and less ideological.) Because there is more self-interest than altruism in most foreign aid, it is not surprising that these three international aid-and-trade agencies have acted in the interests of the continuing economic control of the poor by the rich.[5,6] Since the unfolding of the Third World debt crisis in the early 1980s, poor countries have had to contend with the fact that IMF loans have typically required: (i) scrapping import quotas and tariffs, (ii) devaluing the local currency, (iii) spending money on industrial infrastructure rather than on social welfare, (iv) mechanising agriculture and (v) reducing employment by the State. Through this austere package of economic reform, referred to as 'structural adjustment', the donor seeks to stimulate the recipient country's economic development (and debt repayment) – and, *subsequently*, its social development.

12.1.2 Structural adjustment programmes

From the point of view of public health, adjustment policies are of great and immediate significance because they decrease universality of coverage, lead to rationing of health services by income, and delay or distort the development of primary care at the community level.

A. P. Ruderman, 1991[7]

Structural adjustment, aimed at boosting exports of primary commodities and cutting government expenditures in 'non-productive' areas, seeks to integrate the recipient country into the global market economy, thus reducing the widening gap between rich and poor countries. However, the implementation has usually been less attractive than the theory. Programmes have typically included a regressive cutback in the provision of education, health care and social security – i.e. the 'soft' part of the economy.[8] Structural adjustment has also rendered the expanded, privatised, industrial–agricultural base more accessible to foreign ownership.

In terms of conventional economic criteria, structural adjustment has had mixed success. For example, while incomes in Indonesia rose widely in response, in Brazil wages fell by 5% during 1980–7 and many workers transferred to lower-paid jobs in the informal sector. The World Bank now recognises that these structured austerity measures are taking far longer and proving much more arduous than originally expected.[9,10] In cutting back the 'non-productive' parts of the economy, government health-care expenditure has often been near the front of the sacrificial queue along with food subsidies and other social services. The burden of structural adjustment thus tends to fall most heavily on the poor – but, then, so did the typical antecedent inflation! In the words of a UN report: 'The vulnerable social groups tend to bear a disproportionate share of the burden of adjustment.'[11] Falls in health expenditure and reductions in vaccination in many Latin American and African countries have often meant rises in infant mortality and the incidence of low birthweight babies.[7,12] In some countries such as Korea and Chile, however, improved targeting of health and nutrition expenditures, as part of structural adjustment in a recessionary climate, may have contributed to falls in infant and child mortality.[10]

Despite misgivings, structural adjustment programmes have, so far, not obviously damaged the environment or impaired domestic food production.[13] However, we should heed the warning of the WCED (1987) that economic growth unattuned to environmental constraints will, *in the long term*, endanger population health.[14] In fact, certain structural changes can be beneficial to the environment if they lead to the replacement of expensive imports (such as fertilisers and pesticides) by traditional, less environmentally polluting, technologies. On the other hand, the stimulation of export industry may cause environmental degradation and increased risks to public health – particularly if it occurs in 'export-processing zones' such as in Cubatao, Brazil.[15] During the late 1980s, such adverse consequences of structural adjustment programmes came to be better understood, and, in consequence, recent programmes have increasingly sought to protect the environment.

12.1.3 The control of trade and wealth

Trading is about making money, not about being charitable. Nations have therefore always sought policy and legislative support for their trading interests, and it is no surprise that the international trade system is geared to the monetary interests of politically powerful nations. As Third World debt has grown, so national assets have been sold off. Western banks have acquired equity, and transnational companies have bought out many local industries. The world's largest 500 transnational companies now account for around one-third of global production, and they control most of the world trade in tea, coffee, cocoa, cotton, forest products, tobacco, jute, copper, iron ore and bauxite.[1] Rich industrial countries thus have substantial impact on the environment in poor countries, particularly because of their appetite for fossil fuels, forest products and wildlife products.[11,16] The fastidious Japanese throw away millions of pairs of wooden chopsticks every day.

During the 1980s, the world's poor countries struggled under the continuing burden of expensive oil, falling prices for exported commodities and tariff and quota barriers against exported goods. The Third World faces an uphill battle in a recessionary world. Commodity prices remain low largely because the rich countries, during economic downturn, reduce their discretionary consumption (e.g. of coffee and cocoa) and find affordable, perhaps preferable, substitutes (e.g. optical fibres for copper, and artificial sweeteners for sugar). As a short-term political expedient, tariffs and quotas are raised to protect the jobs of workers in the rich countries – even though the long-term effect may be to inhibit world trade and the general expansion of employment. Since the poor countries have little power to influence this lopsided relationship, they are the perennial losers. As if to rub salt into the wound, an underlying theme during the recent Uruguay Round of GATT was pressure from the rich countries for freer access to the source of raw materials from Third World countries, and to their consumer markets. Such a policy would restrain those countries from, for example, protecting ecologically vulnerable land from exploitative agriculture, since that would restrict the 'freedom' of international trade. Thus, the right of governments to implement environmental and health protection regulations would be seriously compromised, further endangering forest, farmland, water resources and local ecosystems.

It would be naive to imply that the only villains are the rich countries and their transnational accomplices. In most poor countries wealthy urban elites practise similar bias against their own rural poor. Low prices are often paid to farmers, who are a dispersed and politically weak section of the population. Local, often massive, corruption and social inequity have long been compounded by connivance, intended or unintended, between donor countries and political elites in recipient countries. If we cannot reduce this intra-country and intercountry exploitation and inequity, then poverty will remain a fundamental impediment to population control, to ecological sustainability and to the improved health of populations.

In 1990, the UN system established a new fund to help developing countries protect the environment.[16] This fund, the Global Environmental Facility (GEF), is to be administered by the World Bank and is to assist poor countries to pay for new technology to protect the ozone layer, reduce greenhouse gas emissions and sustain forests. Without this type of special provision for poor countries, their governments will continue to ignore environmental considerations through a mix of indigence, local corruption and political deference to their increasingly urbanised domestic constituencies. The Brazilians will continue to destroy rainforest for additional agricultural land and the Chinese will continue to build cheap ozone-endangering refrigeration.

The rich countries are becoming aware of the urgent and fundamental need for a massive, compensatory transfer of wealth to the developing countries, to assist them achieve ecologically sustainable development. This need was reiterated loudly by the poor countries at the 1992 UN Earth Summit. Indeed, those countries objected to the channelling of environmental-aid money through the World Bank, arguing that they, the poor countries, should determine where and how the money should be spent. The politics and ethics of these issues are complex – and militate against the urgent initiatives that the world now needs. Historically, today's rich countries got to the top of the development ladder by taking the cheapest, most direct road and by externalising costs to both environment and the health of working populations. Hence, today's incipient global warming and depletion of the ozone layer are essentially the cumulative consequence of fast-track low-cost development by rich countries. There is therefore both a practical and moral need to make financial redress, so that poorer countries will have a fair chance to undergo development – but using low-impact technology (and slowed population growth) that protects the biosphere.

12.2 Power relations and the Global Commons

12.2.1 Introduction

There is, as we have seen, considerable forward momentum, in the world's population growth, in ozone layer depletion and in the global warming due to already-accumulated excess carbon dioxide. We cannot reverse these processes in less than several decades. Nor should we expect that it will be any easier to change power relations between countries, within countries, or between Third World national governments and transnational companies.

Some commentators foresee the need for an environmental revolution, to reconcile, the forms of human culture and technology that arose from the agricultural and industrial revolutions with the constraints of the biosphere.[17,18] The rich countries would then settle for a lifestyle that is socially richer, but less consumption-dependent. Transfers of wealth and technology, overseen by some form of international governance and tied to substantial fertility reductions, would assist the poor countries to draw alongside the rich. While inequalities and economic exploitation have exacerbated poverty and environmental degradation within the Third World, the environmental and health problems of acid rain, global warming and ozone layer depletion, confront the nations of the world with a 'one in, all in' predicament. The WCED Report speaks of 'Our Common Future'.[14] That word 'common', with its dual meaning, is a key to understanding what, in global terms, is an unprecedented challenge: the need for multinational cooperation within a new sense of shared ecological destiny.

12.2.2 The 'commons'

Historically, the 'commons' were the communal lands in European towns to which all people had access, and upon which anyone's cow could graze. This idea worked well if the commons were large and everyone had just one cow. However, if times changed and everyone had two cows, then the commons would provide less grass per cow. Meanwhile, there was an ethical dilemma for each individual. If he (or, less probably, she) discreetly added a third cow, it would eat a negligible amount of additional grass, and no-one would be noticeably disadvantaged. However, if *he* did that, then why should not *everyone else* do likewise? The result would be that the commons would be overgrazed. This, as Garrett Hardin wrote over two decades ago, is the 'tragedy of the commons'.[19]

The Global Commons are, most obviously, the atmosphere and oceans, into which we dump society's waste products. They belong to, and are depended upon by, all nations of the world. Likewise, the fish and whale stocks in international waters are 'common' to all, and large river systems are often shared by many adjoining countries. Less straightforward are those commodities which are confined within national boundaries, such as oil wells, iron ore and elephants? Are they 'common' too, as part of the world's heritage? From the economist's viewpoint, the commons are a class of goods that are accessible to all and are collectively consumed. This distinguishes them from private goods traded between individuals in the market place. While a bowl of rice (a private good) can only be eaten by one person, everyone can simultaneously enjoy a smog-free day (a public good). The sustaining of public goods requires collective action – and avoidance of free-riders.

The complex political and ethical issues pertaining to reduction of the anthropogenic greenhouse effect were considered in chapter 6. That is clearly a 'commons' issue. An overwhelming 95% of the cumulative total of anthropogenic carbon dioxide emissions comes from the developed countries of the northern hemisphere. By contrast, Bangladesh has more than 2% of the world's population, generates less than 0.1% of the world's greenhouse gas emissions, but will bear much of the brunt of climate change: rising seas, cyclones, storms and flooding. Over the past four decades the average person living in a developed country has dumped about ten times as much carbon in the atmosphere as has the average person living in a developing country. As a result, the developing countries face a world which now has a reduced capacity to absorb further carbon emissions.

It might be argued, in principle, that all nations should have the same entitlement for cumulative per-person carbon emissions (perhaps inversely adjusted for population growth rates). In historical terms, however, the rich countries have already overdrawn their carbon-emissions account to an extent that ought now to preclude the poor countries from exercising their 'entitlement'. A concessionary redistributive formula is therefore needed that allows for those historical inequalities, subject to the curbing of population growth. Although the Montreal Protocol on cessation of chlorofluorocarbon use has granted such concessions, the options are less straightforward for greenhouse emissions.

The world's oceans, covering 70% of the planet's surface, are a major common asset. They are being encroached upon by industrial and domestic waste disposal, by oil spills and leaks, by overfishing and by acts of wanton destruction (such as the recently-outlawed Japanese and Taiwanese drift-net fishing). The 1982 Convention on the Law of the Sea, although not signed by the USA, the former USSR and twenty other countries, was signed by many countries and ratified by some.[16] The convention, which seeks national commitment to reducing pollution, facilitating navigation and controlling coastal-zone fishing, has begun in piecemeal fashion to improve our stewardship of this vast 'common', the world's oceans. We still have far to go. Recently, I was consulted by the waste-water authority in an eastern state of Australia about their plans for reducing coastal pollution (including the contamination of edible fish) by heavy metals and organochlorine chemicals from industrial effluent. They proposed a deep-water offshore outlet, and explained that fortunately it was 'a high-energy coastline' – meaning that wastes would quickly be flushed out to the deep ocean where, they said, 'it is no longer our problem'. That, of course, in microcosm, *is* the problem of the commons.

Managing the global commons will be orders of magnitude more difficult than was, historically, the solving of disputes over the use and overuse of the local village commons.

12.2.3 *War and environment*

Finally, and in a very profound sense, war is an impediment to ecological sustainability and, of course, a scourge of population health. It is one of the age-old Four Horsemen of the Apocalypse. Over many millenia, warfare has been a traditional means of settling disputes, usually on a localised basis. Until recently, war has had its own crude rationality in that one side wins, one side loses and disputes are thus resolved – temporarily if not permanently. However, there has been a dramatic and qualitative shift in the nature and scope of war in recent decades. In particular, the advent of nuclear weapons has moved war to a higher and more dreadful plane. A fundamental discontinuity in the technology of warfare has occurred. Combatants can now expect to decimate their opponent's population and, whether intentionally or unintentionally, to destroy vast swathes of environment and dependent ecosystems. In the short term, this manifest capacity for massive destruction afforded us several decades of mutual deterrence in the Cold War. However, in so doing we have started to play with a new type of fire. Warfare is no longer a behaviour limited to settling disputes; it has become a technology for destroying rather than for conquering.

At first sight, war may seem to belong in a special category within the context of this book. The other global environmental changes that are causing us concern are, in essence, the unintended byproducts of demographic, political and economic processes. War is committed deliberately and with approximate knowledge of consequences. It entails deliberate destruction. However, it is *not* independent of these other problem areas; indeed, it could well become the end-play for many of the ecological pressures that are building up in today's world.

The end-stage of unequal power relations and economic exploitation in the world will be tension and struggle over dwindling life-sustaining resources. Fossil fuels, freshwater, farmland and fishing grounds have already become the foci of armed struggle. Not to put too fine a point on the recent Gulf War, Iraq wanted control of the oil fields of its tiny super-rich neighbour, whereas the Americans could not allow such a destabilising shift in control of the West's economic lifeblood. Iceland and Norway have rattled nautical sabres over fishing rights in the northern oceans. India cannot permit the wheat-rich Punjab to secede. The next war in the Middle East may well involve fighting over access to riverwater – most probably the River Jordan (Jordan and Israel) or the Tigris and Euphrates (Turkey, Syria and Iraq). Today, Israel draws 60% of its water from the river Jordan, mostly via the West Bank territory. An advertisement placed by the Israel Ministry of Agriculture (18 August, 1990) in the *Jerusalem Post* is revealing. It read:

The crucial issue to be considered in any political solution regarding the future of Judea and Samaria is the question of who will have the final authority in resolving issues in dispute. This is especially acute in the case of water resources, as any proposed Palestinian political entity, whether sovereign or autonomous, would have no water resources at all, other than those upon which Israel is so critically dependent for her day-to-day survival.

So, it comes as little surprise that, for the last eight years, one of the largest research projects by the US Central Intelligence Agency in the region has been on water security.[20] In similar vein, the WCED Report said:

The developing and widening environmental crisis presents a threat to national security – and even survival – that may be greater than [that of] well-armed, ill-disposed neighbours and unfriendly alliances. Already in parts of Latin America, Asia, the Middle East, and Africa, environmental decline is becoming a source of political unrest and international tension . . . Climatic change would quite probably be unequal in its effects, disrupting agricultural systems in areas that provide a large proportion of the world's cereal harvest and perhaps triggering mass population movements in area where hunger is already endemic. Sea levels may rise during the first half of the next century enough to radically change the boundaries between coastal nations and to change the shapes and strategic importance of international waterways – effects both likely to increase international tensions.[14]

While poverty, population pressures and ecological decline are rich seed-beds for war, the heightened capacity of modern warfare for environmental destruction can become a form of positive feedback. Environmental deterioration and resource scarcity promotes war which, when it breaks out, further increases environmental deterioration and resource depletion. The jungles of Vietnam, systematically chemically defoliated by the USA, will take several generations to recover. Much of Vietnam's tropical wetlands has since been overrun by monocultures of hardy bamboo. The deliberately oil-fouled Persian Gulf will take years to recover. Extensive beds of sea grass in this shallow, richly-populated, body of water have been damaged and tens of thousands of marine-life deaths occurred within the first weeks after the oil spill – including fish, birds, marine mammals (dugongs and dolphins), sea turtles and countless invertebrates.[21] Long-term damage to fisheries is anticipated, and the survival of several local species is threatened. Civil war in many parts of sub-Saharan Africa, perhaps most chronically in the Sudan, Ethiopia and Somalia, has devastated already hard-pressed farmlands. Soil erosion in food-desperate war-torn Ethiopia has escalated steadily over recent years.

The scale, and therefore the environmental and social impact, of war increases relentlessly. The nuclear arsenal of one Trident submarine is several thousand times more powerful than the Hiroshima bomb. Heavy mechanised war, along with the accompanying years of preparation and simulated war-game exercises, consumes vast amounts of carbon-dioxide-releasing fossil fuels in manufacturing, transportation and operation. Thus does war and its massive infrastructure help to warm the world. Likewise, the civilian consequences of modern warfare are far-reaching. Once the surgically precise Coalition bombing of Iraq ended, the infant and child mortality began inexorably to grow. The destruction of urban habitat – sewage disposal, clean water supply, electricity, housing and health and welfare facilities, along with continuing *de facto* sanctions on medicines and food – resulted, by late 1991, in tens of thousands of extra deaths of children below age five years.[21] The flow of refugees from, and within, war-torn countries takes a mounting toll on the social and physical environment. Currently, tens of millions of war-displaced refugees, from Iraq, Afghanistan, Sri Lanka, Cambodia, non-existent Palestine and many countries in Africa, are crowding into camps and urban slums. The farmlands they leave and the urban environment into which they squeeze thereby become further degraded.

Beyond the dramatic direct environmental costs of warfare, the opportunity cost is staggering. Military expenditure over the past decade has increased three times faster in developing countries than in developed countries; indeed, the aggregate military expenditure in developing countries exceeds their total expenditure on education and health. The world spends nearly US$1 trillion per year on 'military security' – spending in three hours the equivalent of the WHO's yearly budget. A fraction of this money, if spent on reducing world poverty or working more directly for ecological sustainable development, would achieve much – and remove many tensions. Yet, since 1960, military spending by developed countries has been twenty times greater than the total amount given as development assistance. Meanwhile, many poor countries spend 10–20% of their GNP on military training and weapons; an estimated one-fifth of Third World debt is due to military spending.[22] Many similarly obscene summary statistics have been adduced about money spent on 'defence' – but, unrepentant, we continue to make, sell and use hyper-expensive weaponry. War also incurs a huge cost in terms of disrupted trade and the earnings of displaced guest workers who repatriate money to their poorer home countries. The World Bank estimates that, even before international hostilities broke out in the Gulf War, the invasion of Kuwait and the application of sanctions had cost developing countries US$30 billion in lost trade, lost revenue from over two million repatriated guest workers and increased oil prices.

The closure of the Cold War holds promise of a massive swords-into-ploughshares peace dividend. Indeed, military expenditures declined a little in the early 1990s, as nuclear weapons began being dismantled.[23] The need for all nations to work to replace short-sighted, and increasingly irrelevant, notions of military security with far-sighted notions of regional and global environmental security is discussed further in the next chapter.

12.3 Summary

Relations between rich and poor countries, in aid and in trade, continue to set – and to constrain – the possibilities for economic and social development in poor countries. The entrenched mass poverty of the Third World is a prime source of overpopulation and environmental degradation, including increasingly non-sustainable forms of agriculture. While historical precedent and expectations of continuing power and privilege by rich countries impede fairer dealing between countries, it is rapidly becoming clear that some unusually enlightened forms of international cooperation are needed if ecological disasters are to be averted.

The world is largely comprised of 'commons', to which all countries and peoples have a claim. These, in a global sense, are the 'public goods' that require international protection: the climate-setting atmosphere, the oceans, the ozone layer, the rainforests and the supplies of farmland variously suited to different uses. Without that view, countries and groups will persist with warfare – which is increasingly both an exercise in futility in solving environmental disputes and an escalating threat to the world's ecosystems.

References

1. Editorial. The Uruguay round: gunboat diplomacy by another name. *The Ecologist* 1990; **20**: 202–3.
2. Barry M. The influence of the US tobacco industry on the health, economy, and environment of developing countries. *New England Journal of Medicine* 1991; **324**: 917–20.
3. The 'periphery' is a key notion in world-systems theory. On this view, since the rise of Europe as a maritime power 500 years ago, international commerce has been organised and controlled by the rich countries of Europe and, later, North America in ways that ensured the increasing industrial wealth of those 'metropolitan' powers. Initially manned by the local working class, who also became the well-trained consuming class, industrial production has increasingly moved 'off-shore' to seek out cheaper labour in poor countries at the periphery of the system. Those poor countries, burdened by debt, are increasingly under pressure to produce commodities for export and to purchase – whether by trade or as part of an 'aid' package – manufactured goods from the rich countries. The expectations of the rich countries will be hard to change at short notice, as the crisis of international inequality, structural poverty and resultant environmental degradation becomes apparent. For a discussion of this analysis of the origins, evolution and values of the capitalist world economy, see: Taylor PJ. The World-Systems Project. In: Johnston RJ, Taylor PJ. *A World in Crisis? Geographical Perspectives*. Second edition. Oxford: Blackwell, 1989, pp 333–54.
4. Strange S. *Till Debt Us Do Part*. Harmondsworth: Penguin, 1988.
5. McMichael P. Tensions between national and international control of the world food order: Contours of a new food regime. *Sociological Perspectives* 1992; **35**: 343–65.
6. George S. *A Fate Worse Than Debt*. London: Penguin, 1988.
7. Ruderman AP. Economic adjustment and the future of health services in the Third World. *Journal of Public Health Policy* 1990; **11**: 481–90.

8. Tanzania provides an instructive example. Celebrated in the 1960s and 70s as a model of centrally planned economic development in Africa – 'socialism with a human face' – Tanzania ran into economic difficulties around 1980. Following an ill-judged import binge, paid for with coffee-crop profits, and a costly war with Uganda, Tanzania sought money from the IMF. The IMF insisted that any loan be tied to a lowering of import barriers, devaluation of currency and reductions in size of the state bureaucracy. Tanzania resisted; and so no loan was given. Then, in 1986, stricken by the collapse of the world cotton market, Tanzania finally had to agree to IMF conditions. Throughout all of this economic difficulty and the subsequent imposed austerity, Tanzania's health-care and educational systems have deteriorated and maternal and infant mortality has risen. (See also: George S. *A Fate Worse Than Debt*. London: Penguin, 1988, pp 98–101.)

9. World Bank. *World Development Report* 1992. *Development and the Environment*. Oxford: Oxford University Press, 1992.

10. Ribe H *et al. How Adjustment Programs Can Help the Poor. The World Bank's Experience*. World Bank Discussion Paper No. 71. Washington, DC: World Bank, 1990.

11. UN, Department of International Economic and Social Affairs. *World Population Monitoring* 1989. *Population Studies No.* 113. New York: United Nations, 1990.

12. WHO/World Food Programme. *Structural Adjustment, Health, Nutrition and Food Aid in the African Region*, Geneva: WHO, 1988.

13. Cooper Weil DE *et al. The Impact of Development Policies on Health*. Geneva: WHO, 1990.

14. WCED. *Our Common Future*. Oxford: Oxford University Press, 1987.

15. World Bank, Population and Human Resources Division. *Health Policy in Brazil: Adjusting to New Challenges*. Washington DC: World Bank, 1989.

16. IUCN, UNEP, WWF. *Caring For The Earth. A Strategy for Sustainable Living*. Gland, Switzerland: Earthscan, 1991.

17. Lester Brown, of the US-based Worldwatch Institute, predicts that the third great revolution in human history is imminent. The first two – the agricultural and industrial revolutions – happened over millenia and centuries, respectively. This third revolution will have to happen within a decade or two, he argues. (See: Brown LR. Launching the Environmental Revolution. In: *State of the World* 1992. *Worldwatch Institute Report*. New York: Norton, 1992, pp 174–90.)

18. King A, Schneider B. *The First Global Revolution*. New York: Simon & Schuster, 1991.

19. Hardin G. The tragedy of the commons. *Science* 1968; **162**: 1243–8.

20. European Commission. *Europe Without Frontiers: Completing the Internal Market*. Periodical 3/1988, 1988.

21. Lee I, Haines A. Health costs of the Gulf War. *British Medical Journal* 1991; **303**: 303–6.

22. Corson WH (ed). *The Global Ecology Handbook*. Boston: Beacon Press, 1990.

23. Renner M. Military expenditures falling. In: Brown L R, Flavin C, Kane H (eds). *Vital Signs. The Trends That Are Shaping Our Future*. New York: Norton, 1992, pp 84–5.

13

The way ahead

13.1 Science: dealing with uncertainty

Where there are threats of serious or irreversible damage, lack of full scientific
certainty should not be used as a reason for postponing measures to protect
environmental degradation.

Ministerial Communiqué, Bergen Conference on
Environment and Development, 1990

13.1.1 Introduction

The systemic causes of environmental problems in today's world, restated,
are: massive population growth; Third World poverty and a festering
burden of debt between poor countries and rich countries; non-sustainable
consumption and biosphere-damaging waste generation in rich countries;
non-sustainable agricultural practices in many countries; and, in poor
countries struggling for export earnings, environmentally damaging
industrialisation and exploitation of natural resources. The various
impacts of these upon the biosphere pose a range of threats to human
health and survival. This web of problems is summarised in Fig. 13.1.

We should not allow uncertainty about the effects of environmental
change upon human health and wellbeing to induce a 'let's just wait and
see' response. We cannot sensibly await the outcome of further decades of
empirical research before making policy decisions – particularly since the
disruption of some natural systems may eventually, via threshold effects
and feedback loops, trigger unexpectedly rapid destabilisation. How, then,
can we increase the priority given to determining the health impact of
global environmental changes?

13.1.2 The needs of scientific research

Science cannot deliver certainty in knowledge on the global environmental issues,
any more than it can deliver certainty on the moral issues of reproduction

326

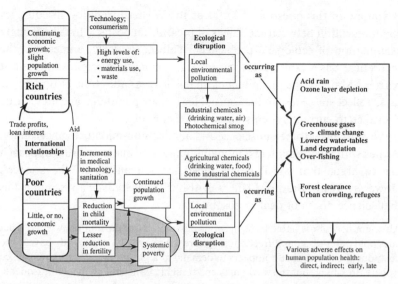

Figure 13.1. Overview of the context, causes and consequences of ecological disruption in rich and poor countries. (The shaded area is described from a First World perspective, while recognising that there are other cultural influences on material expectations and family size.)

engineering. . . . The management of scientific uncertainty has become too big a task for the technical experts alone.

Ravetz, 1990[1]

The widely-held view of science as a set of specialised, highly technical methods for revealing 'the truth' about the natural world is a major obstacle to harnessing science to the study of complex biogeochemical systems. It is both a procedural and a cultural obstacle. We will examine the procedural difficulties further. Since consideration of science as a cultural obstacle exceeds the scope of this book, I simply venture that modern reductionist science, now entrenched in the heartland of Western culture, makes it harder for us to imagine with humility and awe how the human species might be an interdependent part of Nature.

As foreshadowed in chapter 3, assessing the impacts of ecological disruption upon human health will not be tidy science. The task ranges well beyond the classical and secure scientific paradigm of hypothesis formulation, data collection and data analysis. Since the task deals with complex biogeochemical and ecological processes, the research must involve interdisciplinary approaches. Conventional science, which is widely distinguished by compartmentalisation of knowledge, will have to throw open its doors to interdisciplinary collaboration. Earlier, I quoted

Passmore on this question: 'So far as the Western tradition discourages communication between specialists, it presents an obstacle to the adequate examination of ecological problems.'[2] Talbot, similarly, argues that the theoretical bases constructed by separate scientific disciplines present a barrier.[3] Many of the prevailing mechanistic ideas about the biosphere, he says, reflect an older, reductionist Newtonian paradigm and have little relevance to current systems-based understanding.

We must also attune our science and our decision-making to uncertainty. Parties with a vested interest in maintaining a business-as-usual approach like to argue that social policy must not run ahead of 'the science'. However, Austin Bradford Hill, a British medical statistician, reminded us that science does not deal in absolute facts:

All scientific work is liable to be upset or modified by advancing knowledge. That does not confer upon us a freedom to ignore the knowledge we already have, or to postpone the action that it appears to demand at a given time. Who knows, asked Robert Browning, but the world may end tonight? True, but on available evidence most of us make ready to commute on the 8.30 next day.[4]

If the science of global environmental change is to proceed with limited empirical 'facts', then the estimation of population health impacts will require extrapolation and predictive, scenario-based modelling. Such estimates, shrouded in a penumbra of uncertainty, may assail the cherished image of science as the source of certainty. However, Ravetz cautions that: 'The need is not to remove uncertainty (for that is impossible), but to make it open and positive, rather than covert and manipulative.'[1] Now, exhortations to declare their uncertainties may upset classical scientists. Indeed, there were tensions over the fact that, in 1990, the IPCC based its conclusions about the likelihood of greenhouse warming upon a majority view among several hundred climate scientists. The idea that scientific judgement could be based on a majority opinion seemed, to some, to subvert the processes of true science.

Classical rigour is appropriate enough if you are carrying out research to find out, all in good time, how the world works normally. Scientists have been doing that for hundreds of years – often with simplified experimental models of particular components of the real world. But when the research question is more urgent, asking whether the world is going to *stop* working normally, and what the repercussions will be, then step-by-step empirical hypothesis-testing rigour may be an unaffordable luxury. Here, then, is the importance of the 'precautionary principle' when science is being applied to society's urgent and risk-laden problems.[5] That principle exhorts the adoption of prudent social policy ahead of empirical scientific confirmation of 'the facts'. Precaution can be applied in remedial action, moving to reduce environmental damage ahead of full knowledge of the consequences of that existing damage, and in the preemptive assumption that proposed environmental changes are likely to have adverse consequences.

Various governments took the former approach (remedial action) in the late 1970s in deciding to curtail CFC usage, solely on the basis of the theoretical prediction that CFCs would damage stratospheric ozone – and well ahead of any empirical demonstration of ozone damage or adverse biological consequences. While the precautionary principle may draw fire from the classicists, others argue that such an approach can actually *increase* the quality of the science, by liberating the scientist from the claustrophobic confines of idealised reductionist models in dealing with an obviously complex reality. The precautionary approach allows the scientist to say 'I don't know', thus assisting the policy-makers to deal with the uncertainty that permeates this whole field of enquiry. It will also constrain conservative policy-makers from invoking the lack of 'scientific fact' as a justification for doing nothing.

13.1.3 The prospect for 'technical fixes'

Technology, the handmaiden of science, presents a mix of temptations and challenges. The idea that we will find technical fixes for each of our environmental problems is widespread. This, I suppose, is a tribute to human inventiveness and optimism. It is also a tribute to mindless media reporting that hails an endless succession of technical 'breakthroughs' – particularly in the treatment of cancer. (To the gullible lay-person, it must be a wonder that any cancer cell manages to survive in the modern world.) Nevertheless, such things as our success in finding substitutes for CFCs, the genetic engineering of high-yielding cereal plants and the desalination of seawater all give some understandable cause for optimism.

Complex ecological disruptions, however, will almost certainly not be amenable to technical fixes. It is hard to see how some depleted resources – such as the severely lowered, slow-filling, water tables beneath some of the world's major grain-growing lands – can be replenished or replaced by use of technical ingenuity. Unlike CFCs, there is no substitute for water; nor can plants be genetically engineered to need significantly less of it, although there are possibilities for increased efficiency of irrigation. While solving the CFC problem may prove *relatively* easy (although the desired results will not occur for several decades), solving the problem of greenhouse warming will prove very difficult, both politically and technically. Yet still there are enthusiasts who advocate filling the upper troposphere with smoke-enriched jet-airplane exhaust or covering the oceans with floating silver reflectors, each intended to reflect away the Sun's heat. This is the sort of hubris that has helped to *create* our modern ecological predicament.

We should recall that some of the great solutions to population health problems in the past have come from social progress, not from technical breakthroughs. As we saw in chapter 3, the sustained improvements in the population health of industrialising countries last century arose from sanitation, education, improved food supplies and environmental pollution controls. The improvements did not depend on 'magic bullets' from clinical medicine; they mostly occurred before the advent of antibiotics, modern vaccination and technology-assisted clinical rescue. Europe's fertility rates dropped substantially ahead of the popularisation of barrier contraception. Malaria largely disappeared in Europe because of the draining of swamps (local ecological damage notwithstanding!) and improved house design – and well before the advent of chemical pesticides. In contrast, recent reliance on pesticides in tropical countries has had a mixed success, and there now appears to be a widespread resurgence of pesticide-resistant strains of mosquito. In Egypt, although many technical approaches have been tried towards the problem of snail-borne schistosomiasis, success has been very limited – and it is unlikely to improve in the absence of supportive social–political developments.

There will be important contributions to ecological sustainability from biotechnological innovations in food and agricultural production, as well as in renewable energy, waste recycling, pollution control and medical treatment. The US Office of Technological Assessment estimates that, by the year 2000, five-sixths of the annual increase in global agricultural production will result from new biotechnology and genetic engineering while one-sixth will derive from increased land usage.[6] Tissue culture techniques are being used in the development of new plant strains, by transferring genes for resistance to parasites and diseases and for tolerance to salinity and acidity. The genetic engineering of transgenic plants, with desired single-trait genes, will accelerate this directed 'evolution.'[7] Genetic engineering has recently increased bovine milk yields by 20% and may soon yield tomatoes that, once picked, are long-lasting. Other technical advances anticipated in food production include the development of cereals with self-fertilising nitrogen-fixing bacterial nodules on cereal grain roots.

Success in the laboratory, however, is not the same as success in the real world. New plant and animal biotechnologies may have unforeseen adverse effects upon ecological systems. As a simple example, plants bioengineered to have greater pest resistance will produce more natural 'pesticide' toxins; these may also be hazardous to humans. Besides, the mixed success of the Green Revolution – both ecologically and socially – along with the wide-spread loss of fertile land suggest that a reliance on new gene technology may be overly optimistic. 'High technology' is, after all, not new; we have been introducing it for decades – during which time the world's socio-economic disparities and inequities have increased. Context is therefore all-important. Within the prevailing political–economic context, the introduction of the next rounds of new agricultural biotechnology may also cause various social, economic and environmental problems. This is particularly likely in Third World countries. First, foreign-owned bio-technology corporations will set priorities, prescribe high levels of off-farm inputs (new crop varieties, fertilisers, and pesticides), establish a heightened pattern of technical 'dependency' and make most of the profits. Second, traditional high-earning export commodities will be displaced by bio-processed substitutes (e.g. artificial sweeteners, fragrances and spices – and to some extent, perhaps, by 'factory food' synthesised by tissue culture).[8]

With increasing pressure on ocean fish stocks, countries such as Japan must increasingly rely on aquaculture. Japan already obtains over 10% of its fish from fish farms and is developing such techniques as supernutritious algae, produced by cell fusion technology, to boost the efficiency of growing shrimp. The cultivation of seaweed, both in Japan and around coastal Africa, is also developing rapidly as a source of food and of nutrients for fertiliser production. Again, the risk here is that, preoccupied by technical adaptation, we may overlook the more fundamental ecological problem posed by the continuing, excessive human population growth that increasingly strains the carrying capacity of natural systems. Our restless technical prowess is an expression of the unusually large brain that our species inherited from the late stages of primate evolution. That brain-power has subsequently been put to many uses that were not within the original horizons of biological evolution. The real challenge now is to find ways of using that brainpower to diverge from the path leading towards ecological crisis – and not, by adaptive technical ingenuity, merely to defer the crisis.

13.2 Policy initiatives

13.2.1 Introduction

In chapter 11, the arguments for extending the neoclassical economic framework to accommodate full-cost accounting were examined. It will be much more difficult to achieve concerted action towards equitable international and intergenerational sharing of the world's life-supporting and life-enhancing resources. Indeed, this may require an environmental revolution – it is hard to see what else could achieve within decades the capping of world population growth, the reduction of world poverty and hunger, and the elimination of ecologically damaging emissions and wastes from energy-intensive industry, transport and agriculture. These are complex issues that lie beyond the scope of this book, and I will only attempt a quick scan of the major peaks on the policy options skyline.

13.2.2 International aid and trade

In this extraordinary twentieth century, while the global economy has increased from US$1 trillion to US$20 trillion, the control and distribution of the material spoils has left the majority of humanity enmired in environmentally-degrading poverty. Over 40% of the collective GNP of the Third World is now hostage to international debt (US$1.2 trillion). Consequently, in fire-sale fashion, these financially desperate countries are cashing in on their national assets: forests, farmlands, fisheries and waterways. As we have seen in chapter 12, the net flow of money over the past decade, even after allowing for aid given, has been from poor to rich countries. As this wealth gap widens, the consequences of massive Third World environmental damage and impoverishment will almost certainly rebound upon the rich countries in various ways, including greenhouse warming, loss of biodiversity, the drug trade and increasing pressure from refugees. This is the so-called 'debt boomerang'.[9]

The policy options for rich countries are rather basic. Since there is no reasonable prospect of the poor countries achieving economic 'takeoff' and loan repayment within the few decades available before various forms of social and ecological collapse might ensue, debts may have to be partly or wholly cancelled. Debt-for-nature swaps may be of some use – and symbolic significance. But to make real headway, the international community must provide substantial subsidies to poor countries for ecologically sustainable development. This has started to happen – albeit somewhat grudgingly. The Global Environmental Facility (see chapter

12), begun by rich countries in 1990, was substantially upgraded at the 1992 Earth Summit to the tune of US$25 billion. This, however, is only about one-quarter of what many commentators estimate is needed in the first instance.[10] Overall, rich countries give less than 0.5% of their GNP to poor countries, and even this is quite often given in bilateral deals tied to purchase contracts which serve as a form of export promotion for the donor country.

The potentially discriminatory effects of the GATT 'free trade' policies have been discussed in chapter 12. More generally, there is a basic need to dispel the indifference of international trade to questions of population health. Much of the world's coarse grains, soybeans and fish-meal grain – a potentially nutritious food for malnourished human populations – is squandered on feedlot livestock in rich countries (in whose well-fed citizens the resultant high-fat meat contributes to the silting up of coronary arteries). The European Community, in a confusion of politically-motivated subsidisation of surplus production, has produced 'mountains' of butter and warehouses full of dark tobacco. The yellow bricks of saturated fat are dumped on local and international markets. More callously, the carcinogenic high-tar dark tobacco is sold cheaply to northern African countries and to India. As any economic rationalist knows, this production and trade has little to do with international equity, ecological sustainability or population health.

13.2.3 National policies

National governments generally find it easier to adjust their domestic economies than to restructure their international relationships and aid obligations. Money incentives for ecologically sustainable activities are one option. Historically, however, subsidies to private sector activity have often been environmentally damaging. For example, longstanding tax incentives to Australian farmers (discontinued several decades ago) encouraged the indiscriminate 'clearing' of land, often causing soil erosion. In many developing countries, pesticide subsidies to farmers have resulted in excessive and dangerous use and have become a barrier to integrated pest management, while concessions to logging companies have led to quick-profit destruction of forests.

Through 'green taxes', governments working within the existing economic framework at least have an opportunity to restructure the tax base. Ideology aside, the *source* of revenue is not critical, so long as the process is socially equitable and is compatible with desirable economic activity. Green taxes, as a social levy on environmental costs, are an efficient way to adjust for the market's failure to place value on environmental assets. However, since green taxes (like sales taxes) are innately regressive, compensatory adjustments in the income tax formula would be needed. Many countries already have green taxes – on air, water, waste, noise, fertiliser use and so on. In the UK, introduction of an 'environmental' tax on lead in petrol caused sales of unleaded petrol to climb from 4% to 30% in one year, 1989–90. The most widely discussed green tax has been the 'carbon tax'. This seems likely to raise the most revenue and to be the tax that would do the most environmental good. Various European countries are beginning to introduce such taxes; and there may soon be a uniform carbon tax within the European Community. Green taxes in rich countries have another, wider, potential as a source of monetary contributions to international environmental funds.

13.3 Social and political relations

13.3.1 Overview

We seem to be at an historical cross-roads in relation to the locus of political power and the reassertion of cultural identity; there is currently a heightened tension between globalism and localism.[11] This century there have been powerful centrist forces: the rise of two dominant and hegemonic political ideologies within the developed world; the attachment of client states to these blocs; the tentative emergence of world 'government' (albeit of limited powers); the development of trading coalitions and blocs; the growth and spread of transnational corporations; and the spectacular development of technologies that have shrunk the world – radio, air transport, satellites, television, fax machines and more. More recently there have been resurgent forces – often violent – for decentralisation, for political fragmentation, for cultural differentiation and for community autonomy. The patchwork quilt of Eastern Europe has been reestablished; kaleidoscopic nationalism has replaced the monolithic USSR; Yugoslavia has shattered into religious–ethnic factions; overdue concessions are being made to indigenous peoples in Canada and Australia; the worsening plight of traditional jungle-dwelling cultures and other minority groups is beginning to raise concerns; tribal conflict has flared in many parts of Africa (for long the victim of artificially imposed and culturally insensitive 'national' boundaries); and the Kurds are struggling for autonomy. Meanwhile, in rich countries, comfortable middle-class people are creating new, often peri-urban, residential community settings.

The risk to the global environment in this ferment of decentralisation is that systematic decision-making and coordinated action, in relation to regional and global problems, will become harder to achieve. Outweighing this, hopefully, is the new opportunity for culturally and geographically coherent local communities to relate to their environment, to understand the local and regional ecosystems upon which they depend, and to reinvolve themselves in a mix of traditional and modern forms of ecological sustainability. Perhaps we are about to see some change in the world's decision-making structures. One widely remarked aspect of the events of the 1992 Earth Summit was the unprecedented activity of cross-country coalitions at the non-governmental level. International communications within such networks are proliferating, and new populist pressures are being brought to bear upon formal government structures.

At another level – in an age of political realignments, resurgent local culture and autonomy, and global problems that transcend national boundaries – there is a new opportunity for the UN to forge a new coordinating and leadership role in relation to macroenvironmental problems that require supranational policy-making. The embryonic Conventions emerging from the 1992 Earth Summit may yet foreshadow a new global environmental consciousness. Without such an international commitment, it is hard to see how we humans, living in an increasingly overloaded world, can make the necessary transition in awareness, values and collective rational action.

13.3.2 Environmental security

During four Cold War decades we have thought of 'security' primarily as the capacity of superpowers for mutual deterrence and destruction. Indeed, over the millenia, security for many communities has been primarily military security: walled cities, armies, navies and so on. Today, however, concern over long-term security must address both social and environmental security – i.e. relations between categories of people, and relations between people and their environment. These two types of relationship, as I have argued earlier, are intimately related. We will not achieve long-lasting ecological sustainability if we do not understand the ecological context of human existence and the fundamental dependence of human population health upon an intact biosphere and its life-supporting resources. We must learn to live lightly.

Historically, human groups, whether in peaceful coexistence, commercial competition or open war, have assumed that there is a continuing supply of environmental resources. The only things at issue have been gaining access to those resources and developing strategies to exploit them. Notions of long-term environmental security have not been on the agenda. This has resulted in seriously misplaced priorities. For example, despite the continuing difficulties in feeding the world's population, most countries spend several hundred times more on military security than on agricultural research (i.e. food security). As we approach the year 2000 it is rapidly becoming obvious that we need a set of priorities that ensures the transition from cumulative destruction of our environment and its life-support systems to sustainable management of it. If age-old preoccupations with military security were even partially replaced with concerns over environmental security, then massive amounts of environment-sustaining aid would be possible – and the likelihood of future wars over dwindling environmental resources would recede.

13.4 Summary

I have attempted to develop the following three-point argument about the impact of global environmental change on the health of human populations:

1. The implications for human health of some of the emerging global environmental changes are potentially great. They arise from 'planetary overload', entailing circumstances that are qualitatively different from the familiar, localised problem of environmental pollution.

2. These threats to human population health arise within a largescale *ecological* context. Most forms of disease in human populations – other than physical injury, being eaten by predators and occasional climate-induced famine – are a consequence of human cultural evolution since settlements first occurred and population densities increased. This is not to deny the health gains associated with agrarian and industrial settlement, but it emphasises that human cultural evolution has produced distortions of ecological relationships, causing four main types of health hazard. First came infectious diseases. Then came diseases of industrialisation and environmental pollution by toxic chemicals. Simultaneously, in rich populations, various 'lifestyle' diseases of affluence (heart disease, assorted cancers, diabetes, etc.) emerged. Today we face the health consequences of disruption of the world's natural systems.

3. The first three of those health hazards entail direct-acting effects – microbes, chemical toxins and metabolic disturbances. In general, they

impinge immediately on local groups and populations. However, the fourth category of hazard has a *systemic* quality, that reflects the overloading of different aspects of the regenerative and absorptive capacities of our biosphere. The various possible impacts on human health will mostly occur via indirect and gradual (or deferred) mechanisms, and will impinge widely on whole populations.

The sustaining of population health across generations cannot be achieved without the Earth's life-support systems. So, what are the basic conditions necessary for willing global collaboration towards ecological sustainability? For the poor it is to meet their basic needs and aspirations for security, food and access to reasonable levels of material resources, in social conditions that allow participation and fulfilment. Those conditions, supplemented by stringent family planning policies, will enable control of fertility. For the rich, it is to accept lower levels of consumption of energy and materials. The rich 20% of the world currently account for 80% of global consumption and waste production and around three-quarters of all greenhouse gas emissions.

Competitiveness and acquisitiveness may well reflect the survival-enhancing characteristics upon which natural selection, and therefore human biological evolution, is based. However, it is also clear that *Homo sapiens* has a unique capacity for cultural supplementation of basic biology. Indeed, human history can be viewed as a succession of cultural and technological developments enabling us to sidestep the natural ecological constraints on basic human biology. This capacity has led to many gains in human health and security. But, through its reproductive and technological excesses, this same capacity is now creating global ecological problems. The potential losses are piling up, and some of them, unchecked, have the potential to render us an endangered species. Therefore, we now depend on that same cultural ingenuity to find – soon – a path towards an ecologically sustainable, health-supporting, way of life.

References

1. Ravetz J. Knowledge in an uncertain world. *New Scientist* 1990; **1735**: 2.
2. Passmore J. *Man's Responsibility for Nature*. London: Duckworth, 1974 (second edition, 1980).
3. Talbot LM. Man's role in managing the global environment. In Botkin DB, Caswell MF, Estes JE, Orio AA. *Changing the Global Environment. Perspectives on Human Involvement*. London: Academic Press, 1989, pp 17–33.
4. Hill AB. The environment and disease: Association or causation? *Proceedings of the Royal Society of Medicine* 1965; **58**: 295–300.

5. Bodansky D. Scientific uncertainty and the precautionary principle. *Environment* 1991; **33**: 4–5, 43–4.
6. UN. *Global Outlook* 2000. UN Publications ST/ESA/215/Rev 1. New York: UN, 1990.
7. Peacock J. Twenty-first century crops. *Nature* 1992; **357**: 358.
8. Buttel F. Biotechnology and agricultural development in the Third World. In: Bernstein H, Crow B, Mackintosh M, Martin C. *The Food Question: Profits Versus People*. London: Earthscan Publications, 1990, pp 163–80.
9. George S. *The Debt Boomerang: How Third World Debt Harms Us All*. London: Pluto Press, 1992.
10. The UN Environment Program, World Conservation Union and World Wildlife Fund estimate that a graduated increase from US$50 billion to US$150 billion annually would be required throughout this decade. (See: IUCN, UNEP, WWF. *Caring For The Earth. A Strategy for Sustainable Living*. Gland, Switzerland: Earthscan, 1991.)
11. McMichael P. Rethinking comparative analysis in a post-developmentalist context. *International Social Science Journal* 1992; **133**: 351–65.

Glossary

Billion. A thousand million, 10^9.

Carrying capacity. This refers to the population size of a particular species that can be sustained indefinitely by a specified environment. The term is used differently by biologists, ecologists, economists and resource managers. The term originates in theoretical population biology to describe the mathematical properties of a certain type of growth curve – 'carrying capacity' is the asymptotic limit reached when density-dependent mortality comes to equal fertility. Thus, for a particular species, a constant environment would mean a fixed, definable carrying capacity. However, environments vary naturally and steady-state equilibrium systems are uncommon. Further, while the term can be applied relatively easily to describe the carrying capacity of a defined territory for a species of grazing animal, it is less easily applied to the human species. Humans not only remove resources from local environments and return unexpected wastes to it, but they can impose dramatic additional modifications (e.g. agriculture) which continually alter the capacity of the environment to 'carry' humans (at some specified 'standard of living'). Hence, for humans the term is notional rather than formal.

East Asia. This refers to all the countries of Southeast Asia (Malaysia, Thailand, Indonesia, etc.) and China, Taiwan, the Koreas, Hong Kong and Japan.

FAO. The Food and Agriculture Organization of the UN, based in Rome, Italy. FAO compiles international statistics on food production, monitors land use and fisheries, conducts surveys on food consumption and nutritional status, formulates policy advice for the UN and for member nations, and gives field assistance.

Gigatonne. One billion (10^9) tonnes.

339

GNP. Gross National Product. The sum of all 'added value' for goods and services sold within the country (i.e. GDP, the Gross Domestic Product), plus income from international transactions, minus money repatriated by migrant workers.

Hectare. One-hundredth of a square kilometre. A hectare equals approximately 2.5 acres.

IMF. (International Monetary Fund). Established in 1944, alongside the World Bank, as part of the UN's Bretton Woods agreement (see also: World Bank). The IMF oversees international financial statistics and monitors economic trends. It lends money to countries to cover external balance-of-payments deficits, to assist them to pursue economic recovery. The size of each member nation's shareholding, voting and borrowing rights is in proportion to its GNP. Rich countries, who borrow little, have control of loan policy decisions.

IPCC. International Panel on Climate Change. Established in 1988 by UNEP and the World Meteorological Organization.

Latin America. Comprises South America (Bolivia and all countries south) and Central America (all countries from Panama to Mexico).

Millenium. A period of one thousand years.

Nanometre. One-billionth (10^{-9}) of a metre.

South Asia. This refers to the populous region dominated by India, and also includes Sri Lanka, Pakistan, Bangladesh, Bhutan, Nepal, Myanmar and The Maldives.

Sub-Saharan Africa. Comprises all countries south of the Sahara Desert, except South Africa.

UNEP. United Nations Environment Programme. This body was established immediately after the 1972 UN Conference on the Environment, in Stockholm. It was given a broad mandate by the UN to stimulate, coordinate and provide policy guidance for sound environmental action throughout the world. UNEP is based in Nairobi, Kenya.

WCED. World Commission on Environment and Development. Otherwise known as the Brundtland Commission. This body was set up in 1983 by the UN, to which it reported back in 1987. Its report, 'Our Common Future', has been widely quoted and it has influenced national and international policies. Its central ideas helped lay the foundations for the 1992 UN Conference on Environment and Development (UNCED).

West Asia. This is a less familiar term than South Asia and East Asia. It refers to Afghanistan, Iran, Iraq and the other 'Middle East' countries.

(The names 'Middle East' and 'Near East' are Europe-centred terms, still widely used – but are becoming less appropriate as the term 'Far East' is discarded.)

WHO. World Health Organization. WHO was established in 1948 as part of the UN system, and is based in Geneva, Switzerland. UN Members have the option of being 'member states' of WHO, and, in return, pay annual dues. Policies and programmes are formulated by (the annual meeting of) the World Health Assembly, overseen by the Executive Board and implemented via the many Divisions and Special Programmes of WHO, with local participation and delivery via WHO's six regional offices.

World Bank. Established by the UN in 1944 (as part of the Bretton Woods agreement), alongside the International Monetary Fund (IMF – see above). These twin organizations, along with centrally-controlled currency exchange rates and the General Agreement on Tariffs and Trade (GATT), were the core of the First World's strategy for building a stable, expanding global economy. The World Bank makes investment loans to countries for specified developmental projects (e.g. dams, mining, livestock). It also conducts research, carries out reviews and publishes a range of reports. It and the IMF are based in Washington, DC, USA.

Index